AF851041

Birkhäuser

Modeling and Simulation in Science, Engineering and Technology

Series Editor
Nicola Bellomo
Politecnico di Torino
Torino, Italy

Editorial Advisory Board

K.J. Bathe
Department of Mechanical Engineering
Massachusetts Institute of Technology
Cambridge, MA, USA

M. Chaplain
Division of Mathematics
University of Dundee
Dundee, Scotland, UK

P. Degond
Department of Mathematics,
Imperial College London,
London, United Kingdom

A. Deutsch
Center for Information Services
and High-Performance Computing
Technische Universität Dresden
Dresden, Germany

M.A. Herrero
Departamento de Matematica Aplicada
Universidad Complutense de Madrid
Madrid, Spain

P. Koumoutsakos
Computational Science & Engineering
Laboratory
ETH Zürich
Zürich, Switzerland

H.G. Othmer
Department of Mathematics
University of Minnesota
Minneapolis, MN, USA

K.R. Rajagopal
Department of Mechanical Engineering
Texas A&M University
College Station, TX, USA

T.E. Tezduyar
Department of Mechanical Engineering &
Materials Science
Rice University
Houston, TX, USA

A. Tosin
Istituto per le Applicazioni del Calcolo
"M. Picone"
Consiglio Nazionale delle Ricerche
Roma, Italy

More information about this series at http://www.springer.com/series/4960

Alberto d'Onofrio • Alberto Gandolfi
Editors

Mathematical Oncology 2013

Birkhäuser

Alberto d'Onofrio
International Prevention Research Institute
Lyon, France

Department of Experimental Oncology
European Institute of Oncology
Milano, Italy

Alberto Gandolfi
Istituto di Analisi dei Sistemi ed
Informatica "Antonio Ruberti"—CNR
Rome, Italy

ISSN 2164-3679 ISSN 2164-3725 (electronic)
ISBN 978-1-4939-0457-0 ISBN 978-1-4939-0458-7 (eBook)
DOI 10.1007/978-1-4939-0458-7
Springer New York Heidelberg Dordrecht London

Library of Congress Control Number: 2014951354

Mathematics Subject Classification (2010): 92-XX, 92B05, 97M60, 37N25, 35Q92, 00A69, 92C37, 92B99

© Springer Science+Business Media New York 2014
This work is subject to copyright. All rights are reserved by the Publisher, whether the whole or part of the material is concerned, specifically the rights of translation, reprinting, reuse of illustrations, recitation, broadcasting, reproduction on microfilms or in any other physical way, and transmission or information storage and retrieval, electronic adaptation, computer software, or by similar or dissimilar methodology now known or hereafter developed. Exempted from this legal reservation are brief excerpts in connection with reviews or scholarly analysis or material supplied specifically for the purpose of being entered and executed on a computer system, for exclusive use by the purchaser of the work. Duplication of this publication or parts thereof is permitted only under the provisions of the Copyright Law of the Publisher's location, in its current version, and permission for use must always be obtained from Springer. Permissions for use may be obtained through RightsLink at the Copyright Clearance Center. Violations are liable to prosecution under the respective Copyright Law.
The use of general descriptive names, registered names, trademarks, service marks, etc. in this publication does not imply, even in the absence of a specific statement, that such names are exempt from the relevant protective laws and regulations and therefore free for general use.
While the advice and information in this book are believed to be true and accurate at the date of publication, neither the authors nor the editors nor the publisher can accept any legal responsibility for any errors or omissions that may be made. The publisher makes no warranty, express or implied, with respect to the material contained herein.

Printed on acid-free paper

Springer is part of Springer Science+Business Media (www.birkhauser-science.com)

Preface

After a long incubation period in which only sporadic investigations were devoted to the applications of mathematics and physics to the study of tumors (see, e.g., the work of Doll and Armitage on tumorigenesis in the 1950s), in the early 1970s, the papers by Norton and Simon and by Greenspan were a turning point in the scientific interest on tumor modeling.

In this first phase, the majority of models were essentially population-based models. Although many important results were established—for example, the possible onset of a dynamic equilibrium between a tumor and the immune system, only recently experimentally confirmed—the majority of studies did not directly rouse the attention of the oncology world, with the remarkable exception of the papers by Goldie and Coldman on the Darwinian emergence of resistance to chemotherapy and of the above-mentioned works by Norton and Simon.

The pioneering work by Greenspan, where the modeling also involved physico-chemical aspects of tumor growth, remained almost isolated for two decades. Finally, in the 1990s, a large number of mathematical models were devoted to describing the spatial growth of tumors, with approaches ranging from simple diffusive models to complex multiphase mechanical models. Later, the interactions of the tumor with the immune system and with the angiogenesis process also became the object of extensive theoretical research.

The follow-up in the medical world, however, remained rather scarce. This is due mainly to the limited number of joint works between biomedical and biomathematical researchers, although there have been some isolated cases of theoretical scientists—for example, R. Jain—who have been so deeply involved in biological research as to become influential biomedical research leaders.

Things are, however, rapidly changing. In recent years, two major phenomena have given great momentum to research in mathematical oncology.

The first is of a technical nature—that is, the birth of multiscale modeling—where microstructures such as individual cells can be explicitly represented.

The second, and most important from a "sociological" point of view, is that a large number of biomedical scientists are becoming directly involved in quantitative research due to the need to decipher an ever larger mass of "omics" data (mainly from genomics and proteomics).

The increasing number of modeling papers published in journals devoted to oncology, and in particular the opening of the new section on mathematical oncology in *Cancer Research*, one of the most important basic research journals on cancer, is the most important evidence that tumor modeling is slowly but significantly impacting the oncology world.

Moreover, the fact that in these years a number of research groups—often led by biomedical scientists turned to computational sciences—are being "embedded" in research medical centers is further evidence of the scientific interest of the topic.

On the other hand, from the point of view of mathematicians, there is an increasing awareness that the applications of both classical mechanics and nonlinear analysis to the study of tumor growth translate into new challenging problems at the frontier of contemporary mathematics.

This is mirrored by the increasing number of papers on mathematical oncology that are published in mathematical journals.

This book is entirely composed of in-depth contributions reviewing personal research results of outstanding scientists in the field. It is aimed at providing both experienced researchers and the increasing number of newcomers with a careful selection of state-of-the-art results.

Many new researchers who are entering the field of mathematical oncology often experience significant difficulties. Starting work in mathematical cancer modeling, indeed, is a slow and difficult process that requires the acquisition of a special *forma mentis* that goes well beyond that of the usual applications of mathematics and physics, where the learning can be limited to the acquisition of basic concepts and methods of the domain of application. In tumors, on the contrary, many apparently different phenomena are interrelated, and all of them are strictly linked to clinical issues. We believe that in the chapters of this book the authors have successfully transferred not only their results but also, most importantly, their way of seeing and approaching problems.

In order to cope with the state of the art, the book covers different biological subjects and mathematical approaches.

As far as the tumor onset and early phases of tumor growth are concerned, Bortolusso and Kimmel investigate the interplay of spatiality and stochasticity in the process of tumorigenesis by stressing some exquisitely stochastic spatiotemporal phenomena without deterministic counterpart and the role of cellular cooperation. Fasano, Bertuzzi, and Sinisgalli focus on the role of the conservation laws of mathematical physics to decipher the dynamics of the early phases of neoplasias by means of the analysis of some free-boundary problems for partial differential equations. Techniques of nonlinear mathematical physics and statistical mechanics, as well as WKB approximations, are used by Ben Amar to treat two typical features of melanomas: morphological instabilities and phase segregation. In her contribution, Ben Amar also provides an overview of the historical development of "mathematical biophysics" of tumors.

With regard to the intercellular interplay between tumor cells and other cells in the environment, Dyson and Webb extend their models that include the "cell cycle age" of tumor cells by also including the cell-to-cell adhesion theory by Painter and Sherrat and provide a complete mathematical analysis of the resulting model by using the theory of semigroups. Lachowitz, Dolfin, and Szymanska, in the framework of the kinetic theories of tumor-immune system interplay, provide a theoretical framework for the construction of micro- and meso-models that may be related to macroscopic models and that are able to take into account various additional aspects of the microscopic scale. Kareva, Wilkie, and Hahnfeldt review the role of the interplay of tumors with their microenvironment: This includes interplay with endothelial cells in the process of angiogenesis and with the immune system, as well as the role of recycling of nutrients.

Finally, as far as the modeling of antitumor therapies is concerned, de Pillis and Radunskaya review their models of tumor-immune systems and immunotherapies, as well as the effect of chemotherapies on normal, tumor, and immune cells. A variety of approaches are used by de Pillis and Radunskaya, ranging from ordinary differential equations to cellular automata. A hybrid multiscale framework—also including cell cycle dynamics of individual cells—is adopted by Powathil and Chaplain to model the spatiotemporal response of tumors to chemotherapy alone or in combination with radiotherapy. Clairambault reviews the chronobiology of tumor growth and antitumor therapies, i.e., he focuses on how to "exploit" the influence of circadian rhythms on the proliferation of tumors in order to maximize the effects of chemotherapies. Methods and tools of optimal control theories are introduced by Ledzewicz and Schaettler in the final chapter, which reviews the application of these theories to optimize antitumor treatments under various biologically meaningful conditions.

We end the preface with a brief consideration. The field of mathematical modeling of tumor growth and of related therapies has in recent years been named "mathematical oncology." From the point of view of both a layman and of a pure mathematician, this could seem bizarre or as an overstatement. However, as we hope this book will demonstrate, this term is not an exaggeration as all major mathematical tools of analytical and computational mathematics can be fruitfully "exploited" in order to investigate the many problems of oncological interest—methods ranging from ordinary differentialequations to the statistical mechanics of phase transitions, from the theory of semigroups to Gillespie's algorithm, from cellular automata to the geometric theory of optimal control, etc.

However, it is not only a question of adopting analytical theories or computational tools to tackle biological problems from a physical-mathematical point of view. No! The first challenge in mathematical oncology is for biological problems to provide an impetus to develop or substantially improve new mathematical theories and computational algorithms. This is exactly what happened to other related or unrelated branches of mathematical biology, as may be seen by taking a historical perspective of the developments of dynamical systems theory and computational sciences, as well as classical and computational statistics. The second challenge is, of course, to develop and increment the collaboration with biomedical scientists. This point has been considered several times in the past but here we want to also stress that a major effort is needed from the didactic point of view. Indeed, a different and more integrated approach to mathematical oncology might lead to a new generation of theoretical biologists with backgrounds equally divided between quantitative sciences and biomedicine.

Lyon, France
Rome, Italy
Jan 15, 2014

Alberto d'Onofrio
Alberto Gandolfi

Contents

Part I
Cancer Onset and Early Growth

Modeling Spatial Effects in Carcinogenesis: Stochastic and Deterministic Reaction-Diffusion

Roberto Bertolusso and Marek Kimmel

Abstract This paper is a review including some original elements of a family of reaction-diffusion models of carcinogenesis exhibiting diffusion-driven (Turing) instability (DDI), but consisting of a single reaction-diffusion equation coupled with a system of ordinary differential equations (ODE). Such models are very different from the classical Turing-type models in that they exhibit qualitatively new patterns of behavior of solutions, including, in some cases, a strong dependence of the emerging pattern on initial conditions and quasi-stability followed by rapid growth of solutions, which may take the form of isolated spikes, corresponding to discrete foci of proliferation. However, the process of diffusion of growth factor molecules is by its nature a stochastic random walk. An interesting question emerges to what extent the dynamics of the deterministic diffusion model approximates the stochastic process generated by the model. We address this question using simulations with a software tool called sbioPN (spatial biological Petri Nets). The picture emerging suggests that some of the generic features of the deterministic system, such as spike formation, and dependence of the number of spikes on diffusivity, are preserved. However, new elements, such as spike competition and appearance of spikes at isolated random locations, are also present. We also discuss the relevance of the model and particularly the cell cooperativity hypothesis, underlying transition to the DDI.

Keywords Cancer modelling • Deterministic • Stochastic • Reaction-diffusion equations • Pattern formation • Spike solutions

1 Introduction

A spatial model of early carcinogenesis in lung cancer was developed in a series of papers by Marciniak-Czochra and Kimmel [21–23]. The model has the form of a system of two ordinary differential equations (ODE) for cells and bound growth

R. Bertolusso • M. Kimmel (✉)
Department of Statistics, Rice University, 2124 Duncan Hall, 6100 Main St., Houston, TX 77005, USA
e-mail: kimmel@rice.edu

© Springer Science+Business Media New York 2014

A. d'Onofrio, A. Gandolfi (eds.), *Mathematical Oncology 2013*,
Modeling and Simulation in Science, Engineering and Technology,
DOI 10.1007/978-1-4939-0458-7_1

factor molecules and a single reaction-diffusion equation for free growth factor molecules. It has been proved to generate quasi-stochastic patterns of spikes of cell concentration, which might be interpreted as foci of invading early cancer cells. The model offers mathematical challenges, which were addressed in the papers by Marciniak and co-workers [25, 26]. However, it was noticed in Bertolusso and Kimmel [5] that the biologically more realistic system, in which cells and molecules are discrete entities and diffusion is replaced by random walk, behaves in a way that may be quite different. Each single realization produced only some of the spikes observed in the reaction-diffusion approximation. Nevertheless, if averaged over a large number of realizations, the stochastic system became well approximated by the reaction-diffusion system. This behavior is both reassuring and disturbing. In the abstract sense, it seems desirable to know that averaging produces a more regular behavior. However, in the practical sense, each of the realizations is qualitatively different from the average. It is difficult to address this issue analytically. In this paper, we review some of the basic facts concerning the model and contribute some more systematic simulations for two versions of the model. We also discuss the hypothesis of cooperativity embedded in the model in view of recent literature on carcinogenesis. This hypothesis seems to underlie spike formation in the model, one of its unique features.

2 Background on Patterns and Carcinogenesis

2.1 Turing Pattern Formation

One of the most interesting mathematical phenomena arising in models structured by spatial coordinates is pattern formation via diffusion-driven instability (DDI). Discovered by Alan Turing, this effect has been used to explain emergence of biological, physical, and chemical patterns, such as patterns in colonies of microorganisms, embryo segmentation, or dynamics of the Belousov-Zhabotinsky reactions. The usual mathematical framework is that of the system of at least two reaction-diffusion equations, i.e., partial differential equations (PDE) of the form

$$\partial u/\partial t = D_1 \nabla_x u + f(u, v), \tag{1}$$

$$\partial v/\partial t = D_2 \nabla_x v + g(u, v), \tag{2}$$

where $u(x,t)$ and $v(x,t)$ are defined as functions of spatial coordinates x and time t, $\partial(\cdot)/\partial t$ is partial differentiation with respect to time, $\nabla_x(\cdot)$ is the second-order partial differentiation operator with respect to the spatial coordinates x (diffusion operator or Laplacian), nonlinear functions $f(u,v)$ and $g(u,v)$ are reaction terms. Spatial pattern is a stable spatially heterogeneous (nonconstant) equilibrium solution

(DDI or Turing pattern), which arises in the reaction-diffusion system, but which does not exist for the corresponding pure reaction system

$$\partial u/\partial t = f(u, v), \tag{3}$$

$$\partial v/\partial t = g(u, v), \tag{4}$$

for which only spatially homogeneous (constant in x) solution exist. It is known that stable spatial patterns cannot arise in single reaction-diffusion equation with zero-flux boundary conditions on a convex domain, since all stable equilibria are constant. However, in the current paper, we consider systems in which a single reaction-diffusion equation is coupled with an ODE system. These systems exhibit patterns which seem to be inherently unstable in the deterministic case, but which behave somewhat differently in the stochastic case.

2.2 *Field Theory of Carcinogenesis*

Field carcinogenesis states that as a result of exposure to carcinogens and/or of inherited genetic variants (mutations), a substantial portion of an organ (called the field) can be enriched in genetic variants of cells, which then may or may not acquire further genomic modifications. These modifications will result in increased proliferation and invasion of the surrounding tissues. The rearrangements involved may have different nature. They may range from accumulation of point mutations or microsatellite variability, to gene amplification, and to changes in DNA content, number, and organization of chromosomes (aneuploidy and translocation). Therefore, although they may form a linear irreversible progression, it is more likely that they form a more complex network. Because of the spatial dimensions of the field, emerging groups of transformed cells (precancerous and early cancerous tumors; [16, 17]) will represent different levels of transformation, and may exhibit both progression and reversals. They will be frequently multifocal. This viewpoint is in opposition to the clonal theory of carcinogenesis, which assumes linear irreversible progression and generally unifocal lesions. Let us consider a sample scenario of events, which might be partly corroborated by published evidence. Suppose that cells of some type in a region of an organ have (as a result of inheritance or exposure) an ineffective variant of a gene (such as the p53 gene) providing a cell cycle checkpoint. The normal variant of this gene does not allow cells to enter division if unrepaired damage in DNA is detected. However the mutant gene will allow this, and the cells will go through a series of divisions, which may be not exactly symmetric, so that part of the progeny will be deprived of some other genes and part will have these genes amplified (increased number of gene copies; [1,37]). Gradually, this will relax the controls on correct segregation of chromosomes and lead to aneuploidy (irregular amount of DNA in the cells; [32]). Because of increase in the number of divisions, the telomeres in these partly transformed cells will shorten, which will further impair correct segregation [14]. It has to be noted that it is unlikely to expect that in each cell lineage, these sequence of events will be identical. Therefore, the field will frequently give rise to differently transformed (genetically distinct) groups of cells (non-clonality; [27]). The vast majority of studies in cancer research has been done on well-defined tumors in vivo or on discrete neoplastic foci in vitro. Yet, there is evidence that more than 80 % of the somatic mutations found in mutator phenotype human colorectal tumors occur before the onset of terminal clonal expansion, i.e., adenomas were phylogenetically nearly as old as cancers [38]. In other words, the loss of mismatch repair of DNA starts a genetic phase long before the clone size reaches a threshold of clinical detection. The preneoplastic phase likely reflects accumulation of mutations that ultimately confer visible selection. The "occult prologue before visible neoplasia is longer and therefore likely more important than generally realized" [38]. Supportive findings have been made in sporadic colorectal cancers, where there were approximately 10,000 alterations per cell of inter-simple sequence repeats (ISSRs) in early colonic polyps and in carcinomas [36], indicating that most genetic changes occurred in the preneoplastic stages of the disease. Further mutations are required for progression to and through the neoplastic state. This genetic evidence is in accordance with the microscopic examination of the margins of colorectal cancers reported more than 80 years ago, which revealed chronic epithelial hyperplasia that was not grossly visible [20]. Such hyperplasia characterized an extensive field several inches above and below the neoplastic growth, and was considered to be the first stage of neoplastic development, preceding the appearance of adenomas [19]. The term "field cancerization" was first applied to squamous cell carcinomas of the head and neck [7, 34]. Microscopic examination of the grossly normal margin of all excised oral tumors revealed marked hyperplasia. Genetic evidence confirmed that grossly invisible, flat hyperplastic fields precede the appearance of neoplasia. Field cancerization has been described for a wide variety of tissues [7] and may be characteristic of all cancers [8].

In a commentary in Nature Reviews Cancer, Seton-Rogers [33] discusses a paper by Hu et al. [12]: "Field cancerization is primarily attributed to changes in the epithelium, but whether changes in mesenchymal cells might also have a primary role in the establishment of a field effect is not clear". Hu et al. [12] employed mouse model to explore cancerization field mechanisms in squamous epithelial tumors. Quoting from this paper: "One of the questions asked was whether tissue changes surrounding multifocal epithelial tumors are a cause or consequence of cancer. They employed mice carrying a mesenchymal-specific deletion of CSL/RBP-Jk, a Notch effector, which exhibit spontaneous multifocal keratinocyte tumors that develop after dermal atrophy and inflammation. CSL-deficient dermal fibroblasts promote increased tumor cell proliferation through upregulation of c-Jun and c-Fos expression and consequently higher levels of diffusible growth factors, inflammatory cytokines, and matrix-remodeling enzymes. These changes in gene expression are also induced by UVA irradiation, which is a known cancerization field-promoting agent."

Rubin [31] proposes an experimental model of field cancerization which may be developed to understand and apply the principles to carcinogenesis in the organism. He also proposes that mitigation of selective conditions to prevent or delay carcinogenesis can be developed in culture [30] and then adapted to the organism.

Another important aspect of early carcinogenesis is signaling and production of growth factors. It has been proposed [2] that early carcinogenesis may be a result of cooperative interactions between partly progressed cell populations, which produce growth factors in a complementary way, for example, cells of type 1 produce growth factor needed by cells of type 2 and vice versa. Although this theory has experimental foundations, it is missing the dynamic transformation element prominent in field carcinogenesis. One of the important elements to be taken into account is the spatial diffusion of growth factors and other signaling molecules. As it is known, diffusion in connection with nonlinear kinetics and growth regulation may lead to instability and pattern formation. This is an important prediction, since it suggests that in field carcinogenesis, diffusion of free growth factor molecules leads to diversified multifocal lesions arising on the background of primed tissue, the exact pattern of which depends on chance and small fluctuations of initial conditions.

2.3 *The AAH-BAC: Adenocarcinoma Sequence in Lung Cancer*

A specific biological example motivating our model is the atypical adenomatous hyperplasia (AAH), which is considered a candidate precursor lesion for adenocarcinoma of the lung [15]. As illustrated in Fig. 1A in [15], also reproduced in [21], the AAH lesions are placed along the walls of smallest branchings of the bronchial tree (the alveoli). Citing from [15]:

> From a morphological perspective there seems little doubt that AAH might progress and develop, through a stage of bronchoalveolar adenocarcinoma (BAC), into invasive adenocarcinoma. AAH lesions are usually invisible to the naked eye, but the larger ones or those with more fibrosis in their walls may be visible on the cut surface of well inflated and formalin fixed lung. In the process of progression, the tubular (or approximately linear) AAH structures thicken in places and invade the surrounding tissue.

3 Reaction-Diffusion Model of Early Carcinogenesis

3.1 Hypotheses of the Deterministic Model

The model of a precancerous cell population is as in Marciniak-Czochra and Kimmel [23], which adopts elements of the models previously proposed in [21,22]. The present model is based on the following hypotheses (Fig. 1):

- Precancerous cells c, existing in a spatial domain, proliferate at a rate $a(b,c)$, which is reduced by cell crowding but enhanced in a paracrine manner by a hypothetical biomolecular growth factor b bound to cells.
- Precancerous cells are supplied at a constant rate μ by mutation of normal cells.
- Free growth factor g is secreted by the cells at the rate $\kappa(c)$, and then it diffuses among cells with diffusion constant $1/\gamma$ and binds to cell membrane receptors at a rate $\alpha(c)$, becoming the bound factor b. It then dissociates at a rate d, returning to the free factor pool.
- Free and bound growth factor particles decay at rates d_g and d_b, respectively.

Discussion of possible geometries for the spatial variable x can be found in previous papers (Marciniak-Czochra and Kimmel [21, 22]). One natural geometry is that of a line of cells, occupying the interval $x \in [0, 1]$. There are three substances distributed over the line's length: cells and free and bound growth factor molecules, with densities $c(x,t)$, $g(x,t)$, and $b(x,t)$, respectively. The resulting equations are as follows:

$$\frac{\partial c}{\partial t} = (a(b,c) - d_c)\,c + \mu \tag{5}$$

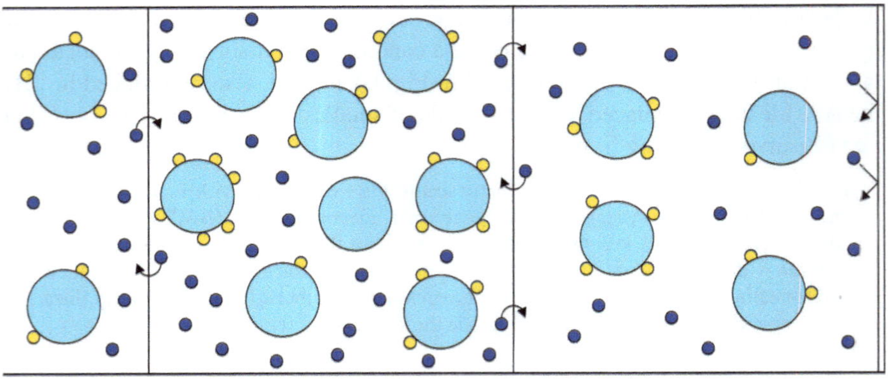

Fig. 1 Graphical depiction of the model. The system is composed of cells (*light blue*), free growth factor molecules (*blue*), and bound growth factor molecules (*yellow*). Three compartments next to a boundary are depicted. Each compartment is considered a perfectly mixed system. Free growth factor diffuses among compartments and is subject to Neumann boundary conditions. Cells and bound growth factor do not diffuse. Free growth factor binds to receptors on cell membrane (receptors not shown)

$$\frac{\partial b}{\partial t} = \alpha(c)g - d_b b - db \tag{6}$$

$$\frac{\partial g}{\partial t} = \frac{1}{\gamma}\frac{\partial^2 g}{\partial x^2} - \alpha(c)g - d_g g + \kappa(c) + db \tag{7}$$

with homogeneous Neumann boundary conditions for g:

$$\partial_x g(0,t) = \partial_x g(1,t) = 0$$

Figure 1 presents a schematic of the model. The kinetics were derived from the stochastic model describing the transitions between different states of the growth factor molecules [21, 22]. Diffusion equation is a macroscopic approximation of the microscopic process of growth factor binding to membrane receptors, under homogeneity hypotheses. A deterministic derivation based on homogenization methodology has been published in [24]. Coefficient $1/\gamma$ is a composite parameter including the diffusion constant and scaling parameters, $\gamma = 1/D$. Proliferation rate has the Hill function form

$$a(b,c) = \frac{a_1(b/c)^m}{1+(b/c)^m}$$

where $a_1 = (2p-1)a_0$ and p is the efficiency of divisions. We will consider the special case $m = 1$. Production of free growth factor by cells has the Michaelis-Menten form

$$\kappa(c) = \kappa_0 + \frac{\kappa_1 c}{1+c}.$$

Supply of Cells Versus Supply of Growth Factor

We consider two versions of the model, which differ with respect to the interactions with environment:

Model 1 Influx of growth factor, no influx of mutated cells = 0, $\kappa(0) > 0$.
Model 2 Influx of mutated cells, no influx of growth factor > 0, $\kappa(0) = 0$.

3.2 *Conditions for Turing Instability*

Theorem 1 (Proposition 3.1 in [23]). *Let us consider the system*

$$\frac{\partial c}{\partial t} = F_c(c,b,g) \tag{8}$$

$$\frac{\partial b}{\partial t} = F_b(c,b,g) \tag{9}$$

$$\frac{\partial g}{\partial t} = \frac{1}{\gamma}\frac{\partial^2 g}{\partial x^2} + F_g(c, b, g) \tag{10}$$

with $A = \{a_{ij}\}$, $i = 1, 2, 3$, $j = 1, 2, 3$ *being the Jacobian matrix of the kinetics system of the system, computed at a positive spatially homogeneous steady state,* $(\bar{a}, \bar{b}, \bar{c})$, *and such that (i) the diagonal elements of A are negative, i.e.,* $a_{ij} < 0$ $i = 1, 2, 3$, *and (ii)* $a_{12}a_{21} > 0$. *There is a DDI for the system if and only if the following conditions are fulfilled*

$$-tr(A) > 0$$

$$-tr(A)\sum_{i<j}|A_{ij}| + |A| > 0$$

$$-|A| > 0$$

$$|A_{12}| > 0$$

where A_{ij} *is a submatrix of A consisting of the i-th and j-th column and i-th and j-th row.*

The first three conditions are necessary for the kinetics system (system without diffusion) to be stable. The last condition is the condition of destabilization of the system with diffusion under (i) and (ii). Let us notice that it is equivalent to the condition of instability for the first two equations, which constitute the ODE part of the system. When applied to our model (Model 1 or Model 2), the DDI condition assumes the following special form.

Theorem 2 (Theorem 3.2 in [23]). *The DDI occurs if and only if there exists a nonnegative steady state* $(\bar{a}, \bar{b}, \bar{c})$, *which is stable to the spatially homogeneous perturbations (i.e., the three first conditions of the proposition hold), and*

$$a_1\bar{c}^2\bar{b}(\alpha(\bar{c}) - \bar{c}\alpha'(\bar{c})) + \mu(\bar{b} + \bar{c})^2\alpha(\bar{c}) < 0.$$

This latter condition is satisfied only if

$$\alpha(\bar{c}) - \bar{c}\alpha'(\bar{c}) < 0.$$

One example of the function $\alpha(c)$ satisfying the above condition is $\alpha(c) = \alpha c^{s+1}$, with $s > 0$. Let us notice that this assumption may be considered "non-physical" (however, see Discussion). For Model 1, this form of $\alpha(c)$ automatically guarantees DDI (of course if the kinetics are stable).

3.3 Perturbation of the Spatially Homogeneous Steady State

Cosinusoidal perturbation is applied to resulting in initial condition of the form

$$c_0 = \bar{c} + \epsilon cos(2n\pi x), 0 < x < 1$$

where $\epsilon = 0.05$ is the amplitude used and n is the number of peaks. If the spatially homogeneous steady state is unstable, this is sufficient for the numerical solutions to be repelled.

3.4 Spike Instability in the Deterministic Model

Existence of the Inhomogeneous Steady States

We will outline a technique in the paper by Marciniak-Czochra and Kimmel [23], i.e., analysis of existence of periodic solutions of a second-order inhomogeneous steady-state ODE, satisfying the zero-flux boundary conditions. This equation emerges from a mixed-type system and therefore its explicit solution is difficult to find. Marciniak-Czochra and Kimmel [23] found sufficient conditions for the existence of periodic solutions and then verified they were satisfied in our system, under certain conditions on coefficients. The method hinges on the following theorem.

Theorem 3 (Theorem 3.6 of [23]). *A general solution of differential equation*

$$g''(x) = f[g(x)] \tag{11}$$

with a C^N function f, defined on an appropriate domain in $\mathbf{R}$, *is given by a $C^2[0, 1]$ function $g(x)$, implicitly defined by the relationship:*

$$\int \frac{1}{\sqrt{C_2 + 2\int f(g)dg}} dg = C_1 \pm x.$$

The existence of two branches of the solution reflects the fact the second derivative is invariant under sign change of x. Moreover, the sufficient condition for the solution of the "+" branch to satisfy the boundary conditions $g'(0) = g'(1) = 0$ is that there exist $g^{**} > g > 0$ such that

$$\int_{g^*}^{g} f(u)du \geq 0, g \in (g^*, g^{**}); \int_{g^*}^{g^{**}} f(u)du$$

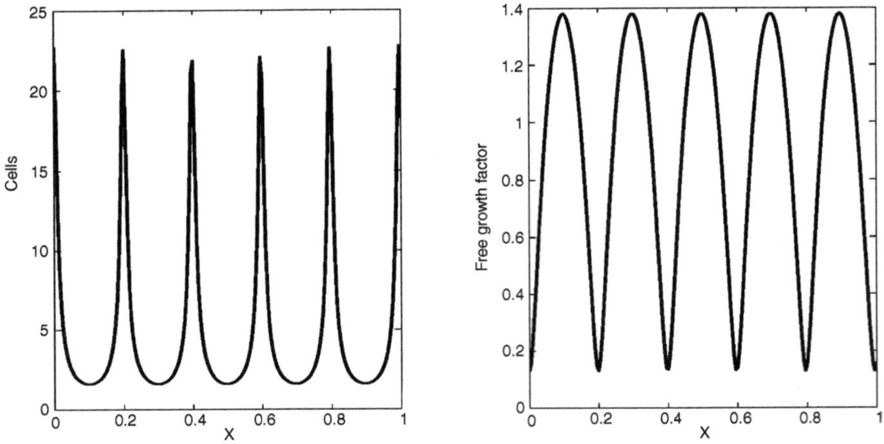

Fig. 2 Spatial profile of the periodic equilibrium spatially inhomogeneous solutions c (*left panel*) and g (*right panel*) of model with parameters $a_1 = 1/12$, $d_c = 0 : 0.05$, $d = d_b = d_g = 0.1$, $\kappa = 1, \alpha = 0.1, \mu = 0.01$, and $\gamma = 1000$. The graph was obtained as a solution of the equilibrium equation (after Fig. 4 of [23])

and

$$\int_{g^*}^{g^{**}} F(u)du$$

where $F(u) = (\int_{g^*}^{u} f(v)dv)^{-1/2}$.

Corollary. *Consider the Eq. (11) of the theorem with the right-hand side defined as*

$$f_a(g) = af(g), a > 0.$$

If there exists an interval $(g_+, g_.)$ *such that* $f_a(g_+) > 0$ *and* $f_a(g_.) < 0$, *then there exists an* $a > 0$ *such that the assertion of the theorem holds.*

Definition 1 (After Hillen [11]). Assume $f \in C_5$, and x_0 x0 is an extremum with $f''(x_0) \neq 0$ and $f^{IV}(x_0) \neq 0$. Then, the extremum is a spike (respectively, a plateau) I f $f^{IV}(x_0)$ has a different (respectively, the same) sign as $f''(x_0)$.

The definition above allows relating the intuitive notion of a "spiky" (sharp) or "plateau-like" extremum to verifiable mathematical conditions. This is one of the ways in which numerical simulations can be mathematically verified.

Based on the last theorem and its corollary, it is possible to prove the following version of the existence result for the periodic inhomogeneous equilibria, which is used to guarantee the existence of periodic spatially inhomogeneous equilibria, such as the one depicted in Fig. 2.

Theorem 4 (Theorem 3.10 in [23]). *Suppose that $g(x)$ is the increasing branch of the solution to Eq. (11), on the interval $x \in [0,1]$, satisfying conditions $g'(0) = g'(1) = 0$ and $g(0) = g^* < g(1) = g^{**}$. Let us choose natural number n, such that $n \geq 1$. Then, the function $h(x)$, defined on $x \in [0,1]$, and such that*

$$h|_{(k-1)/(2n),k/(2n)} = h_k., k = 1,2,\ldots,2n,$$

where

$$h_k(u) = g(k - 2nu), u \in [(k-1)/(2n), k/(2n)], \ k \text{ odd}$$

$$h_k(u) = g(2nu - k + 1), u \in [(k-1)/(2n), k/(2n)], \ k \text{ even}$$

is of class $C_5[0,1]$ and it satisfies the following equation:

$$\frac{1}{4n^2} h''(u), u \in [0,1].$$

Moreover,

$$h(k/2n) = g^*, h'(k/2n) = 0, h^{II}(k/2n) > 0,$$

$$h^{III}(k/2n) = 0, h^{IV}(k/2n) = f'[h(k/2n)]h^{II}(k/2n), \ k \text{ odd}$$

$$h(k/2n) = g^{**}, h'(k/2n) = 0, h^{II}(k/2n) < 0, h^{III}(k/2n) = 0,$$

$$h^{IV}(k/2n) = f'[h(k/2n)]h^{II}(k/2n), \ k \text{ even}.$$

This theorem implies that if $f'[h(k/2n)]$ is positive (respectively, negative), then the solution h has an extremum at $u = k/(2n)$, which is a plateau (respectively, a spike).

We will now apply the results stated above to demonstrate that there exist coefficient values such that inhomogeneous steady states exist. In summary, we have the following corollary.

Corollary (Corollary 3.12 of [23]). *For any $n \geq 1$ there exists a value of diffusion constant γ^{-1}, such that inhomogeneous periodic steady states with period $1/n$ exist, if μ and d_c are small enough and κ is large enough. Estimates for μ, d_c, and κ can be provided, based on the detailed derivations, which are notationally complicated.*

Numerical studies (not shown) indicate that the sufficient conditions provided in the corollary are by far not necessary. However, the number of special cases to consider seems to make a more accurate analytical result impractical.

Figure 2 depicts one example of the spatially inhomogeneous periodic steady states, obtained using the method of the theorem above. The curves are smooth, and spikes are alternating with plateaus. Please notice the superficial resemblance of the

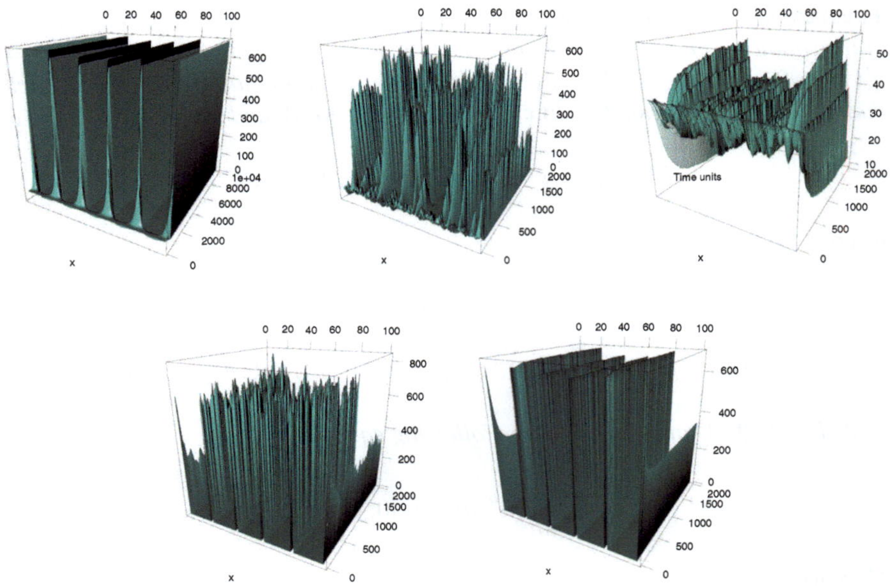

Fig. 3 *Top left*: deterministic evolution (10,000 time units) of the perturbed system. *Top right*: sample stochastic evolution (2,000 time units) of a system starting from deterministic spatially homogeneous steady state. *Central panel*: average of 1,000 stochastic runs starting from deterministic spatially homogeneous steady state. *Bottom left*: sample stochastic evolution (2,000 time units) of a system starting from final spike solutions (at time equal to 10,000 units) of the deterministic system. *Bottom right*: average of 1,000 stochastic runs starting from final spike solutions. Model 1, $\gamma = 1$

profiles to the deterministic profile in Fig. 3 obtained by numerical simulation of dynamics. What is surprising is that the profiles obtained using the method of the theorem are always unstable. This has been demonstrated in a series of papers by [25, 26]. The limit profiles seem to be (in the truly continuous system) Dirac-type distributions. This convergence was be approached numerically using computations with adjustable grid density (Marciniak-Czochra, personal communication).

4 Stochastic Version of the Model

It has been always puzzling, what is the relationship of any time- and space-continuous reaction-diffusion system and the underlying discrete stochastic particle system? In the case of stable patterns, it is expected that the deterministic case is approximated by the averages of the stochastic realizations. However, the system considered in this paper does not exhibit stable patterns.

This problem has been tackled by Bertolusso and Kimmel [5]. Diffusion has been considered as a succession of two first-order stochastic chemical reactions each with reaction constant [9, 13], $j = 1/(\gamma h^2)$ where h is the distance between the centers of two contiguous compartments. Compartments are illustrated in Fig. 1. The model was simulated with an optimized variant of the stochastic simulation algorithm known as constant-time composition-rejection algorithm [35], part of the developed library sbioPN [4], developed in the doctoral thesis of Roberto Bertolusso (available electronically at scholarship.rice.edu/handle/1911/70209). Cells and bound growth factor particles are immobile, but free growth factor molecules are mobile. Cell proliferation, binding and unbinding of growth factor particles, as well as diffusion are modeled as stochastic chemical reactions in space represented by a periodic grid. System trajectories are paths of the free growth factor molecules and counts of the cells and of the bound growth factor molecules. Output plots depict statistics, specifically spatial histograms derived from the trajectories.

Stochastic simulations are started from two different sets of initial conditions:

1. Using the non-perturbed deterministic and spatially homogeneous steady state
2. Using the spiky pattern evolved by the deterministic model

The model without diffusion has been simulated using the bioPN software package [4] in the deterministic and stochastic case.

4.1 Comparison of the Spatial Model in the Deterministic and Stochastic Case

Results of the simulations of the spatial carcinogenesis models in deterministic and stochastic version are presented in Figs. 3–8. In each figure, five panels depict (row by row, from left to right): deterministic simulations starting from a spatially homogeneous steady state perturbed by a cosine function, a single stochastic simulation starting from a non-perturbed deterministic and a spatially homogeneous steady state, an average over 1,000 such stochastic simulations, a single stochastic simulation starting from the spikey pattern evolved by the deterministic model, and an average over 1,000 such stochastic simulations. Figures depict the following variants of the two models: Fig. 3, Model 1, $\gamma = 1$, Fig. 4, Model 1, $\gamma = 5$, Fig. 5, Model 1, $\gamma = 50$, Fig. 6, Model 2, $\gamma = 1$, Fig. 7, Model 2, $\gamma = 5$, and Fig. 8, Model 2, $\gamma = 50$, with the other parameters as listed in Table 1.

Relationship between the deterministic and stochastic simulations is different for Models 1 and 2. In Model 1, the deterministic system does not become extinct, but the solution converges to a number of spikes (this number depends on diffusivity and initial conditions). The stochastic system, when started from the discretized deterministic spatially homogeneous steady state, tends to produce persistent spikes, the number and location of which are comparable to those in the deterministic case. However, the spikes in different simulation runs are out of phase, and when averaged, cancel each other (see the mean plot in Figs. 3, 4, and 5). If the initial conditions for stochastic simulations are set equal to discretized final pattern of

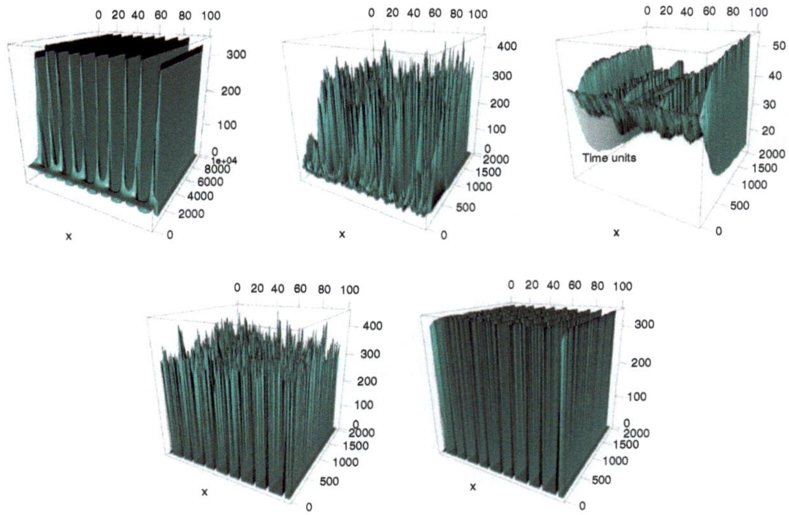

Fig. 4 *Top left*: deterministic evolution (10,000 time units) of the perturbed system. *Top right*: sample stochastic evolution (2,000 time units) of a system starting from deterministic spatially homogeneous steady state. *Center*: average of 1,000 stochastic runs starting from deterministic spatially homogeneous steady state. *Bottom left*: sample stochastic evolution (2,000 time units) of a system starting from final spike solutions (at time equal to 10,000 units) of the deterministic system. *Bottom right*: average of 1,000 stochastic runs starting from final spike solutions. Model 1, $\gamma = 5$

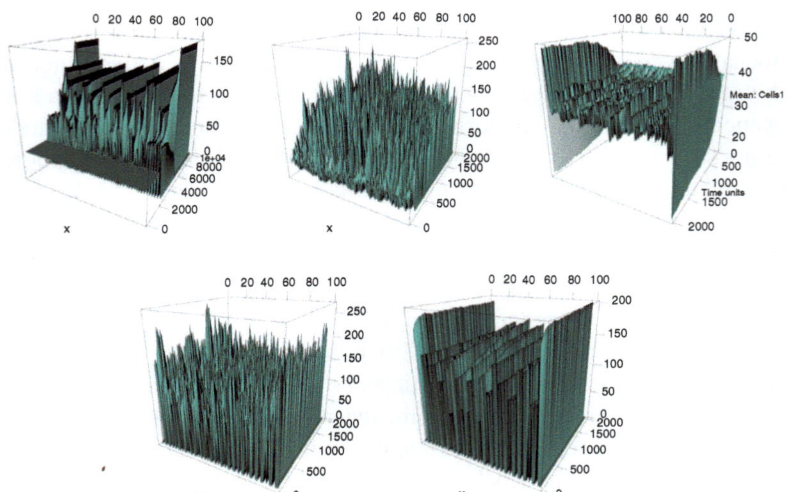

Fig. 5 *Top left*: deterministic evolution (10,000 time units) of the perturbed system. *Top right*: sample stochastic evolution (2,000 time units) of a system starting from deterministic spatially homogeneous steady state. *Center*: average of 1,000 stochastic runs starting from deterministic spatially homogeneous steady state. *Bottom left*: sample stochastic evolution (2,000 time units) of a system starting from final spike solutions (at time equal to 10,000 units) of the deterministic system. *Bottom right*: average of 1,000 stochastic runs starting from final spike solutions. Model 1, $\gamma = 50$

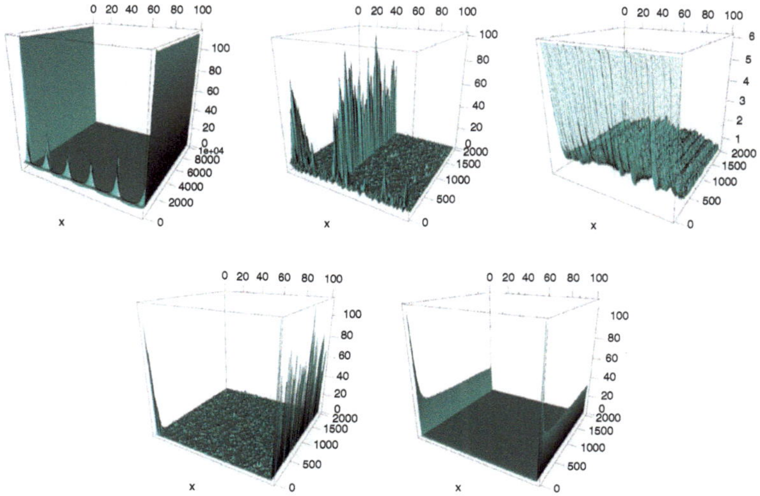

Fig. 6 *Top left*: deterministic evolution (10,000 time units) of the perturbed system. *Top right*: sample stochastic evolution (2,000 time units) of a system starting from deterministic spatially homogeneous steady state. *Center*: average of 1,000 stochastic runs starting from deterministic spatially homogeneous steady state. *Bottom left*: sample stochastic evolution (2,000 time units) of a system starting from final spike solutions (at time equal to 10,000 units) of the deterministic system. *Bottom right*: average of 1,000 stochastic runs starting from final spike solutions. Model 2, $\gamma = 1$

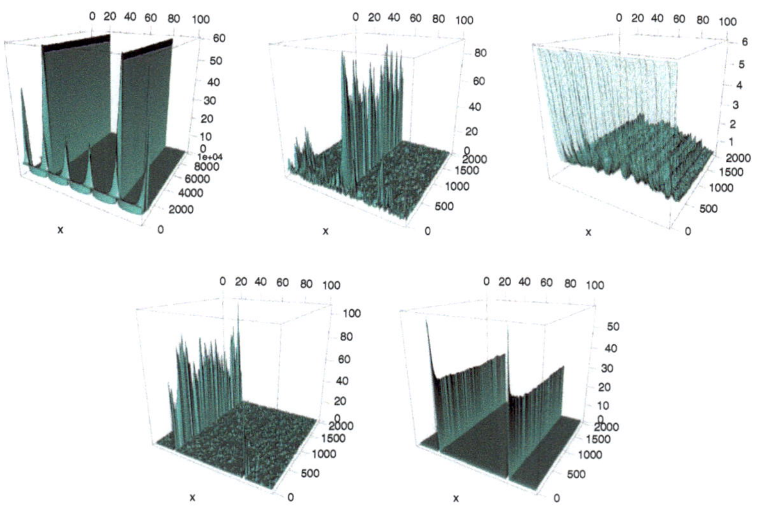

Fig. 7 *Top left*: deterministic evolution (10,000 time units) of the perturbed system. *Top right*: sample stochastic evolution (2,000 time units) of a system starting from deterministic spatially homogeneous steady state. *Center*: average of 1,000 stochastic runs starting from deterministic spatially homogeneous steady state. *Bottom left*: sample stochastic evolution (2,000 time units) of a system starting from final spike solutions (at time equal to 10,000 units) of the deterministic system. *Bottom right*: average of 1,000 stochastic runs starting from final spike solutions. Model 2, $\gamma = 5$

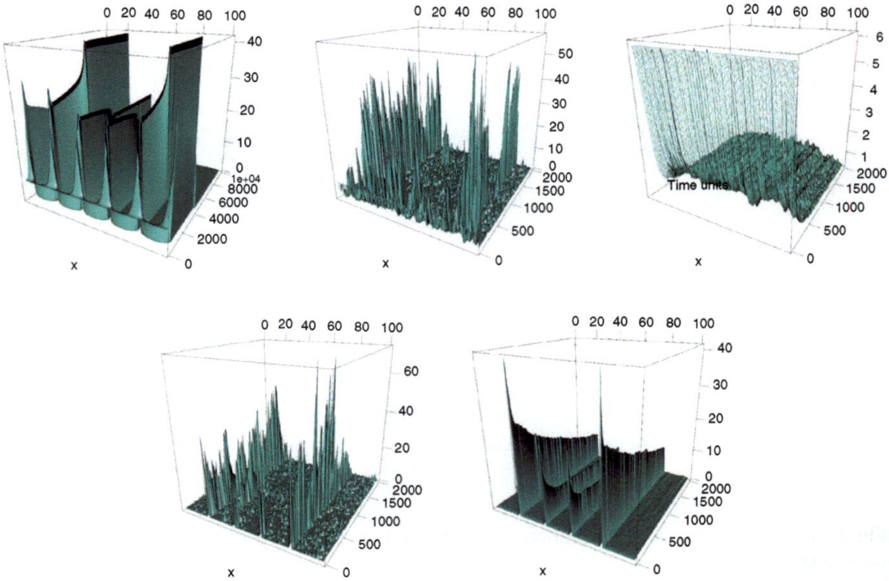

Fig. 8 *Top left*: deterministic evolution (10,000 time units) of the perturbed system. *Top right*: sample stochastic evolution (2,000 time units) of a system starting from deterministic spatially homogeneous steady state. *Center*: average of 1,000 stochastic runs starting from deterministic spatially homogeneous steady state. *Bottom left*: sample stochastic evolution (2,000 time units) of a system starting from final spike solutions (at time equal to 10,000 units) of the deterministic system. *Bottom right*: average of 1,000 stochastic runs starting from final spike solutions. Model 2, $\gamma = 50$

Table 1 Values of the parameters used in all studied systems

Parameter	Value
a_1	1/12
α	10^{-1}
κ_0	5
κ_1	1
d_c	5×10^{-2}
μ	10^{-2}
$d = d_b = d_g$	10^{-1}

the deterministic simulations, the spikes are preserved, but their magnitude changes because of the effect known as stochastic focusing [28]. The positions and heights of spikes are preserved, which becomes clear by inspection of the mean plot. When γ is varied, the frequency of spikes increases, quite similarly in the deterministic and stochastic variant of the model.

As for Model 2, in the deterministic case, for high diffusivity values, most spikes occur only transiently, whereas, as diffusivity decreases, some of them stabilize. In the stochastic case, with initial conditions proportional to the discretized deterministic spatially homogeneous steady state (including spikes), the model tends to produce a single spike at a random, which is selected out at the initial time period. It is usually maintained until the end of simulation. When averaged, the pattern emerging includes a weak effect of the boundary conditions, favoring the formation of spikes in the central part of the unit interval. If the initial conditions for stochastic simulations are set equal to discretized final pattern of the deterministic simulations, then the effect depends on diffusivity. For high diffusivity only one of the spikes is preserved, with no preference, as it can be observed in the averaged values. For low diffusivity, more than one spike can be maintained, but the averaged values show that this effect is transient.

Dynamics of the deterministic and stochastic RD displays general qualitative consistency, notably tendency towards spike formation, with the number of spikes depending on diffusivity $(1/\gamma)$ and initial conditions (compare stochastic simulations starting from equilibrium to those starting from spikes). Stochastic trajectories exhibit new modes of spike competition. Model 2 (more prone to extinction for large diffusivity or small γ) seems stabilized by stochasticity at least for a range of parameters.

4.2 Behavior of the Model Without Diffusion

The results of simulations of the deterministic version of Models 1 and 2 are presented in Figs. 9 and 10. Panel A of Fig. 9 includes trajectories of the deterministic and stochastic versions of Model 1, starting from $c(0) = 1$, $c(0) = b(0) = 40$. The stochastic simulations either oscillate around a value close but lower than the deterministic nontrivial equilibrium (because of stochastic focusing) or become extinct. In the scenario depicted, 65 % of simulations end with extinction of cells. Panel B includes trajectories of the deterministic and stochastic versions of Model 2, also starting from $c(0) = 1, c(0) = b(0) = 40$. The oscillatory pattern is analogous as for Model 1, although the extinct and non-extinct trajectories are less clearly resolved. In the scenario depicted, 62.3 % of simulations end with extinction of cells. Panel A of Fig. 10 depicts dependence of the cell extinction probability on initial conditions, based on 96 cases of initial conditions. Dependence on $c(0)$ is substantial, as opposed to almost no dependence on $b(0)$ and $g(0)$.

5 Discussion

5.1 Role of Stochastic Effects

Reaction-diffusion is a continuous approximation of an essentially discrete (finite object) phenomenon, involving stochastic chemical reactions as well as random walk of free growth factor particles. Continuity "corset" superimposed on nonlinear

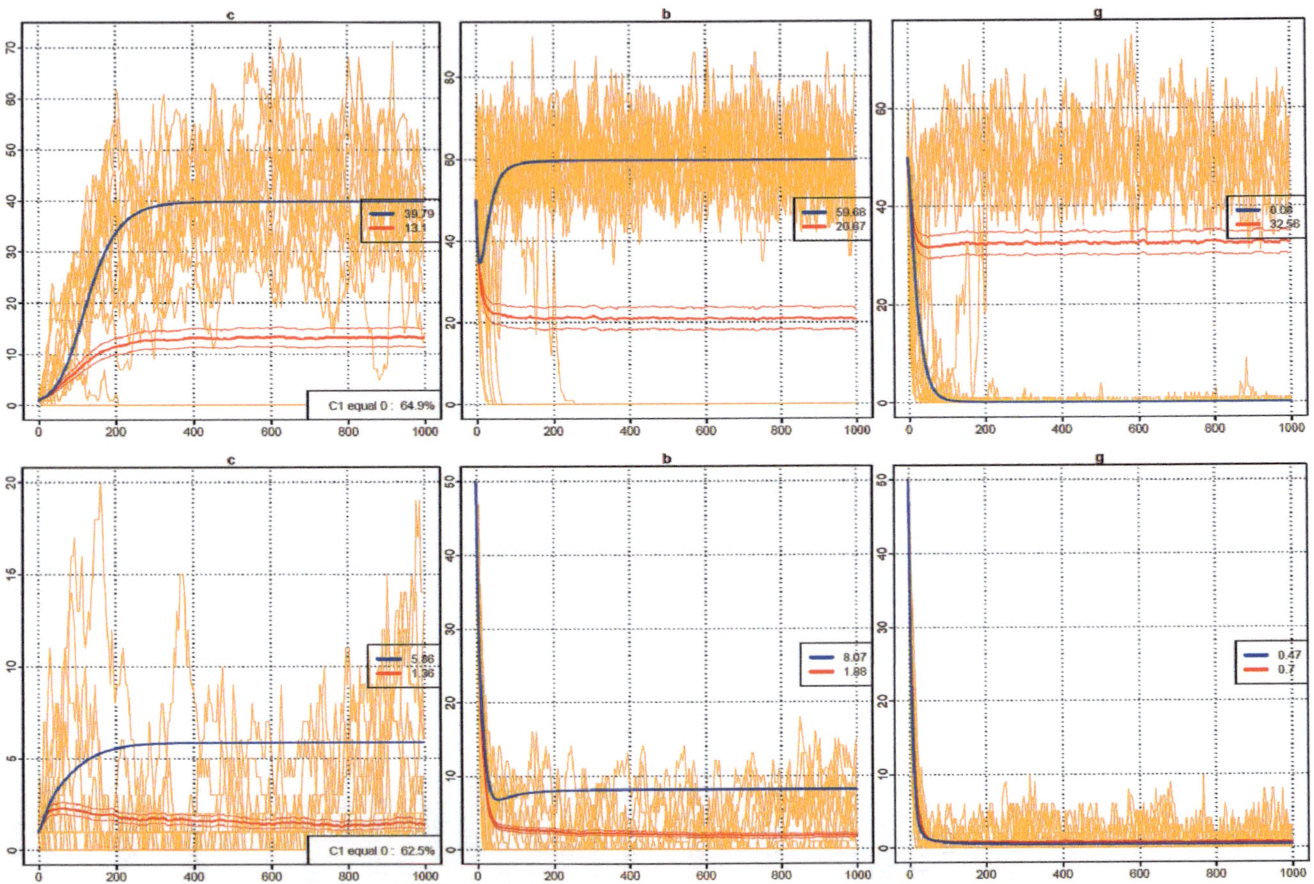

Fig. 9 Reaction-only (non-spatial) system. *Blue lines*, deterministic solutions; *orange lines*, single stochastic runs; *red lines*, averages 3 standard errors of 1,000 stochastic runs). *Insets* depict values of c(1,000) and percentage of cells with value zero (extinction for Model 1) at time 1,000. *Top*, Model 1; *bottom*, Model 2. In both cases the initial conditions are $c(0) = 1$, $b(0) = 50$, $g(0) = 50$

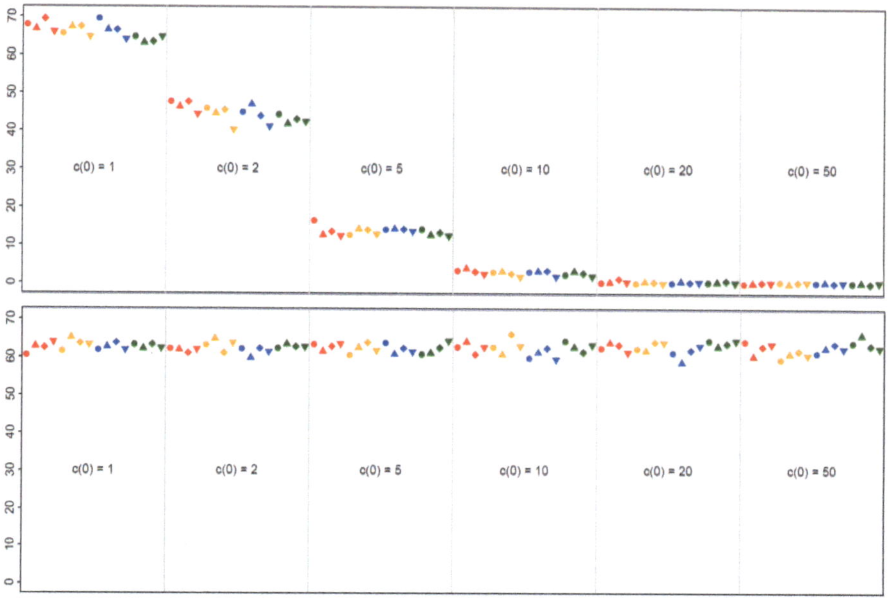

Fig. 10 *Top*, Model 1 percentage of extinct cells at time 1,000, starting from different initial conditions. *Bottom*, Model 2 percentage of cells at value 0 at time 1,000, starting from different initial conditions. Color coding: *red*, $b(0) = 1$, *orange*, $b(0) = 5$, *blue*, $b(0) = 10$, and *green*, $b(0) = 50$. Symbols: *circle*, $g(0) = 1$, *upper triangle*, $g(0) = 5$, *rhomboid*, $g(0) = 10$, and *lower triangle*, $g(0) = 50$. $c(0)$ values are shown in the plot, separated by *gray vertical lines*

kinetics may generate patterns which are absent in finite-object systems. However, as detailed in the results section, the main qualitative feature, which is spike development, remains unchanged. Stochastic versions of the system display new features such as spike competition. Interestingly, no spatial perturbation is needed in the stochastic systems to initiate spike formation, and the intrinsic randomness is sufficient.

We carried out a large number of simulations, which address the behavior described under two different specific assumptions: (i) no influx of mutated cells but external influx of growth factor molecules and (ii) the opposite, external influx of mutated cells but no influx of growth factor molecules. In both cases, we start the stochastic model either from (a) discretized deterministic space-homogeneous equilibrium, or (b) from discretized final "spiky" profiles produced by the deterministic system. We compare these two scenarios to their deterministic counterparts and to each other and find striking differences in dynamics. Under assumption (i), in scenario (a), each stochastic run produces a spikey profile similar to the deterministic one, but with a random number and positions of spikes, so that when these are averaged, they produce a profile which is much less variable. In scenario (b), the opposite is true, while single realizations produce only some of the spikes, when averaged they approximated deterministic solutions. Under assumption (ii), in scenario (a), each stochastic run usually produces one spike at a random position, so that when these are averaged, they produce a homogeneous profile, partially affected by boundary conditions. In scenario (b), for high diffusivity only one of the spikes is preserved, with no preference. For lower diffusivity, more than one spike can be preserved, but the averaged values show that this is a transient effect.

It is seen in the simulations without diffusion that there is no ultimate extinction ($c(\infty) = 0$) for Model 2 ($\mu > 0$). However, at any specific time point, $Pr[c(t) > 0]$ is higher in Model 1, than in Model 2, except for a subset of initial conditions. In other words, the effect of external supply of free growth factor is stronger than that of supply of cells. This is confirmed in the simulations with diffusion.

5.2 Cooperativity Condition and Communication of Cancer Cells

One of the interesting features of the model is that DDI and spike formation require the apparently nonphysical assumption of cooperativity, $\alpha(c) = \alpha_1 c^{1+s}$, where $s > 0$, i.e., that the process of binding free growth factor particles to cells is super-linear. However, this assumption becomes more plausible, if we accept that cells may communicate with each other. In other words, cells accepting free growth factor particles transmit this behavior to their neighbors. The biological mechanism may, speculatively, involve increasing the number of receptors on cell membrane. Information sharing of this type has been known for a long time in radiation effect research, where it is called the bystander effect [6]. Under different exposure scenarios, irradiated cells can communicate with each other or with nonirradiated cells. As a result, the "bystander" cells display responses similar to those of irradiated cells.

Interestingly, the role of communication among cancer cells has been recently emphasized in a model that compares the process of tumor progression to the dynamics of organized microbial communities [3]. Tumor is viewed as a community of interacting cells, and the first stage of metastasis involves outward cell migration in the form of path-finder and path-generator cell types. Even more complex behavior is proposed. In view of this, the cooperativity hypothesis, which is sufficient for cell growth to become organized in discrete foci, represented by cell density spikes in our model, seems justified.

Literature describing tumor invasion along linear structures includes [29] who used advanced microscopy to determine the organization of specific collagen structures around ducts and tumors in mammary glands. They observed and defined three tumor associated collagen signatures (TACS) that provide novel markers to locate and characterize tumors. In particular, local cell invasion was found predominantly to be oriented along certain aligned collagen fibers, suggesting that radial alignment of collagen fibers relative to tumors facilitates invasion. Consistent with this observation, primary tumor explants cultured in a randomly organized collagen matrix realigned the collagen fibers, allowing individual tumor cells to migrate out along radially aligned fibers. This is one more example of tumor invasion along linear structures.

5.3 *More Models*

Spike solutions in the reaction-diffusion systems are not unknown. Iron et al. (2001) and then among other Kolokolnikov et al. [18] established stability conditions for the spikes in the solutions of the one-dimensional reaction-diffusion systems with two diffusions describing the Gierer-Meinhardt activator-inhibitor model of Hydra. In particular, [18] constructed three types of solution: (i) an interior spike; (ii) a boundary spike; and (iii) two boundary spikes. They found that an interior spike is always unstable; a boundary spike is always stable. The two boundary spike configuration can be either stable or unstable. They carried out numerical simulations consistent with the theoretical results.

Foo et al. [10] recently developed a spatiotemporal stochastic model of epithelial carcinogenesis, related to the spatial Moran model, combining cellular reproduction and death dynamics with genetic progression. They investigate how the size and geometry of premalignant fields depend on tissue renewal rate, mutation rate, selection, and so forth. They also investigate the time to emergence of multiple primary tumors. This approach uses the apparatus of contact processes such as voter models but also interesting simulations. In contrast to the stochastic model of Bertolusso and Kimmel, Foo et al. [10] do not seem to model explicitly growth factor diffusion. In many other respects their model seems more detailed.

The microscopic (non-mean field) model pursued in the present paper is different from the microscopic deterministic model of [24] who used homogenization techniques. In general, the relationship between averaging of stochastic trajectories and homogenization of microscopic deterministic models remains unclear.

References

1. H. Awaya, Y. Takeshima, O. Furonaka, N. Kohno, K. Inai, Gene amplification and protein expression of EGFR and HER2 by chromogenic in situ hybridisation and immunohistochemistry in atypical adenomatous hyperplasia and adenocarcinoma of the lung. J. Clin. Pathol. **58**(10), 1076–1080 (2005)
2. R. Axelrod, D.E. Axelrod, K.J. Pienta, Evolution of cooperation among tumor cells. Proc. Natl. Acad. Sci. **103**(36), 13474–13479 (2006)
3. E. Ben-Jacob, D.S. Coffey, H. Levine, Bacterial survival strategies suggest rethinking cancer cooperativity. Trends Microbiol. **20**(9), 403–410 (2012)
4. R. Bertolusso, Computational models of signaling processes in cells with applications: influence of stochastic and spatial effects. Ph.D. thesis, Rice University, Houston, TX, 2011
5. R. Bertolusso, M. Kimmel, Modeling spatial effects in early carcinogenesis: stochastic versus deterministic reaction-diffusion systems. Math. Model. Nat. Phenom. **7**(01), 245–260 (2012)

6. B.J. Blyth, P.J. Sykes, Radiation-induced bystander effects: what are they, and how relevant are they to human radiation exposures? Rad. Res. **176**(2), 139–157 (2011)
7. B.J.M. Braakhuis, M.P. Tabor, J.A. Kummer, C.R. Leemans, R.H. Brakenhoff, A genetic explanation of Slaughter's concept of field cancerization evidence and clinical implications. Cancer Res. **63**(8), 1727–1730 (2003)
8. J. Califano, P. van der Riet, W. Westra, H. Nawroz, G.Clayman, S. Piantadosi, R. Corio et al. Genetic progression model for head and neck cancer: implications for field cancerization. Cancer Res. **56**(11), 2488–2492 (1996)
9. R. Erban, J. Chapman, P. Maini, A practical guide to stochastic simulations of reaction-diffusion processes. arXiv preprint arXiv:0704.1908 (2007)
10. J. Foo, K. Leder, M. D. Ryser, Multifocality and recurrence risk: a quantitative model of field cancerization. J. Theor. Biol. **355**, 170–184 (2014)
11. T. Hillen, A classifycation of spikes and plateaus. SIAM Rev. **49**(1), 35–51 (2007)
12. B. Hu, E. Castillo, L. Harewood, P. Ostano, A. Reymond, R. Dummer, W. Raffoul, W. Hoetzenecker, G.F.L. Hofbauer, G.P. Dotto, Multifocal epithelial tumors and field cancerization from loss of mesenchymal CSL signaling. Cell **149**(6), 1207–1220 (2012)
13. S.A. Isaacson, C.S. Peskin, Incorporating diffusion in complex geometries into stochastic chemical kinetics simulations. SIAM J. Sci. Comput. **28**(1), 47–74 (2006)
14. T. Kawai, S. Hiroi, K. Nakanishi, A.K. Meeker, Telomere length and telomerase expression in atypical adenomatous hyperplasia and small bronchioloalveolar carcinoma of the lung. Am. J. Clin. Pathol. **127**(2), 254–262 (2007)
15. K.M. Kerr, Pulmonary preinvasive neoplasia. J. Clin. Pathol. **54**(4), 257–271 (2001)
16. K. Kishi, S. Homma, A. Kurosaki, S. Tanaka, H. Matsushita, K. Nakata, Multiple atypical adenomatous hyperplasia with synchronous multiple primary bronchioloalveolar carcinomas. Int. Med. **41**(6), 474–477 (2002)
17. H. Kitagawa, A. Goto, T. Niki, M. Hironaka, J. Nakajima, M. Fukayama, Lung adenocarcinoma associated with atypical adenomatous hyperplasia. A clinicopathological study with special reference to smoking and cancer multiplicity. Pathol. Int. **53**(12), 823–827 (2003)
18. T. Kolokolnikov, J. Wei, M. Winter, Existence and stability analysis of spiky solutions for the Gierer-Meinhardt system with large reaction rates. Phys. D Nonlinear Phenom. **238**(16), 1695–1710 (2009)
19. M. Lipkin, Phase 1 and phase 2 proliferative lesions of colonic epithelial cells in diseases leading to colonic cancer. Cancer **34**(S3), 878–888 (1974)
20. J.P. Lockhart-Mummery, C. Dukes, The precancerous changes in the rectum and colon. Surg. Gynecol. Obstet. **46**(5), 591–596 (1928)
21. A. Marciniak-Czochra, M. Kimmel, Dynamics of growth and signaling along linear and surface structures in very early tumors. Comput. Math. Meth. Med. **7**(2–3), 189–213 (2006)
22. A. Marciniak-Czochra, M. Kimmel, Modelling of early lung cancer progression: influence of growth factor production and cooperation between partially transformed cells. Math. Model Meth. Appl. Sci. **17**(Suppl 01), 1693–1719 (2007)
23. A. Marciniak-Czochra, M. Kimmel, Reaction-difusion model of early carcinogenesis: the effects of influx of mutated cells. Math. Model. Nat. Phenom. **3**(07), 90–114 (2008)
24. A. Marciniak-Czochra, M. Ptashnyk, Derivation of a macroscopic receptor-based model using homogenization techniques. SIAM J. Math. Anal. **40**(1), 215–237 (2008)
25. A. Marciniak-Czochra, G. Karch, K. Suzuki, Unstable patterns in reaction-diffusion model of early carcinogenesis. J. Math. Pures Appl. **99**(5), 509–543 (2013)
26. A. Marciniak-Czochra, S. Härting, G. Karch, K. Suzuki, Dynamical spike solutions in a nonlocal model of pattern formation. Arxiv:1307.6236 (2013, preprint)
27. L. Morandi, S. Asioli, A. Cavazza, A. Pession, S. Damiani, Genetic relationship among atypical adenomatous hyperplasia, bronchioloalveolar carcinoma and adenocarcinoma of the lung. Lung Cancer **56**(1), 35–42 (2007)
28. J. Paulsson, O.G. Berg, M. Ehrenberg, Stochastic focusing: fluctuation-enhanced sensitivity of intracellular regulation. Proc. Natl. Acad. Sci. **97**(13), 7148–7153 (2000)

29. P.P. Provenzano, K.W. Eliceiri, J.M. Campbell, D.R. Inman, J.G. White, P.J. Keely, Collagen reorganization at the tumor-stromal interface facilitates local invasion. BMC Med. **4**(1), 1–15 (2006)
30. A.L. Rubin, Suppression of transformation by and growth adaptation to low concentrations of glutamine in NIH-3T3 cells. Cancer Res. **50**(9), 2832–2839 (1990)
31. H. Rubin, Fields and field cancerization: the preneoplastic origins of cancer. Bioessays **33**(3), 224–231 (2011)
32. T. Sano, Y. Kitayama, H. Igarashi, M. Suzuki, F. Tanioka, K. Chida, K. Okudela, H. Sugimura, Chromosomal numerical abnormalities in early stage lung adenocarcinoma. Pathol. Int. **56**(3), 117–125 (2006)
33. S. Seton-Rogers, Tumour microenvironment: field effect. Nat. Rev. Cancer **12**(8), 508–509 (2012)
34. D.P. Slaughter, H.W. Southwick, W. Smejkal, "Field cancerization" in oral stratified squamous epithelium. Cancer **6**, 963–968 (1953)
35. A. Slepoy, A.P. Thompson, S.J. Plimpton, A constant-time kinetic Monte Carlo algorithm for simulation of large biochemical reaction networks. J. Chem. Phys. **128**, 205101 (2008)
36. D.L. Stoler, N. Chen, M. Basik, M.S. Kahlenberg, M.A. Rodriguez-Bigas, N.J. Petrelli, G.R. Anderson, The onset and extent of genomic instability in sporadic colorectal tumor progression. Proc. Natl. Acad. Sci. **96**(26), 15121–15126 (1999)
37. K. Takamochi, T. Ogura, K. Suzuki, H. Kawasaki, Y. Kurashima, T. Yokose, A. Ochiai, K. Nagai, Y. Nishiwaki, H. Esumi, Loss of heterozygosity on chromosomes 9q and 16p in atypical adenomatous hyperplasia concomitant with adenocarcinoma of the lung. Am. J. Pathol. **159**(5), 1941–1948 (2001)
38. J.-L. Tsao, Y. Yatabe, R. Salovaara, H.J. Järvinen, J.-P. Mecklin, L.A. Aaltonen, S. Tavaré, D. Shibata, Genetic reconstruction of individual colorectal tumor histories. Proc. Natl. Acad. Sci. **97**(3), 1236–1241 (2000)

Conservation Laws in Cancer Modeling

Antonio Fasano, Alessandro Bertuzzi, and Carmela Sinisgalli

Abstract We review mathematical models of tumor growth based on conservation laws in the full system of cells and interstitial liquid. First we deal with tumor cords evolving in axisymmetric geometry, where cells motion is simply passive and compatible with the saturation condition. The model is characterized by the presence of free boundaries with constraints driving the free boundary conditions, which in our opinion are particularly important, especially in the presence of treatments. Then a tumor spheroid is considered in the framework of the so-called two-fluid scheme. In a multicellular spheroid, on the appearance of a fully degraded necrotic core, the analysis of mechanical stresses becomes necessary to determine the motion via momentum balance, requiring the specification of the constitutive law for the "cell fluid." We have chosen a Bingham-type law that presents considerable difficulties because of the presence of a yield stress, particularly with reference to the determination of an asymptotic configuration. Finally, we report some recent PDE-based models addressing complex processes in multicomponent tumors, more oriented to clinical practice.

Keywords Conservation laws • Cancer modeling • Tumor cords • Tumor spheroids • Multicomponent tumors

A. Fasano (✉)
Dipartimento di Matematica "U. Dini", Universita' di Firenze,
Viale Morgagni 67/A, 50134 Firenze, Italy

Istituto di Analisi dei Sistemi ed Informatica "A. Ruberti" - CNR,
Viale Manzoni 30, 00185 Roma, Italy
e-mail: fasano@math.unifi.it

A. Bertuzzi • C. Sinisgalli
Istituto di Analisi dei Sistemi ed Informatica "A. Ruberti" - CNR,
Viale Manzoni 30, 00185 Roma, Italy
e-mail: bertuzzi@iasi.cnr.it; carmela.sinisgalli@iasi.cnr.it

© Springer Science+Business Media New York 2014

A. d'Onofrio, A. Gandolfi (eds.), *Mathematical Oncology 2013*,
Modeling and Simulation in Science, Engineering and Technology,
DOI 10.1007/978-1-4939-0458-7_2

1 Introduction

The literature on cancer modeling has been rapidly increasing during the present century, paralleling the remarkable intensification and diversification of the research in this field. The last International Conference in Industrial and Applied Mathematics (ICIAM 2011, Vancouver) hosted a surprisingly large number of talks on that subject, emphasizing many new and important areas of investigation, including the rather new subject of the role of cancer stem cells (see, e.g., [42]). Among the recent review papers we quote [4, 7, 25, 26, 34, 51].

A feature that has been treated differently in many growth models is the one of conservation laws, accompanying the choice of the phenomena to be included in the model, such as cell displacement mechanisms (whether totally passive or with a chemotactic or haptotactic component), drug actions, angiogenesis, and so on. One of the main issues is mass conservation, which in several instances has been disregarded with the aim of producing a treatable model. Raising a far too obvious criticism may be simply not constructive, because the target of simplifying a subject whose nature is so tremendously complicated, trying to preserve the basics of biological behavior, has often proved to be useful and has to be rather assessed on the basis of the results. Take for instance the paper [35] on the acid-mediated invasion of healthy tissue by tumor cells, where only three species are present: tumor cells, normal cells, and the H^+ ions produced by tumor cells and attacking the other species. The model does not specify how the acidity is produced (thus glucose metabolism is completely absent), it does not consider any interstitial fluid carrying nutrients, and not even oxygen consumption. Nevertheless it reproduces at least qualitatively the main biological phenomenon, emphasizing the presence of a gap between the advancing and the receding species, represented in a one-dimensional geometry by travelling waves, for appropriate values of the parameters. The richness of that oversimplified model has been further clarified in the paper [27]. A nontrivial extension has been presented very recently in [36]. In our opinion this is a remarkable example of how effective a lean model can be, provided it is constructed assembling the essential elements. Without entering the elegant and appealing subject of travelling waves in tumors, we quote the recent paper [56] in which a three-species model (tumor, normal, and dead cells) described by a treatable system of PDEs describes the spread of an aggressive glioma in the form of a diffusion-dominated spherical expanding wave.

"Completeness" remains of course a legitimate aspiration, worth to be pursued with some caution. The quotation marks allude to the extremely hard task not only of putting all the relevant ingredients but then of specifying how they mutually interact. Such an attempt necessarily calls for choices, which in most cases have some degree of arbitrariness: cell–cell and cell–matrix interactions, active or passive cell displacement, cytoskeleton and membrane mechanics, cells electrochemistry, signaling, cell metabolism (aerobic or anaerobic), proliferation, death and degradation, mutations, interstitial fluids, angiogenesis (including vessels sprouting, leaking and occlusion), metastatic processes, dormancy, etc. For each element the model formulation requires the selection of constitutive laws, containing parameters of various nature, whose numerical value has to be provided, at least in a reasonable range. In some cases the latter task may be prohibitive. When we introduce the cell-killing action of drugs, we are faced with the problem of transport through vasculature and through the interstitial liquids, with the interaction with the cell membrane and with the various cell components. Not to speak of the mechanics of the tumor mass as a whole and its modification following treatments it is, we believe, sufficiently clear that completeness is an objective which makes sense only leaving an exceedingly complex reality and restricting the area to specific aspects. What is reasonable is trying to attain some degree of accuracy focusing on the most relevant elements in the particular process considered. For sure in this framework conservation laws are invariably the backbone of any mathematical schematization.

In this paper we want to go through the literature of the last decade or so taking conservation laws as the leading subject, trying to emphasize the diversity of various approaches, both in the modeling and in the scopes.

A striking example is the case we are going to deal with in Sect. 2, the so-called tumor cords, for which we summarize a two-fluid model keeping the mechanics at a very simple level but emphasizing some nontrivial mathematical aspects arising in the presence of massive cell death due to treatments. In our opinion this crucial feature has not received enough attention in the literature.

The other class of problems in cancer modeling with a simple geometry is the one of multicellular spheroids, which, in the framework of continuum mechanics, are treated as spheres with all quantities depending just on time and on the radial coordinate. This kind of symmetry is even more treatable than the not-so-nice axisymmetric geometry of tumor cords, since for instance all fluxes are just radial, though it does not always allow to skip the analysis of the stress field. The literature is huge and deals with many fundamental subjects. In the recent review paper [24] the important question of incorporating glucose metabolism has been discussed at length, together with the noticeable consequence of pH decrease. Therefore we will not insist here on such a question. Instead in Sect. 3 we analyze a refinement of the two-fluid scheme in which the "cell fluid" exhibits a yield stress, i.e., a Bingham fluid. This approach, first presented in [2] and further developed in [57], has been studied extensively in [15]. Occasionally we will point out some basic differences accompanying the switch from axisymmetric geometry (cords) to spherical symmetry, which for instance deeply influences the structure of the necrotic region. In both cases it turns out that the mechanical behavior of the necrotic region has a crucial influence on the evolution of the tumor, a fact that deserves to be greatly emphasized and that has been discussed in the review paper [25].

The first two sections, based on our work, are mainly concentrated on mass and momentum balance and its implications on the mechanical behavior of systems possessing an idealized geometry. The models considered there incorporate the continuous mass exchange between cells and interstitial fluid. In such a framework it is possible to analyze the role of mechanics in full detail and to carry out a complete investigation of the mathematical structure of the problem. The reader may object that, in spite of its mathematical complexity, such an approach has a rather limited target in the much wider horizon of malignancies and of their treatments. Indeed, great efforts are being made to produce mathematical models for the growth of specific tumors and related therapies in the presence of concurrent phenomena. For this reason in the last section we report some recent researches, taken from the literature, addressing complex processes in multicomponent tumors, confining our attention to the class of models expressed through systems of partial differential equations. Though obviously not exhaustive, the choice is certainly quite representative of the positive trend to get modeling closer to clinical practice.

2 Tumor Cords and the Doubly Constrained Boundary Conditions

In the three papers [8–10] some peculiar features that may occur in cancer growth have been pointed out for the first time in the framework of the microstructures, mainly observed in experimental tumors, where tumor cells proliferating around a blood vessel form an approximately axisymmetric aggregate called tumor cord. Generally tumor cords have different sizes and orientation and may or may not be surrounded by necrosis [44, 53, 65]. Despite its complexity, the model described in [8, 9] was based on a rather naïve description of cell metabolism, since oxygen was the only "nutrient" considered. A novel feature at that time was instead the analysis of the flow of the interstitial fluid from the central vessel to the cord periphery [10]. The introduction of free boundaries, such as the sharp interface between viable and necrotic tissue, led to the formulation of boundary conditions depending on the system evolution and regulated by suitable constraints. A fundamental simplification was provided by the assumption that the cords are arranged in a regular array of parallel identical elements, so that, because of symmetry, the cords are separated by no-flux boundaries forming a bee nest structure. As a consequence, an individual cord can be studied, approximating the hexagonal boundary with a cylinder, so that the whole system is axisymmetric (Fig. 1).

Though computationally complicated, the partition of the cord (Fig. 2) by means of cylindrical interfaces separating homogeneous species has the advantage of facilitating the calculation of the cell and fluid velocity fields. In the quoted papers the analysis of well posedness was fully performed, along with the study of important qualitative properties. Concerning our principal theme, namely conservation laws, *the interstitial liquid plays the basic role of allowing the fulfillment of mass balance*. Indeed, the fluid provides the material for the production of new cells and receives the material released by the degradation of dead cells in the necrotic region. The cell-liquid mass exchange looks quite natural, but it requires the analysis of fluid motion, which is not trivial. As a matter of fact, often in the literature the shortcut is taken by simply ignoring the liquid. For example, in the early age of cancer modeling, Greenspan [38] introduced some volume loss rate after necrosis and his choice was adopted by many authors, even recently, though the actual removal mechanism has never been specified. Such a particular aspect in cancer modeling has been extensively discussed in [25] with special attention to the consequences for the attainment of a steady state. It has to be said that, however strange it

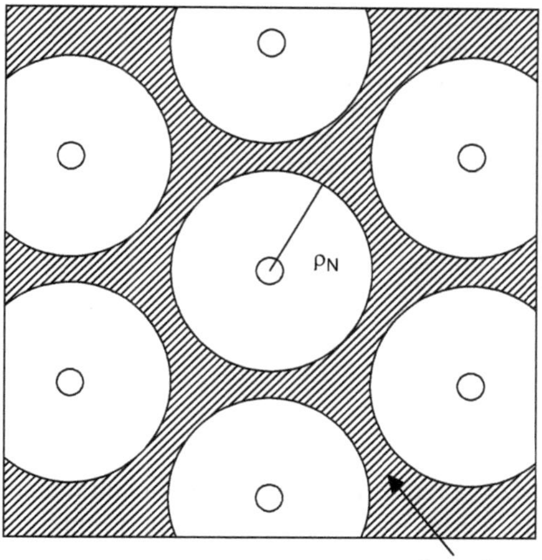

Fig. 1 Sketch of the cross section of a regular array of tumor cords exhibiting the presence of a necrotic region (at distance ρ_N from the cord axis). *Small circles* are cross sections of vessels. Each cord is surrounded by a no-flux surface, approximated by a *cylinder*

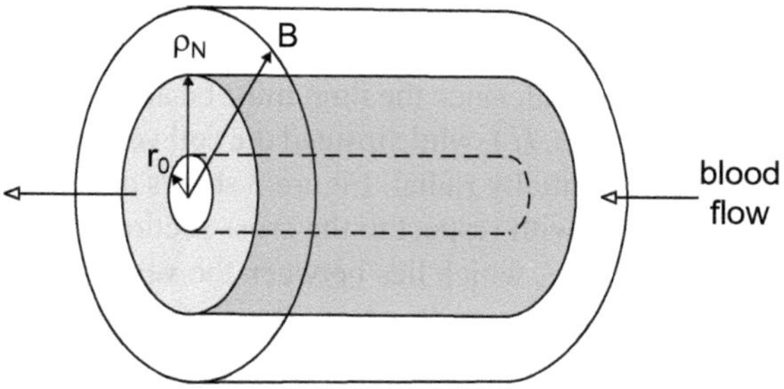

Fig. 2 Sketch of a single cord, externally bounded by the no flux surface $r = B$ and showing the viable region (in *grey*)

may look, models not including a basic element like the interstitial fluid can be meaningful if, as we said in the introduction, such a simplification is introduced in a suitable way, so as to capture anyway some essential phenomena. Moreover, in the scheme below, the cell motion is completely passive, generated by proliferation, thus disregarding the autonomous motility (haptotaxis, chemotaxis) as well as the random motility (diffusion). Nevertheless, for the case of cords, and even more for the case of multicellular spheroids that will be considered later, neglecting such causes of motion looks reasonable.

Still considering the kinematic aspect, we note that once the tumor has spread longitudinally along the vessel, the average motion of cells takes place in the radial direction. The question of tumor progression along the vessel has been considered in [5] in a different framework (cord expansion against a host tissue).

Coming back to the system sketched in Fig. 2, we treat the cell-liquid system as a mixture, adopting the scheme of continuum mechanics. Clearly, this is not the only possible choice. Discrete or hybrid models have been used extensively and with some success (see, e.g., [1, 43]). Continuum models are justified in the presence of a sufficiently large number of cells. It could be objected that for the typical cord

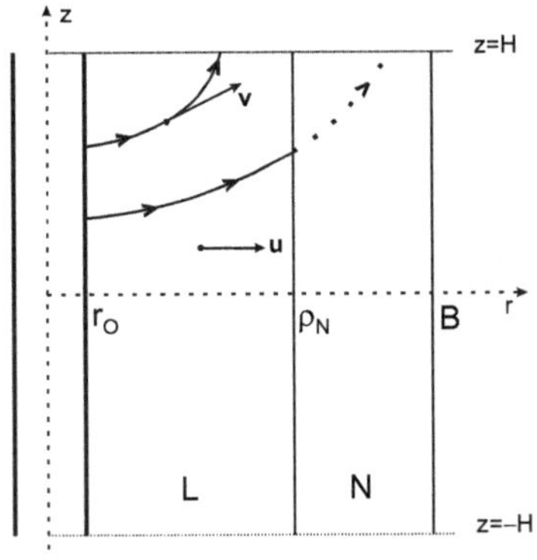

Fig. 3 Sketch of velocity fields: cellular velocity **u**; radial, fluid velocity **v**. The liquid enters the cord crossing the wall of the axial blood vessel. Viable and necrotic regions are indicated by L and N, respectively

size (a radius of the order of ten cell diameters) the latter requirement is not met. However, we must consider that a representative volume element is a cylindrical shell of radius smaller than a cell diameter, extending over the whole length of the vessel (up to 1 mm), actually intersecting a large number of cells, so that the continuum approach ultimately makes sense.

A glance to Fig. 3 clarifies why the liquid velocity field can be taken axisymmetric, but not radial, since the fluid must be allowed to leave the cord from the extreme sections ($z = \pm H$), while instead the cell velocity field (in the continuum sense) can be taken essentially radial. Figure 3 shows also the spatial coordinates. The system is symmetric with respect to the cross section $z = 0$. Living cells occupy the region L: $r_0 < r < \rho_N$, which lies between the vessel and the necrotic region N, and when developed: $\rho_N < r < B$, where B is the outer radius of the cord. Living cells can be either proliferating (volume fraction ν_P)[1] or quiescent (volume fraction ν_Q). Dead cells are actually disseminated in the cord (their volume fraction is denoted by ν_A), but the region N is exclusively necrotic. We consider three possible causes of cell death:

(i) extreme hypoxia, occurring when the oxygen concentration σ does not exceed a viability threshold σ_N,
(ii) the action of a cytotoxic drug (or radiation), taking place with a known kinetics, and
(iii) apoptosis, also regulated by some kinetics.

As a consequence the region N is normally characterized by the inequality $\sigma \leq \sigma_N$, with the exception that will be clarified later.

A basic hypothesis is that the medium is saturated, so that the volume fraction of the extracellular liquid ν_E is complementary to the total volume fraction of the

[1]Some authors adopt the extreme view point that proliferation takes place only at the tumor surface because of contact inhibition (e.g., [17]) and then migrate, driven by the surface curvature. Here we stick to the experimental observation that in the tumors we are talking about proliferation occurs in the tumor mass, whenever enough oxygen is available.

cells, $\nu = \nu_P + \nu_Q + \nu_A = 1 - \nu_E$. All the components are assumed to have the same density. Another fundamental assumption is that ν_E is constant.

One of the most difficult aspects in modeling a growing mass is the mechanical behavior. Mixture theory (see, e.g., [58]) seems a perfect tool, since we are dealing with a multicomponent system, but, independently of the constitutive law attached to each component, the presence of mass conversion processes is a substantial difficulty. Several papers have addressed this issue in various ways. One of the first papers adopting mixture theory was [19]. In the recent paper [2] a very complex analysis of the mechanics of the cellular component has been performed justifying the presence of a yield stress and therefore of a Bingham-like constitutive law.

We will return to such a question in the next section. Here we are pursuing the aim of keeping the mechanics as simple as possible. For this reason we perform some operations that ultimately will circumvent the dynamical problem almost completely, exploiting symmetry to bypass the necessity of writing down the momentum balance equations. First of all, we adopt the so-called two-fluid approach [18], assimilating the cell aggregate to a Newtonian fluid and considering the interstitial fluid as inviscid. There are several conceptual limitations in the two-fluid scheme that have been illustrated in [25], but the advantage of dealing with clearly well-defined constituents, with the possibility of making suitable assumptions on their mutual interaction, makes it extremely valuable. As we pointed out in many instances, modeling cancer growth is a compromise between accuracy and simplicity, so that the goal is to keep as much as possible of the biology, limiting at the same time the number of physical parameters involved. Of course the hypothesis attributing no viscosity to the interstitial fluid is justified by the fact that its viscosity (comparable to the one of water) is by many orders of magnitude smaller than the viscosity of the "cellular fluid." Nevertheless, the interaction with the cells is taken into account assuming that the flow relative to the cells obeys Darcy's law:

$$\nu_E(\mathbf{v} - \mathbf{u}) = -\kappa \nabla \hat{p}\,, \tag{1}$$

where κ is the hydraulic conductivity of the cell aggregate experienced by the liquid and $\hat{p}$ is the liquid pressure.

The way of bypassing a finer description of the mechanics of the system consists in averaging two quantities in the longitudinal direction, namely the radial component of the liquid velocity $v_r(r,z,t)$ and pressure $\hat{p}(r,z,t)$, thus defining

$$v(r,t) = \frac{1}{2H}\int_{-H}^{H} v_r(r,z,t)\,dz\,, \tag{2}$$

$$p(r,t) = \frac{1}{2H}\int_{-H}^{H} \hat{p}(r,z,t)\,dz\,. \tag{3}$$

The first quantity, multiplied by $2\pi rH$, provides the total radial discharge through a surface coaxial with the central vessel. The mass balance of the components in the region L is expressed by the system

$$\frac{\partial \nu_P}{\partial t} + \frac{1}{r}\frac{\partial}{\partial r}(ru\nu_P) = \chi\nu_P + \gamma\nu_Q - \lambda\nu_P - \mu_P\nu_P\,,$$
$$\frac{\partial \nu_Q}{\partial t} + \frac{1}{r}\frac{\partial}{\partial r}(ru\nu_Q) = -\gamma\nu_Q + \lambda\nu_P - \mu_Q\nu_Q\,,$$
$$\frac{\partial \nu_A}{\partial t} + \frac{1}{r}\frac{\partial}{\partial r}(ru\nu_A) = \mu_P\nu_P + \mu_Q\nu_Q - \mu_A\nu_A\,,$$
$$\nu_E\nabla\cdot\mathbf{v} = \mu_A\nu_A - \chi\nu_P\,, \tag{4}$$

which, together with the saturation assumption, yields the global mass balance

$$\nabla\cdot[\nu\mathbf{u} + \nu_E\mathbf{v}] = 0\,. \tag{5}$$

In (4) χ is the proliferation rate, and the coefficients γ, λ are the transition rates from the class Q (quiescent cells) to P (proliferating cells) and vice versa. They are functions of the oxygen concentration σ (γ is nondecreasing and λ nonincreasing). μ_P, μ_Q are death rates in the respective classes possibly depending on the concentration of a cytotoxic drug; μ_A is the mass conversion rate of apoptotic cells (class A) to liquid, due to degradation. If we take the longitudinal average of the longitudinal component of (5), exploiting the assumption that $\mathbf{u}$ is $u(r,t)$ times the outward directed radial unit vector, we get an equation containing the difference $v_z(r,H,t) - v_z(r,-H,t)$, whose product with ν_E gives the local liquid efflux rate from the cord, for which we assume that

$$\nu_E[v_z(r,H,t) - v_z(r,-H,t)] = 2\zeta_{out}(p(r,t) - p_\infty)\,. \tag{6}$$

In other words, the liquid loss rate is regulated by the pressure excess with respect to a far-field pressure p_∞, established by the lymphatic system. The coefficient ζ_{out} may in principle depend on r, but is taken constant for simplicity. The procedure just described leads to the equation

$$\frac{1}{r}\frac{\partial}{\partial r}(rv) = -\frac{1}{\nu_E}\Big(\chi\nu_P - \mu_A\nu_A + \frac{\zeta_{out}}{H}(p - p_\infty)\Big)\,. \tag{7}$$

The longitudinal average of the radial component of (5) leads finally to express the pressure in terms of the relative liquid-cell velocity:

$$p(r,t) = p_0(t) - \frac{1-\nu}{\kappa}\int_{r_0}^{r}[v(r',t) - u(r',t)]dr'\,. \tag{8}$$

An important feature of the model is the appearance of the new unknown $p_0(t)$, i.e., the pressure at the blood vessel wall. Indeed, equation (7) requires the inlet boundary condition

$$(1-\nu)v(r_0,t) = \zeta_{in}(p_b - p_0(t)), \tag{9}$$

meaning that the liquid inflow rate from the blood vessel is proportional to the pressure jump across the vessel wall.[2] The coefficient ζ_{in} is a positive constant, and p_b is the blood pressure in the specific vessel considered. It has to be said that actually blood pressure decreases along the flow, so that p_b is a function of z (averaging out time dependence, since the time scale of heart pulsation is much smaller than the scale of tumor evolution). The feasibility of taking constant both p_b and, later on, the oxygen concentration in blood σ_b along the cord has been discussed in the paper [10], concluding that for vessels shorter than a millimeter (the case of capillaries in a vascular tumor) the approximation is compatible with the many other sources of error included in the scheme. The equation governing the cell velocity field can be derived in the form of a mass balance by summing the first three equations in (4):

$$\frac{1}{r}\frac{\partial}{\partial r}(ru) = \frac{1}{\nu}(\chi\nu_P - \mu_A\nu_A)\,, \tag{10}$$

to which the boundary condition

$$u(r_0,t) = 0 \tag{11}$$

must be associated.

Concerning oxygen, because of its large diffusivity ($D_{O_2} \simeq 10^{-5}\,\mathrm{cm}^2/\mathrm{s}$) and easy penetration through cell membrane, at each time its concentration is assumed to be at the equilibrium profile satisfying the equation

$$D_{O_2}\,\Delta\sigma = f_P(\sigma)\nu_P + f_Q(\sigma)\nu_Q\,, \tag{12}$$

where the consumption rates $f_P \geq f_Q$ are of Michaelis-Menten type.

Oxygen concentration at the blood vessel wall is taken constant (see [10] for a justification):

$$\sigma(r_0,t) = \sigma_b\,. \tag{13}$$

---

[2]Here we neglect osmotic pressure.

The conditions for σ at the necrotic interface (when present) are far less obvious. Considering the role of the threshold σ_N, the natural conditions for $r = \rho_N$ would be

$$\sigma(\rho_N, t) = \sigma_N \tag{14}$$

$$\left.\frac{\partial \sigma}{\partial r}\right|_{r=\rho_N} = 0\,. \tag{15}$$

The second condition is a consequence of the absence of consumption in the necrotic core and it implies that the oxygen profile in the N zone is flat. However, a model accounting for treatments must be able to describe a sudden rise in oxygen concentration due to massive cell death. When such a phenomenon takes place (an increase in σ_b would have a similar effect), the necrotic interface cannot always be identified as the level set by (14), for the simple reason that if σ rises above threshold at the necrotic interface, dead cells will not return to life. In that case σ must be left free to evolve, while the necrotic interface becomes a material surface, moving with the velocity of the viable cells that stay on it:

$$u(\rho_N(t), t) = \dot{\rho}_N(t)\,. \tag{16}$$

The state described by (16) lasts as long as $\sigma(\rho_N, t)$ remains above threshold. When the threshold is recovered, the cells resume entering the necrotic region. In summary, we conclude that:

(i) the following unilateral constraints have to be satisfied at each time instant:

$$\sigma(\rho_N, t) \geq \sigma_N \tag{17}$$

$$u(\rho_N(t), t) \geq \dot{\rho}_N(t)\,; \tag{18}$$

(ii) when one constraint is satisfied in the strict sense, the other has to hold as an equality;
(iii) equation (15) always holds true.

This very peculiar structure of the boundary conditions at the necrotic interface makes the mathematics considerably difficult.

Unfortunately, when we come to modeling the necrotic region we find more complications, because we realize that the necrotic region can be "prevalently solid" or "prevalently liquid." In the first case the degrading cells are in mutual contact and can bear an external stress, while in the opposite case the task of sustaining external stress is given to the liquid. Thus we have two regimes that we call N-solid and N-liquid, respectively. The basic time-dependent quantities to be considered, besides the interface $r = \rho_N(t)$, are

- The cord outer boundary $r = B(t)$
- The volume occupied by the degrading cell $V_N^c(t)$

- The volume occupied by the liquid $V_N^l(t)$
- The liquid pressure $p_N(t)$

The total volume $V_N(t)$ of the necrotic region is

$$V_N(t) = V_N^c(t) + V_N^l(t) = 2\pi H(B^2 - \rho_N^2). \tag{19}$$

The two partial volumes evolve according to the equations

$$\dot{V}_N^c = 4H\pi\rho_N(1-\nu_E)[u(\rho_N,t) - \dot{\rho}_N] - \mu_N V_N^c, \tag{20}$$

$$\dot{V}_N^l = 4H\pi\rho_N\nu_E[v(\rho_N,t) - \dot{\rho}_N] + \mu_N V_N^c - q_{out}(t). \tag{21}$$

In (20) the first term on the RHS is the contribution of cells entering the necrotic region (always nonnegative, as we know); the second term is the rate of conversion into liquid. In the second equation the only term to be explained is the last one:

$$q_{out} = \frac{\zeta_{out}^N}{H} V_N^l(p_N - p_\infty), \tag{22}$$

namely the liquid efflux rate from the cord ends $z = \pm H$, where ζ_{out}^N is a positive coefficient.

The solid volume fraction is subject to the constraint

$$\frac{V_N^c}{V_N} \le \nu_N < 1, \tag{23}$$

(for the sake of generality ν_N is distinguished from ν, but in practice they can be taken equal). Now, in the N-solid regime, the above constraint is at work, so we can take

$$V_N^c = \nu_N V_N \quad \Leftrightarrow \quad V_N^l = (1-\nu_N)V_N, \tag{24}$$

where $V_N(t)$ has the expression (19). Thus (20) yields a differential equation for the difference $B^2 - \rho_N^2$:

$$\frac{d}{dt}(B^2 - \rho_N^2) = 2\frac{1-\nu_E}{\nu_N}\rho_N[u(\rho_N,t) - \dot{\rho}_N] - \mu_N(B^2 - \rho_N^2), \tag{25}$$

and operating with (21) it is possible to express q_{out} in terms of geometrical and kinematical unknowns, which in turn allows to derive $p_N(t)$.

Let us now discuss the question of how to detect the transition to the N-liquid regime and vice versa. The key point is to monitor the pressure $p_N(t)$, comparing it with the pressure exerted on the cord by the surroundings, i.e., by the neighboring cords. Such an external pressure is due to the reaction of the host tissue to the expansion of the cord cohort and therefore is ultimately related with the size of the individual cords, that is, with their equivalent radius B. We denote it by $\Psi(B)$, a continuous increasing function. Clearly, the N-solid regime is characterized by the fact that this pressure is fully sustained by the solid component, while $p_N(t) < \Psi(B(t))$. Therefore the N-liquid regime sets in when

$$p_N(t) = \Psi(B(t)), \tag{26}$$

which replaces (24), no longer valid as an equality but as the inequality $V_N^c < \nu_N V_N$. When (26) is enforced, $V_N^c(t)$ evolves according to (20) from which it can be deduced as a functional of $\rho_N(t)$ and of $u(t)$.

In conclusion, we are back to a doubly constrained problem:

$$V_N^c \leq \nu_N V_N, \quad p_N \leq \Psi(B), \tag{27}$$

with at least one of the two constraint being active.

We remark that the fact that the boundary conditions are actually selected by the constraints, which come into play depending on the evolution of the system, has deep consequences on the mathematical structure. In particular, one has to be aware that not necessarily the various types of boundary conditions alternate over finite time intervals, since it cannot be excluded a priori that there are accumulation points of switching times. This circumstance requires the adoption of particular techniques in the existence proof (see the reference papers [9, 10]). It is really surprising that even at the level of a model including just some minimal requirements and with substantial simplifications, the corresponding mathematical structure is necessarily quite complicated. It seems to us that the presence of the constraints here emphasized is a major feature of the model. The fact that they actually come into play is unquestionably put in evidence by numerical simulations, as illustrated by Fig. 4 taken from [26], which shows an example of the time evolution of the cord in case of a drug affecting mainly the proliferating cells. Panel A illustrates the evolution induced by the treatment of the viable cell population, showing the ratio between the total volume per unit cord length of viable cells (P+Q) and its value at $t = 0$. The decrement of the amount of viable cells reduces oxygen consumption and thus causes a transient increase of the mean oxygen concentration (panel B). The re-oxygenation of the cord produces a recruitment of quiescent cells into proliferation. Thereafter, the populations P and Q tend to the stationary value (panel A). The radius ρ_N shows an initial shrinkage [52] followed by a regrowth (panel C). The interface ρ_N quickly becomes a material boundary, so satisfying (16), and remains material until, at about $t\chi = 3$, it becomes nonmaterial again, an event marked by a slope discontinuity. In the same panel, the time course of the boundary B is plotted. Panel D shows the time evolution of the pressure p_N and of the cellular fraction in the necrotic region. In the initial state the constraint (23) is satisfied with the equality sign and $p_N < \Psi(B)$. Due to the increased influx of liquid caused by cell death, p_N increases reaching $\Psi(B)$ at $t\chi \simeq 0.5$. At this point the regime changes, with $p_N = \Psi(B)$, and the cellular fraction goes below ν_N. During the cord regrowth, the

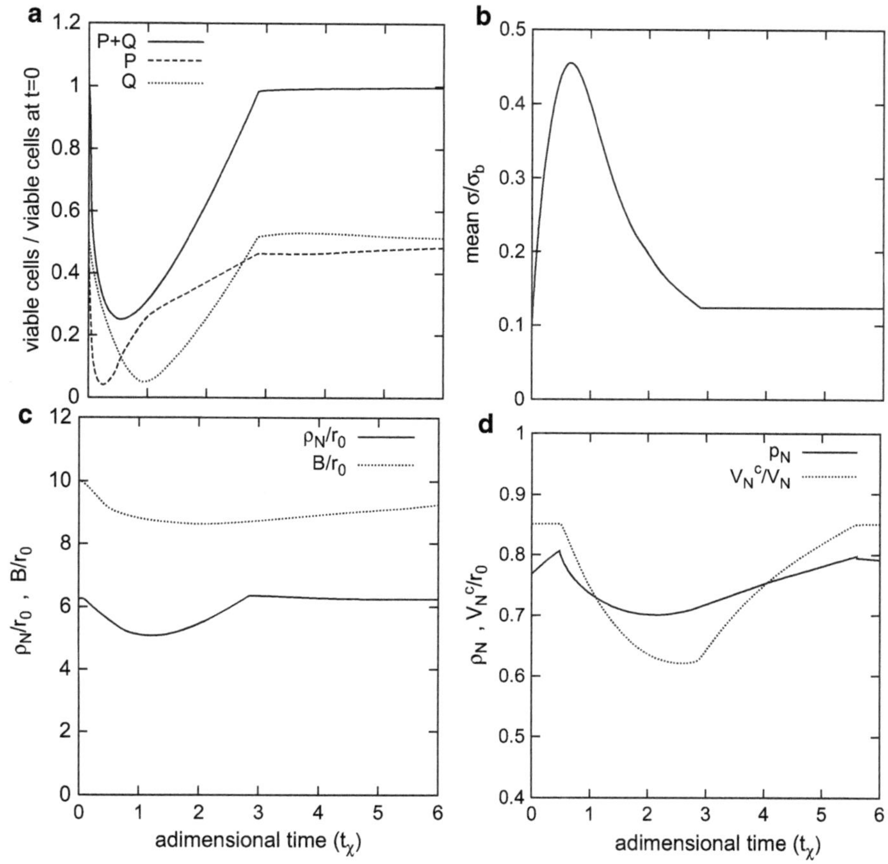

Fig. 4 *Panel A*: time course of the viable cell subpopulations after a single-dose treatment; *P* proliferating cells, *Q* quiescent cells. *Panel B*: mean oxygen concentration. *Panel C*: cord radius ρ_N and outer boundary B. *Panel D*: pressure and cell fraction in the necrotic region. Parameters values given in [26]

influx of liquid decreases and the system switches again to the regime characterized by a cellular fraction equal to ν_N ($t\chi \simeq 5.5$).

We conclude this short review about the tumor cords recalling that besides the existence and uniqueness theorem and some qualitative analysis for the doubly constrained free boundary problem, much more has been done in the wake of the papers [9, 10]. A specific analysis of interstitial pressure has been performed in [12]. One of the main aspects to be kept into account in modeling treatments is the re-oxygenation of the tumor following massive cells destruction [8]. The effects of delayed exit from quiescence after re-oxygenation were considered in [11]. The consequence of re-oxygenation on chemo- and radiotherapy has been investigated in [13, 14], including the analysis of the possible advantages of dose splitting, a subject treated in the literature in the context of the optimal fractionation of radiation dose [16, 66]. For the more general context of the application of optimization in cancer treatment, see, e.g., [49].

3 What Kind of a Fluid Can the "Cell Fluid" Be? A Model for Bingham-Like Spheroids

We have mentioned the advantages of the two-fluid approach, which, despite all the internal contradictions, have been responsible for its success. It is quite evident that the choice of identifying cells with a Newtonian fluid, which is frequent in this context, is arbitrary. In such a fluid the shear stress is generated by viscosity, whose physical origin in the case of cells should be a kind of mutual "friction." However, cells do not just slide one upon the other. Their mutual adherence is due to bonds that can resist some traction and that, when destroyed, can be restored in a different configuration. Such a situation is much more similar to what happens in "fluids" possessing some fragile internal structure, breakable by a stress beyond a threshold. This is precisely the main characteristic of Bingham fluids. Cell-exerted tractions have been measured [55] and can be surprisingly strong. Therefore it seems that resistance to motion comes primarily from the necessity of overcoming mutual bonds and then from the membrane-to-membrane friction. Accordingly, we may associate to these phenomena a nonnegligible yield stress and a viscosity, namely the two quantities intervening in the constitutive law of a Bingham fluid. The problem of finding possible steady states for a spheroid with an inner liquid core originated by dead cells degradation has been addressed in [29, 30], both in the Newtonian and in the Bingham framework, imposing the continuity of the normal stress throughout the system. Fasano et al. [29] follows a previous attempt [28] to solve the same problem imposing energy balance, stemming from the approximation that proliferating cells produce a known amount of mechanical power.

The passage from Newtonian to Bingham is by no means trivial. It is well known that defining a Bingham fluid is relatively easy in 1-D cartesian flows, but it offers different options in higher dimensions. In [15] the difficulty of selecting a constitutive law compatible with the radial motion to be found in spheroids has been emphasized, showing for instance that the definition proposed in [6] leads to a contradiction, which would make impossible to describe the early stage of the spheroid evolution. It was found in [15] that the following way of defining the stress tensor of the cell component of the mixture suites our purposes. The cellular Cauchy stress tensor in the viable region is given by

$$\mathbf{T}_{\mathrm{C}} = -\nu \left(p_C + \frac{2}{3}\eta_C \nabla \cdot \mathbf{u} \right) \mathbf{I} + \nu \boldsymbol{\tau} , \tag{28}$$

where ν is the cell volume fraction, p_C is the cell pressure, η_C is the so-called Bingham viscosity, and $\boldsymbol{\tau}$ is the deviatoric stress tensor defined as

$$\boldsymbol{\tau} = \left(2\eta_C + \frac{\tau_0}{\sqrt{II_D}} \right) \mathbf{D} \tag{29}$$

in which, as usual, $\mathbf{D}$ is the strain rate tensor and its second invariant II_D is given in the form

$$II_D = \frac{1}{2} \mathrm{Tr} \mathbf{D}^2 , \tag{30}$$

provided II_τ exceeds the threshold τ_0. In the opposite case we have $\mathbf{D} = 0$. We refer to [15] for additional comments on the definition (28).

The definition above applies to an aggregate of cells whose membrane is integer. We will return to the case of degrading cells later on.

The task of describing the liquid is much simpler. We put (inviscid fluid)

$$\mathbf{T}_E = \nu_E \big[- p_E \mathbf{I} \big] \tag{31}$$

with $\nu_E = 1 - \nu$, and we adopt Darcy's law for its motion relative to the cells:

$$\nu_E (\mathbf{v} - \mathbf{u}) = -\kappa \nabla p_E \tag{32}$$

thus treating the cells aggregate as a porous medium.

Remark 1. It is important that p_C and p_E be kept distinct. As a matter of fact, at the surface of the spheroid there may be forces acting differently on the two components. For instance the so-called tumor surface tension, attributable to stretched intercellular bonds, and the resistance of an external medium, like the polymer network in a gel hosting the spheroid, are applied to cells only. The concept that p_C and p_E are separate quantities has received some attention in the literature. For instance, in [48] the inequality $p_C > p_E$ was taken as a condition for cells viability.

Living cells can be proliferating or quiescent. Instead of introducing transition rates from one class to another, in [15] the two species have been separated by a sharp interface. Of course this is an extreme schematization, but, in our opinion, it does not make much difference, also owing to the uncertainty about the definition of the transition rates. The separation of a proliferating region P and a quiescent region Q, possibly surrounding a necrotic core, is very helpful to simplify the computation of the velocity field.

Again, the structure of the necrotic core N is going to play a basic role in the evolution of the system. Introducing a deterministic degradation time after which a dead cell is turned into a material mechanically behaving like a liquid, the core N is in turn divided into a liquid (NL) and a solid (NS) region. Despite the fact that this approach is rather extreme, it finds some support in experimental observations, based on NMR techniques, pointing out a prevalence of free water in the spheroid core [54], suggesting that the NS→NL transition is due to membrane degradation.

Still under the assumption that the two components have the same density and that ν is constant, the mass balance in the various regions is expressed by the system

$$\begin{aligned} &\nabla \cdot \mathbf{u} = \chi, && \text{in } P, \\ &\nabla \cdot \mathbf{u} = 0, && \text{in } Q \cup NS, \\ &\nabla \cdot \mathbf{v} = -\chi \frac{\nu}{1-\nu}, && \text{in } P, \\ &\nabla \cdot \mathbf{v} = 0, && \text{in } Q \cup NS \cup NL \end{aligned} \tag{33}$$

where χ is the cell proliferation rate. The supposed constancy of ν eventually provides global mass conservation in the form of a relationship between the two radial velocity fields:

$$\nu \mathbf{u} + \nu_E \mathbf{v} = 0\,. \tag{34}$$

The fact that inertia is absolutely negligible and the analysis of the liquid-cell interaction forces lead to express the momentum balance equation for the cell component in the form

$$\nabla \cdot \mathbf{T_C} = \frac{\nu_E}{\kappa}\mathbf{u} = -\nabla \cdot \mathbf{T_E}\,. \tag{35}$$

Recalling (28), the discontinuity of $\nabla \cdot \mathbf{u}$ across the P/Q interface creates a singularity in (35), which however is easily overcome imposing the *continuity of the normal stress throughout the system.* As a matter of fact, this is one of the main assumptions in the model.

The unknown interfaces

- $r = \rho_P(t)$ between the regions P and Q
- $r = \rho_N(t)$ between the regions Q and N

are defined implicitly via the system

$$\begin{aligned} &D_{O_2}\Delta\sigma(r,t) = f(\sigma(r,t))\nu, && \text{in P}, \\ &D_{O_2}\Delta\sigma(r,t) = \frac{1}{m} f(\sigma(r,t))\nu, && \text{in Q}, \\ &\sigma(R,t) = \sigma^*\,, \\ &\sigma(\rho_P,t) = \sigma_P\,, \\ &[\sigma_r(\rho_P,t)] = 0\,, \end{aligned}$$

$$\sigma(\rho_N, t) = \sigma_N ,$$
$$\sigma_r(\rho_N, t) = 0 , \tag{36}$$

e.g., with the choice

$$f(\sigma)\nu = nM\frac{\sigma}{H+\sigma} ,$$

(n cell number density, M maximum consumption rate per cell, H Michaelis constant, $m > 1, \sigma^* > \sigma_P > \sigma_N$). It can be proved that, for a given R sufficiently large, problem (36) is uniquely solvable. Actually the qualitative nature of the solution depends on the spheroid size. The fully developed structure with the two interfaces $\rho_P(t)$, $\rho_N(t)$ is observable only for R exceeding some increasing function $R_N(\sigma^*)$, with $\sigma \equiv \sigma_N$ throughout the region N. For R below this threshold and greater than another increasing function $R_P(\sigma^*)$ there is no necrotic core and the last two conditions in (36) must be replaced with $\sigma_r(0,t) = 0$. The latter condition is still operating if $R < R_P(\sigma^*)$, in which case only the region P is present.

Unlike the previous section, here we just consider Cauchy-type boundary data on the necrotic interface, since we are only interested in the tumor growth towards its possible steady state, with no cell death cause other than hypoxia. Thus, the switch to a freely evolving oxygen concentration on the moving necrotic interface never occurs. The necrotic core will appear at the time the spheroid radius reaches the value $R_N(\sigma^*)$. The structure of the necrotic region, though not simple because of the ongoing degradation, is less complicated than in cords, for which there are constraints to be satisfied. As we said, in the approach of [15] a deterministic degradation time is introduced. Once the necrotic (hypoxic) region is formed, it keeps being fed by cells arriving from the region Q. If the necrotic core is old enough transition to "liquid" will take place, and dead cells are pushed inwards, while degrading. Of course the assumption of a fixed degradation time is artificial, not differently from all other thresholds that have been introduced. All phenomena going on in the spheroid are characterized by some degree of stochasticity; therefore all interfaces are just mathematical tools, approximating transition regions by means of sharp surfaces. Therefore a question arises very naturally: why bother with unreal pictures? The question does not apply just to the specific case at hand, but it involves the whole domain of cancer modeling, since this particular branch of biomathematics (like many others) inevitably goes through strongly simplifying assumptions and compromises. A model with no interfaces would not necessarily be more accurate, since it would anyway contain some gross approximation. Interfaces simplify the computation of the velocity fields, and this largely compensates for the mathematical complications connected to the presence of free boundaries.

If θ_D is the degradation time and t_N is the (unknown) appearance time of the necrotic core, the liquid necrotic region NL will appear at time $t_D = t_N + \theta_D$. From that time on, the NL/NS interface $r = \rho_D(t)$ will be present. Its evolution depends on the feeding rate of the region NS, which is of course one of the unknowns.

Before we move to illustrating the spheroid evolution, we have to deal briefly with the boundary conditions on the outer moving surface $r = R(t)$.

The liquid pressure equals some given external pressure (typically atmospheric pressure):

$$p_E(R,t) = p_{ext} \tag{37}$$

irrespective of whether the spheroid is grown in a water suspension or in a gel.

The computation of the cellular normal stress leads to the following equation, expressing that a jump of normal stress across the boundary is produced by surface tension. The resulting boundary condition is the following:

$$p_C(R,t) = -\frac{2}{3}\eta_C\chi + \left(2\eta_C + \frac{\tau_0}{\sqrt{II_D(R,t)}}\right)u'(R,t) + p_{ext} + \psi(R) + \frac{2\gamma}{R}. \tag{38}$$

where $\frac{2\gamma}{R}$ is the pressure exerted by surface tension, $\psi(R)$ is the one produced by the reaction of the host medium (gel) to the spheroid expansion, and the symbol u' is an abbreviation of $\frac{\partial u}{\partial r}$.

It has to be stressed that, even if the steady state is reached, the velocity gradient u' will not vanish at the outer surface, according to the first equation in (33). The function $\psi(R)$ is assumed to be nonnegative and Lipschitz continuous. Since it is originated by the gel deformation, namely by the displacement of the gel polymer network, it is expected to increase up to a certain value and then to stabilize.

A spheroid growing from an initial size so small that it consists entirely of the region P, evolves through the following stages:

- Stage I: fully proliferating. It ends at the time t_P at which R attains the value $R_P(\sigma^*)$.
- Stage II: the interface $r = \rho_P(t)$ appears, enclosing a quiescent core. It ends at the time t_N.
- Stage III: the region NS appears, with the boundary $r = \rho_N(t)$, but conversion to liquid is not achieved yet. It ends at time t_D.
- Stage IV: the interface $r = \rho_D(t)$ is present.

We report the main results concerning the four stages, addressing the reader to [15] for their derivation, which is definitely long and not simple. Useful pieces of information when performing the calculations are:

(i) $\nabla \cdot \mathbf{u}$ constant $\Rightarrow$ $\nabla \cdot \mathbf{D} = 0$ (however, care is needed when crossing the discontinuity of $\nabla \cdot \mathbf{u}$);

(ii) where $\nabla\cdot\mathbf{D} = 0$ the only nonzero component of the vector $\nabla \cdot \left[\left(2\eta_C + \frac{\tau_0}{\sqrt{II_D}}\right)\mathbf{D}\right]$ is the radial one, which takes the form $\frac{\tau_0}{\sqrt{II_D}}\left(-\frac{1}{2II_D}\frac{\partial II_D}{\partial r}\right)\frac{\partial u}{\partial r}$.

3.1 Stage I

All quantities can be computed explicitly:

$$u(r,t) = \frac{\chi}{3} r\,, \tag{39}$$

implying that the ODE providing the outer radius is readily solvable, leading to

$$R(t) = R_0 \exp(\frac{\chi}{3} t)\,,$$

from which we get

$$t_P = \frac{3}{\chi} \ln(\frac{R_P}{R_0})\,.$$

Since

$$\mathbf{D} = \frac{\chi}{3} \mathbf{I}\,, \qquad II_D = \frac{\chi^2}{6}\,,$$

$$\nabla \cdot \boldsymbol{\tau} = \nabla \cdot \mathbf{D} = \mathbf{0}\,,$$

the integration of (35) with conditions (38) and (37) yields

$$p_C(r,t) = \tau_0 \sqrt{\frac{2}{3}} + p_{ext} + \psi(R) + \frac{2\gamma}{R} + \frac{\chi \nu_E}{6 \nu \kappa}(R^2 - r^2) \tag{40}$$

$$p_E(r,t) = p_{ext} - \frac{\chi}{6\kappa}(R^2 - r^2)\,. \tag{41}$$

3.2 Stage II

Now we have to distinguish region P from region Q, which at this stage is an immobile core. In region P we have

$$u(r,t) = \frac{\chi}{3} r \big[1 - (\frac{\rho_P(t)}{r})^3\big]\,, \tag{42}$$

where ρ_P is a functional of R and $R(t)$ is found by integrating the equation

$$\dot{R}(t) = \frac{\chi}{3} R(t) \big[1 - (\frac{\rho_P(t)}{R(t)})^3\big]\,.$$

Equation (35) now gives

$$-p_C' + \tau_0 u' \Big(\frac{1}{\sqrt{II_D}}\Big)' - \frac{\nu_E}{\nu\kappa} u = 0$$

with

$$II_D = \frac{\chi^2}{3}\Big[\frac{1}{2} + \Big(\frac{\rho_P}{r}\Big)^6\Big],$$

whose integration yields

$$p_C(r,t) = p_C(R,t) - \frac{\tau_0}{\sqrt{3}}\Big\{\frac{\sqrt{2}[1 + 2(\frac{\rho_P}{s})^3]}{[1 + 2(\frac{\rho_P}{s})^6]^{1/2}}\Big|_r^R - 2\ln\Big[\Big(\frac{\rho_P}{s}\Big)^3 + \Big[\frac{1}{2} + \Big(\frac{\rho_P}{s}\Big)^6\Big]^{1/2}\Big]\Big|_r^R\Big\}$$
$$+ \frac{\chi\nu_E}{3\nu\kappa}\Big[\frac{R^2 - r^2}{2} + \rho_P^3\Big(\frac{1}{R} - \frac{1}{r}\Big)\Big], \tag{43}$$

where

$$p_C(R,t) = \frac{4}{3}\eta_C \chi \left(\frac{\rho_P}{R}\right)^3 + \tau_0 \sqrt{\frac{2}{3}} \frac{1 + 2\left(\frac{\rho_P}{R}\right)^3}{\left[1 + 2\left(\frac{\rho_P}{R}\right)^6\right]^{1/2}} + p_{ext} + \psi(R) + \frac{2\gamma}{R}. \tag{44}$$

In the region Q, owing to the absence of motion, p_C turns out to be uniform. The discontinuity of the proliferation rate, i.e., of $\nabla \cdot \mathbf{u}$, through the P/Q interface produces a jump of p_C which can be computed imposing the continuity of the normal stress:

$$p_C(\rho_P^+, t) - p_C(\rho_P^-, t) = \frac{4}{3}\eta_C \chi + \sqrt{2}\tau_0. \tag{45}$$

No discontinuity is experienced by p_E which is found to be

$$p_E(r,t) = \begin{cases} p_{ext} - \dfrac{\chi}{3\kappa}(R - r)\Big(\dfrac{R + r}{2} - \dfrac{\rho_P^3}{rR}\Big), & \rho_P \le r \le R, \\ p_E(\rho_P, t), & 0 < r < \rho_P. \end{cases} \tag{46}$$

3.3 *Stage III*

The necrotic region is now present without "liquid" core. From the dynamical point of view the situation is very similar to the previous stage, the only difference being the presence of the necrotic interface $r = \rho_N(t)$, a known functional of R.

3.4 Stage IV

Stage IV evolves for $t > t_D$. As it frequently happens in tumor modeling, the structure of the necrotic region has a decisive influence on the system evolution. Here too, when entering the stage in which the liquid necrotic region appears, the mathematical nature of the problem is deeply affected. The point is that, while the liquid core is immobile, cells in its vicinity are moving inwards and we no longer have information on their velocity at the interface, while in Stage III we knew that the velocity was zero in N∪Q, hence also at $r = \rho_P(t)$. This fact introduces a very substantial complication. We emphasize the fact that the need of studying the whole stress field is originated by the necessity of determining precisely that unknown velocity.

It was therefore natural in [15] to introduce a new unknown $\omega_N(t)$ in the physical range $\omega_N(t) \leq 0$, namely the velocity of the cells crossing the interface $r = \rho_N(t)$, so that the quantity $4\pi\rho_N^2\nu(\dot{\rho}_N - \omega_N)$ represents the feeding rate of the necrotic region. At the beginning of Stage IV $\omega_N(t_D) = 0$.

Since for $\rho_D < r < \rho_P$ the cell velocity field in that region is divergence free, we have

$$u(r,t) = \frac{1}{r^2}\rho_N^2(t)\omega_N(t)\,, \qquad \rho_D \leq r \leq \rho_P\,, \tag{47}$$

and

$$u(r,t) = \frac{1}{r^2}\rho_N^2(t)\omega_N(t) + \frac{\chi}{3}\left(r - \frac{\rho_P^3(t)}{r^2}\right), \qquad \rho_P \leq r \leq R\,. \tag{48}$$

From the latter we deduce the differential equation for $R(t)$, namely

$$R^2(t)\dot{R}(t) = \rho_N^2(t)\omega_N(t) + \frac{\chi}{3}(R^3(t) - \rho_P^3(t))\,. \tag{49}$$

Of course the equation contains the new unknown $\omega_N(t)$. Following the motion of a dead cell through the region NS during the degradation time θ_D, we find that $\rho_D(t)$ is expressed in terms of $\omega_N(t)$ as follows:

$$\frac{1}{3}[\rho_D^3(t) - \rho_N^3(t-\theta_D)] = \int_{t-\theta_D}^{t} \rho_N^2(\tau)\omega_N(\tau)\,d\tau\,. \tag{50}$$

At this point it is clear that the determination of the kinematic unknown $\omega_N(t)$ relies on the stress analysis. In order to proceed further, we have to say something more on the degradation process, which affects the dynamical behavior. In [15] it was assumed that *membrane degradation is accompanied by a reduction of the yield stress*. This requires monitoring the age from death, $\theta(r,t)$, of cells located at a given point of NS. Proceeding as in the derivation of (50), the latter quantity is found to have the implicit expression

$$\frac{1}{3}\left[r^3 - \rho_N^3(t-\theta(r,t))\right] = \int_{t-\theta(r,t)}^{t} \rho_N^2(\tau)\omega_N(\tau)\,d\tau\,. \tag{51}$$

Now the yield stress varies in NS according to the formula

$$\tilde{\tau}_0(r,t) = \begin{cases} \tau_0, & \rho_N \le r \le R, \\ \tau_0\left[1 - \frac{\theta(r,t)}{\theta_\tau}\right]_+, & \rho_D < r < \rho_N, \end{cases} \tag{52}$$

i.e., it reduces linearly to zero in the time θ_τ, which is between 0 and θ_D. This generates a new free boundary $r = \rho_\tau(t)$, tending to ρ_N or to ρ_D in the respective limits $\theta_\tau \to 0$ or $\theta_\tau \to \theta_D$. In any case, the yield stress is gone when the interface ρ_D is reached. Once more, the choice was to introduce a deterministic law, extrapolating from the biological randomness of the phenomenon in agreement with the general deterministic setting of the entire model.

The new condition at our disposal, which closes the model, is the continuity of the normal stress across the interface ρ_D. Concerning p_E, we know that it is continuous everywhere, while the just-mentioned condition of normal stress continuity eventually yields the following limit for p_C when approaching ρ_D from the "solid" side:

$$p_C(\rho_D^+,t) - 2\eta_C u'(\rho_D^+,t) = p_E(\rho_D,t)\,. \tag{53}$$

In view of Remark 1 about the comparison between the two pressures p_C, p_E, the following result from [15] is of some interest: $\omega_N < 0 \Rightarrow p_C > p_E$, stressing the physical relevance of the orientation of the cell motion at the necrotic interface.

We just report the full expression of (53) in the case $\theta_\tau \to 0$, which requires a special procedure, referring to [15] for the more general case:

$$\begin{aligned} \frac{2\gamma}{R} = &- \psi(R) + 4\eta_C\rho_N^2\omega_N\left(\frac{1}{R^3} - \frac{1}{\rho_D^3}\right) + \frac{4}{3}\eta_C\chi\left[1 - \left(\frac{\rho_P}{R}\right)^3\right] \\ &- \frac{1}{\nu K}\left\{\rho_N^2\omega_N\left(\frac{1}{\rho_D} - \frac{1}{R}\right) + \frac{\chi}{3}(R-\rho_P)^2\left(\frac{1}{2} + \frac{\rho_P}{R}\right)\right\} \\ &+ \frac{2}{\sqrt{3}}\tau_0 \ln \frac{\left(\frac{\rho_P}{\rho_N}\right)^3\left[1 + \sqrt{\frac{\chi^2\rho_P^6}{6\left(\frac{\chi}{\sqrt{3}}\rho_P^3 - \sqrt{3}\rho_N^2\omega_N\right)^2} + 1}\,\right]}{\left(\frac{\rho_P}{R}\right)^3 + \sqrt{\frac{\chi^2\rho_P^6}{6\left(\frac{\chi}{\sqrt{3}}\rho_P^3 - \sqrt{3}\rho_N^2\omega_N\right)^2} + \left(\frac{\rho_P}{R}\right)^6}}\,. \end{aligned} \tag{54}$$

In (54), the fractions having ρ_D in the denominator are potentially singular at the beginning of Stage IV. The presence of the term ω_N/ρ_D^3, which cannot be balanced by any other term, says in particular that, when $t \to t_D^+$, $\omega_N(t)$ is infinitesimal (as predicted) of the same order as ρ_D^3. Moreover, owing to the sign restriction over ω_N, the limit of the ratio above has to be nonpositive. This requires a condition that, for convenience, in [15], has been imposed in the strict sense. Again we write it only with reference to the special case $\theta_\tau \to 0$, namely

$$\frac{2\gamma}{R} + \psi(R) - \frac{4}{3}\eta_C\chi\Big[1 - \Big(\frac{\rho_P}{R}\Big)^3\Big] + \frac{1}{\nu\kappa}\frac{\chi}{3}(R-\rho_P)^2\Big(\frac{1}{2} + \frac{\rho_P}{R}\Big)$$
$$- \frac{2}{\sqrt{3}}\tau_0 \ln\left[\Big(\frac{\rho_P}{\rho_N}\Big)^3 \frac{1+\sqrt{\frac{3}{2}}}{\Big(\frac{\rho_P}{R}\Big)^3 + \sqrt{\frac{1}{2} + \Big(\frac{\rho_P}{R}\Big)^6}}\right] > 0\,. \tag{55}$$

This inequality is actually a limitation on the choice of the two quantities γ and τ_0, but a biological explanation is missing.

Another assumption is made on the function f expressing the oxygen absorption rate in (36) that we write in the form

$$f(\sigma) = \tilde{f}(\sigma) + (m-1)\tilde{f}(\sigma_P)H(\sigma - \sigma_P)\,, \tag{56}$$

where H is the Heaviside function and $\tilde{f}(\sigma)$ is a continuously differentiable function of Michaelis-Menten type. It is required that for all $R > R_D$,

$$R^2\Big[\frac{1}{6} - \frac{1}{2}\Big(\frac{\rho_N}{R}\Big)^2 + \frac{1}{3}\Big(\frac{\rho_N}{R}\Big)^3\Big] \sup_{\sigma\in(\sigma_N,\sigma^*)} |\tilde{f}'| < 1\,, \tag{57}$$

which makes sense because the quantity $R^2\Big[\frac{1}{6} - \frac{1}{2}\Big(\frac{\rho_N}{R}\Big)^2 + \frac{1}{3}\Big(\frac{\rho_N}{R}\Big)^3\Big]$ is bounded for $R > R_D$. In addition, $(m-1)\tilde{f}(\sigma_P)$ is supposed to be sufficiently small (for the details see [15]).

The stated conditions on ψ, f, together with (55), have been employed in [15] to show existence and uniqueness for Stage IV, first in a neighborhood of t_D and then extended by means of a standard argument.

The proof is very long and it goes through the study of how σ and the interfaces ρ_P, ρ_N depend on R. More precisely a priori estimates of the derivatives $\frac{\partial\sigma}{\partial R}$, $\frac{\partial\rho_P}{\partial R}$, $\frac{\partial\rho_N}{\partial R}$ have been obtained in terms of the data, which are instrumental in the fixed point argument employed in the proof.

Besides the well-posedness analysis, in [15] numerical simulations have been performed, dealing with the nontrivial question of selecting appropriate values for the parameters and investigating the possible attainment of a steady state. The problem is characterized by a rather large uncertainty about some critical parameters in the model, a constant obstacle in this kind of research. The tumor hydraulic conductivity κ appearing in Darcy's law (32) is certainly one of such parameters. A chosen value was $\kappa = 4 \cdot 10^{-8}\,\mathrm{cm}^3{\cdot}\mathrm{s/g}$ (or alternatively $4 \cdot 10^{-9}\,\mathrm{cm}^3{\cdot}\mathrm{s/g}$), much larger than the values usually reported for solid tumors in consideration of the relatively low value of the cell volume fraction ($\nu \sim 0.6$) in spheroids. Concerning the rheological parameters of the Bingham-like cell fluid, the value $\eta_C = 10^4$ g/(cm·s) is acceptable, in view of the results of [45]. The determination of the yield stress τ_0 can be deduced on the basis of the measurements of the force F able to detach two adhering cells [55], according to the formula $\tau_0 = Fn^{2/3}$. The so-called tumor surface tension γ is apparently related to the upper bound of adhesion forces, hence to τ_0, as it was already observed in [29].

The interplay between τ_0 and γ is crucial for the existence of a steady state. This fact emerges very clearly when looking for the spheroid size at a possible steady state. The investigation of the possible equilibrium can be performed with the help of equation (54), in which ω_N has to be replaced with

$$\omega_N = -\frac{\chi}{3\rho_N^2}\left(R^3 - \rho_P^3\right) \tag{58}$$

which is the cell velocity at the necrotic interface corresponding to a steady spheroid of radius R. Indeed, for a spheroid at the steady state the cell velocity vanishes at the outer surface, providing the information that allows the computation of the whole velocity profile. The radius ρ_D of the inner liquid core and the age from death $\theta(r)$ can be found in terms of R:

$$\rho_D^3 = \rho_N^3 - \chi\theta_D\left(R^3 - \rho_P^3\right), \tag{59}$$

$$\theta(r) = \frac{\rho_N^3 - r^3}{\chi(R^3 - \rho_P^3)}, \qquad \rho_D \le r \le \rho_N\,. \tag{60}$$

Thus all elements are available to formulate the steady-state version of normal stress continuity condition, which, still in the limit $\theta_\tau \to 0$, turns out to be

$$2\gamma + R\psi(R) = \frac{4}{3}\eta_C\chi R\frac{R^3 - \rho_P^3}{\rho_D^3} + \frac{\chi R^3}{3\nu\kappa}\left\{\frac{R}{\rho_D}\left[1 - \left(\frac{\rho_P}{R}\right)^3\right] - \frac{3}{2}\left[1 - \left(\frac{\rho_P}{R}\right)^2\right]\right\}$$
$$+\frac{2}{\sqrt{3}}\tau_0 R\ln\left(\frac{\rho_P}{\rho_N}\right)^3\frac{\left(\frac{R}{\rho_P}\right)^3 + \sqrt{\frac{1}{2} + \left(\frac{R}{\rho_P}\right)^6}}{1 + \sqrt{\frac{3}{2}}}, \tag{61}$$

where ρ_D is expressed by (59) and ρ_P, ρ_N are known functionals of R.

The above formula allows to establish the conditions ensuring the existence of a steady state. For instance, in the case $\psi = 0$, the plot in Fig. 5 compares the

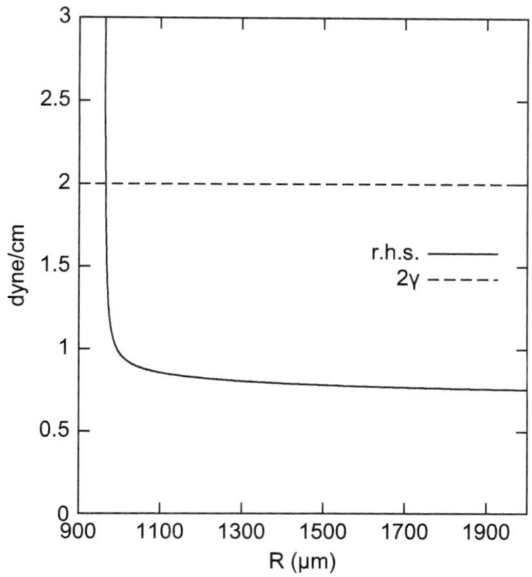

Fig. 5 Profile of right-hand-side (*solid*) and left-hand-side (*dashed*) of (61) as a function of R. $\gamma = 1$ dyne/cm, $\tau_0 = 10$ dyne/cm^2. Other parameters given in [15]. The predicted stationary radius is $R = 966\,\mu\text{m}$

LHS 2γ with the RHS in a case exhibiting an intersection. From (61) it is clear that for a solution to exist it is necessary that γ is large enough. Of course a mechanical reaction from the environment, meaning $\psi > 0$, helps in reaching the equilibrium. Since the RHS of (61) can be shown to go to infinity as $R \to \infty$, the equation can have two roots. The physical one is the smaller, since it is reached as an asymptotic state. The larger one is normally too big and it looks nonphysical.

The whole evolution model has been numerically simulated in [15] with the same data as in Fig. 5. The results are shown in Fig. 6, in which the moving boundaries entering the problem are followed from their origin to their asymptotic value (upper panel) and the velocity ω_N (appearing simultaneously with the interface ρ_D) is shown to reach the final value (58).

As we said, equation (61) may have no solution at all, meaning that the spheroid grows to infinity. Such a possibility has been numerically investigated too, but the values to be attributed to γ and τ_0 were out of the expected physical range.

It is legitimate to ask whether a tumor spheroid reaching equilibrium is actually observable. In principle the answer is positive, and it has to be said that experimental measurements (see, e.g., [33]) may suggest that this is indeed the case. However, it is impossible from a few experimental points to infer more than a trend to reach equilibrium, and experiments performed over a very long time show that spheroids may go into a state of senescence [31] no longer describable with a model for a viable system.

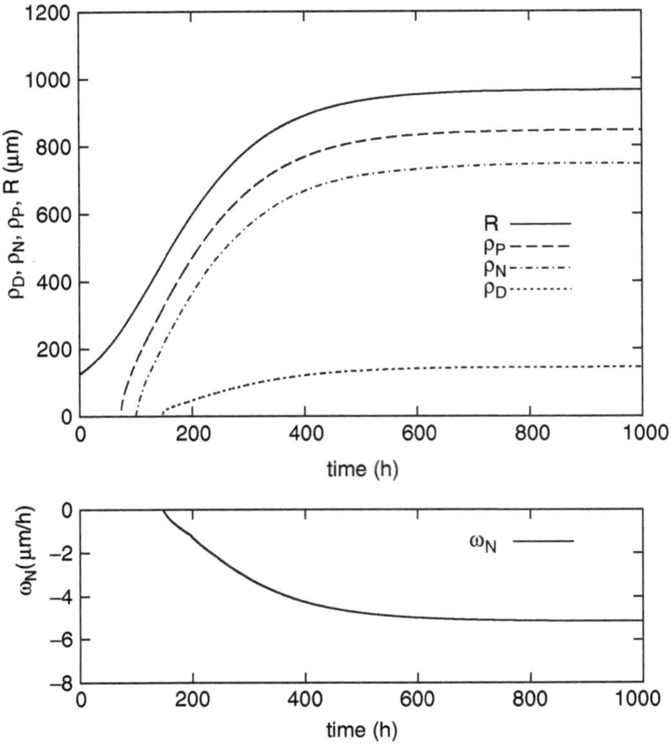

Fig. 6 *Upper panel*: Time evolution of the external radius and of all the interfaces. The velocity ω_N is reported in the *lower panel*. $\gamma = 1$ dyne/cm, $\tau_0 = 10$ dyne/cm^2. Other parameters given in [15]

4 Recent Models for Multicomponent Systems

In this final section, we review some very recent cancer growth models taken from the literature that include more elements than just a species of tumor cells. Active cell displacement (chemotaxis and haptotaxis) along with diffusion transport is considered, e.g., in the context of angiogenesis. Chemotaxis and reaction-diffusion are very large subjects entering those kinds of models. For them we refer to the important review papers [40,41], without mentioning their general features. Here we are more concerned in highlighting the efforts in cancer modeling to tackle problems having an immediate clinical impact. We confine our attention to models describing the spatial structure of the tumor and utilizing partial differential equations. Among compartmental models we quote for instance the paper [47], dealing with the treatment of prostate cancer by means of the so-called androgen ablation therapy, which is a very good example of how to keep the level of complexity within manageable limits, still retaining the essential pieces of information, and reaching conclusions of theoretical and practical interest. We summarize here just two models concerning systems with many cellular and molecular species. We remark that a common trend is to ignore the contribution of interstitial fluids to the mass balance and that in both instances the model setting is such to avoid the analysis of stresses. In this sense the present section is complementary to the previous ones.

4.1 Gliomas Invasion and Angiogenesis: Diffusion Driven Processes

Since the discovery of the tumor angiogenic factor (TAF) [32] (see [59] for the historical aspects), the phenomenon of angiogenesis has been investigated for long time in two main directions: modeling its development [3, 20, 50] and modeling the action of antiangiogenic drugs [22, 39, 62]. Recently a new subject emerged, related to the chaotic structure of tumor vascularization, partly immature and leaking and with numerous loops, slowing down blood circulation in vascularized tumors and consequently reducing the efficiency of drugs delivery within the tumor mass. In [37, 46] the so-called pruning procedure was illustrated, consisting in a partial destruction of the vasculature preceding drug administration (see also [22]), so to facilitate the tumor perfusion. Here we have no space to deal with the huge topics of angiogenesis, and we concentrate on one very specific theme: the recent claim that high-grade gliomas receive their aggressiveness from angiogenesis. The reference paper is [64] (see also the literature quoted therein), which is in the wake of an earlier model proposed in [63] where diffusion was assumed to be an important (though slow) transport mechanism for cells. In the model of [64] the tumor cells can be in two states (normoxic and hypoxic), depending on the vasculature density, considered to be the direct oxygen source. Only normoxic cells proliferate. Death rates are different for the two species. The vasculature development is driven by VEGF (vascular endothelial growth factor), produced by living tumor cells. The unknowns are:

- c, normoxic cells concentration
- h, hypoxic cells concentration
- n, necrotic cells concentration
- v, vascular endothelial cells concentration
- a, VEGF concentration

Moreover, the following fractions have some role:

- $V = v/(c+h+v)$ (endothelial cells vs. the total oxygen consuming population)
- $T = (c+h+v+n)/k$ (total number of cells compared to the carrying capacity k).

The model is summarized as follows (we do not comment on the meaning of the parameters):

$$\frac{\partial c}{\partial t} = \nabla \cdot [D(1-T)\nabla c] + \rho c(1-T) + \gamma h V - \beta c(1-V) - \alpha_n n c, \tag{62}$$

$$\frac{\partial h}{\partial t} = \nabla \cdot [D(1-T)\nabla h] - \gamma h V + \beta c(1-V) - \alpha_n n h - \alpha_h h(1-V), \tag{63}$$

$$\frac{\partial n}{\partial t} = \alpha_h h(1-V) + \alpha_n n(c+h+v), \tag{64}$$

$$\frac{\partial v}{\partial t} = \nabla \cdot [D_v(1-T)\nabla v] + \mu \frac{a}{K_m + a} v(1-T) + \gamma h V - \alpha_n n v, \tag{65}$$

$$\frac{\partial a}{\partial t} = \nabla \cdot (D_a \nabla a) + \delta_c c + \delta_h h - q \frac{a}{K_m + a} v(1-T) - \omega a v - \lambda a\,. \tag{66}$$

Thus cell diffusion is limited by a crowding effect, expressed by the factor $(1-T)$, which also limits proliferation of normoxic cells and of endothelial cells; the factors V and $(1-V)$ control the $c \leftrightarrow h$ transitions, and all the rest needs no explanations. The model does not consider extracellular matrix and the consequent occurrence of haptotaxis nor the chemotactic motion of endothelial cells.

The paper [64] contains an interesting discussion about the combined influence on growth of angiogenesis, diffusivity, proliferation rate, and transitions $h \to c$ and $c \to h$. The interesting conclusion is that increased aggressiveness is not necessarily originated by mutations affecting D and ρ. Even if the invasion rate is known to be strictly related to the parameters D and ρ, the final outcome is strongly conditioned by the tumor ability of developing a vascularization.

As a final remark to this subsection, we stress the fact that cell motility is a subject that has been approached in many different ways in the literature. For instance, in connection with the role of diffusivity in cancer invasion, it is worth mentioning the paper [43], preceding [64], which adopts a similar scheme in a larger context, since, besides the equations for the three species of tumor cells, and the equations for endothelial cells and for VEGF, it includes the equation for oxygen diffusion-consumption and for the development of the extracellular matrix (inducing haptotaxis of normoxic tumor cells). In that paper cell motility is treated in a very different way, since hypoxic and apoptotic cells are considered immobile and normoxic cells have a diffusivity made of two terms: a background random diffusivity D, like in (62), and an additional term, which, with the same symbols used above, is expressed by $D_c \max[c - c^*, 0]$, where c^* is some threshold concentration. The meaning of this extra term is a "pressure-driven" motility due to crowding, thus representing an opposite point of view with respect to (62), where crowding was opposing diffusion. More specifically, both D and D_c are taken to be of the same order $10^{-9}\,\text{cm}^2{\cdot}\text{s}^{-1}$. Moreover, in [43] the equation governing vascularization includes the usual chemotactic term, differently from the simpler process described by (65).

4.2 The Anti-angiogenic Role of Macrophages During Cancer Growth

We are now back to angiogenesis, but in a different perspective. In the paper [21] a mathematical model has been developed accounting for an important action of tumor-associated macrophages (TAMs). In hypoxic conditions such cells are chemically induced by tumor cells to produce VEGF, precisely as tumor cells do. However, when treated with another growth factor, namely GM-CSF (granulocyte/macrophage colony stimulating factor), TAMs are strongly stimulated to produce the VEGF inhibitor sVEGFR-1, namely the soluble VEGF receptor-1, which neutralizes VEGF by binding to it [23]. The paper [21] is based on the experimental work illustrated in [60, 61] about the influence of the transcription factors HIF-1α, HIF-2α on the production of VEGF and sVEGFR-1, respectively, studied by comparing the tumor development in normal mice and in mice with genetically induced deficiency of either factors. Interestingly enough, in view of the discussion in the previous subsection, cells diffusivity is totally disregarded, as it is very small in most tumors (here we are dealing with breast cancer). For the reader's convenience we try to preserve, as long as possible, the symbols already adopted in the previous section. Differently from that approach, the model of [21] does not deal with normoxic and hypoxic tumor cells as different species.

Thus, from the list of the previous section we remove the unknown h. Now c represents the concentration of living tumor cells with no further specification, and to the symbol list we add:

- m, macrophages concentration
- p, q, g, concentrations of specific cytokines (to be explained soon)
- s, sVEGFR-1 concentration
- w, oxygen concentration

The cytokines entering the model are:

- MCP-1/CCL2 (monocyte chemoattractant protein-1: p), produced by TAMs in response to M-CSF (q), acts as a chemoattractant to recruit more macrophages
- M-CSF (macrophage colony stimulating factor: q), produced by tumor cells, stimulates the secretion of MCP-1/CCL2 by TAMs
- GM-CSF (g), already mentioned

The transcription factors HIF-1α, HIF-2α intervene by regulating the production rates of VEGF, sVEGF-1 (without a kinetics of their own).

Let us write down the governing differential system, according to [21], which includes also the tumor velocity field $\boldsymbol{v}$:

$$\frac{\partial c}{\partial t} + \nabla \cdot (c\,\boldsymbol{v}) = \lambda_1(w)c(1 - \frac{c}{k}) - \lambda_2(w)c - \mu_c c, \tag{67}$$

where $\lambda_1(w)$ (proliferation rate) is a piecewise linear, increasing function connecting zero (for low w) to the value λ_1, through two thresholds $w_h < w_0$, and $\lambda_2(w)$ (hypoxia-induced necrosis) is a piecewise linear, decreasing function connecting the value λ_2 (for low w) to zero across two thresholds $w_n < w_h$. Then we have

$$\frac{\partial n}{\partial t} + \nabla \cdot (n\,\boldsymbol{v}) = \lambda_2(w)c + \mu_c c - \mu_n \frac{w}{w_0} mn, \tag{68}$$

where the last term describes clearance by macrophages. Equations (67) and (68) are to be solved in the unknown domain $\Omega(t)$ occupied by the tumor and embedded in a larger domain D, in which the complement to $\Omega(t)$ is occupied by the host tissue. Thus the problem contains the free boundary $\partial\Omega(t)$. Moreover,

$$\frac{\partial m}{\partial t} + \nabla \cdot (m\,\boldsymbol{v}) = -\nabla \cdot (k_p m \nabla p) - \nabla \cdot (k_g m \nabla g), \tag{69}$$

expressing mass balance of macrophages under the two chemotactic motions induced by p and g, to be solved in the whole domain D, as well as all the remaining equations for the molecular components p, q, a, s, g. The mass balance equation for the last cellular species is the one for endothelial cells:

$$\frac{\partial v}{\partial t} + \nabla \cdot (v\,\boldsymbol{v}) = -\nabla \cdot (k_a v \nabla a)\,. \tag{70}$$

Next we write the mass balance equations for the cytokines, VEGF and its inhibitor, and the drug.

$$\frac{\partial p}{\partial t} + \nabla \cdot (p\,\boldsymbol{v}) = \nabla \cdot (D_p \nabla p) + \lambda_4(w)\frac{q}{q+q_0} m - \mu_p p \tag{71}$$

with $\lambda_4(w)$ a stepwise increasing function with values $(0, 0.4\lambda_4, \lambda_4)$ through the same thresholds as $\lambda_2(w)$,

$$\frac{\partial q}{\partial t} + \nabla \cdot (q\,\boldsymbol{v}) = \nabla \cdot (D_q \nabla q) + \lambda_3 c - \mu_q q \tag{72}$$

$$\frac{\partial a}{\partial t} + \nabla \cdot (a\,\boldsymbol{v}) = \nabla \cdot (D_a \nabla a) + \lambda_5(w)c + \theta_1 \lambda_6(w)\frac{q}{q+q_0} m - \bar{\mu}_s s a - \mu_a a \tag{73}$$

where both $\lambda_5(w)$ and $\lambda_6(w)$ are proportional to another piecewise linear function $\phi(w)$ increasing from 0 to 0.3 through three thresholds $w_n < w^* < w_0$. We remark the appearance of the coefficient θ_1, linked to the level of the transcription factor HIF-1α (set equal to 1 for normal level). For the sVEGFR-1 and the oxygen we have:

$$\frac{\partial s}{\partial t} + \nabla \cdot (s\,\boldsymbol{v}) = \nabla \cdot (D_s \nabla s) + \theta_2 \lambda_7 \frac{g + \beta g_0}{g + g_0} m - \bar{\mu}_a a s - \mu_s s \tag{74}$$

$$\frac{\partial w}{\partial t} + \nabla \cdot (w\,\boldsymbol{v}) = \nabla \cdot (D_w \nabla w) + \lambda_8 v - \lambda_9 m w - \lambda_{10} c w \tag{75}$$

where the coefficient θ_2 in (74) has for HIF-2α a role parallel to the one of θ_1 for HIF-1α and the term $\lambda_8 v$ in (75) represents oxygen delivery by endothelial cells.

The mass balance equation introducing the treatment by GM-CSF is

$$\frac{\partial g}{\partial t} + \nabla \cdot (g\,\boldsymbol{v}) = \nabla \cdot (D_g \nabla g) + f(t) - \mu_g g \tag{76}$$

with $f(t)$ expressing the drug injection rate.

All transport terms contain the velocity field $\boldsymbol{v}$ and, as we have seen in several instances, its determination is a quite delicate issue, since in general it calls for the investigation of the whole mechanics of the system (constitutive equations, momentum balance, etc.). As in the case of Sect. 2, imposing saturation and dealing with a simple geometry reduces that complex dynamical problem to a much simpler kinematical condition, since transport velocity is ultimately forced by the ideal arrangement imposed to the set of volume occupying components. Thus, if one considers just a spherical geometry, with the tumor occupying a sphere $r < R(t)$, the velocity being purely radial, and takes the sum $c + n + m + v$ constant, the equation for the only scalar component of $\boldsymbol{v}$ is derived summing up the equations for all cells mass balance. Then imposing that the boundary $r = R(t)$ moves with the cell velocity provides the necessary free boundary condition. After a long discussion on the selection of the parameters, numerical simulations show the consistency of the model with the experimental results of [60, 61].

5 Conclusions

We have reviewed a few mathematical models of tumor growth based on conservation laws and pursuing different targets. In Sects. 2 and 3 we consider tumors with only two components: cells and extracellular fluid, taking advantage of the small number of constituents to carry out the analysis of mass and momentum balance, as well as of the mathematics involved to a full extent.

The model in Sect. 2 deals with tumor cords evolving in axisymmetric geometry, where the cells motion is simply passive and compatible with the saturation condition. The main difficulty there consists in the presence of free boundaries with constraints driving the free boundary conditions, with severe mathematical and numerical implications which in our opinion are particularly important. Though those results are not extremely recent, we decided to include them anyway in our review in order to reaffirm the crucial role of constraints, particularly in the presence of treatments. All the other material here exposed is taken from quite recent publications.

In Sect. 3 a tumor spheroid is considered in the framework of the so-called two-fluid scheme. In a multicellular spheroid, unlike the previous case, on the appearance of a fully degraded necrotic core the analysis of mechanical stresses becomes necessary to determine the motion via momentum balance, requiring the specification of the constitutive law for the cell "liquid." The case in which such a liquid is of Bingham type presents considerable difficulties linked to the presence of the yield stress that have been described, particularly with reference to the determination of an asymptotic configuration.

Despite the formidable mathematical complexity of the models treated in Sects. 2 and 3, the fact remains that two-component tumors are in a sense too schematic structures. Therefore, in the last section we illustrated two very recent studies dealing with multicomponent tumors, in order to give at least a feeling of the trends in the mathematical modeling of complex tumor structures, based on conservation laws, though in a perspective rather different from the one pursued in the models of Sects. 2 and 3. The specific studies considered are (i) Gliomas invasion and angiogenesis (reference paper [64]) and (ii) the anti-angiogenic role of macrophages during cancer growth (reference paper [21]). These are extremely interesting cases both for the modeling technique and under the perspective of their practical implications.

We regret that, because of space limitations, we could only illustrate a limited number of models. A typical feature of this research field is that it is expanding at an impressive rate, and as mathematicians come closer to the clinical practice their models become oriented to more specific targets. Despite its conciseness, we hope that our exposition can be stimulating.

References

1. T. Alarcón, H.M. Byrne, P.K. Maini, A cellular automaton model for tumour growth in inhomogeneous environment. J. Theor. Biol. **225**, 257–274 (2003)
2. D. Ambrosi, L. Preziosi, Cell adhesion mechanisms and stress relaxation in the mechanics of tumours. Biomech. Model. MechanoBiol. **8**, 397–413 (2009)
3. A.R.A. Anderson, M.A.J. Chaplain, Continuous and discrete mathematical models of tumor-induced angiogenesis. Bull. Math. Biol. **60**, 857–999 (1998)
4. R.P. Araujo, D.L.S. McElwain, A history of the study of solid tumour growth: the contribution of mathematical modelling. Bull. Math. Biol. **66**, 1039–1091 (2004)
5. S. Astanin, A. Tosin, Mathematical model of tumour cord growth along the source of nutrient. Math. Model. Nat. Phenom. **2**, 153–177 (2007)
6. I.V. Basov, V.V. Shelukhin, Generalized solutions to the equations of compressible Bingham flows. Z. Angew. Math. Mech. **79**, 185–192 (1999)

7. N. Bellomo, N.K. Li, P.K. Maini, On the foundations of cancer modelling: selected topics, speculations, and perspectives. Math. Mod. Meth. Appl. Sci. **18**, 593–646 (2008)
8. A. Bertuzzi, A. d'Onofrio, A. Fasano, A. Gandolfi, Regression and regrowth of tumour cords following single-dose anticancer treatment. Bull. Math. Biol. **65**, 903–931 (2003)
9. A. Bertuzzi, A. Fasano, A. Gandolfi, A free boundary problem with unilateral constraints describing the evolution of a tumour cord under the influence of cell killing agents. SIAM J. Math. Anal. **36**, 882–915 (2004)
10. A. Bertuzzi, A. Fasano, A. Gandolfi, A mathematical model for tumor cords incorporating the flow of interstitial fluid. Math. Mod. Meth. Appl. Sci. **15**, 1735–1777 (2005)
11. A. Bertuzzi, A. Fasano, L. Filidoro, A. Gandolfi, C. Sinisgalli, Dynamics of tumour cords following changes in oxygen availability: a model including a delayed exit from quiescence. Math. Comput. Model. **41**, 1119–1135 (2005)
12. A. Bertuzzi, A. Fasano, A. Gandolfi, C. Sinisgalli, Interstitial pressure and extracellular fluid motion in tumour cords. Math. Biosci. Eng. **2**, 445–460 (2005)
13. A. Bertuzzi, A. Fasano, A. Gandolfi, C. Sinisgalli, Cell resensitization after delivery of a cycle-specific anticancer drug and effect of dose splitting: learning from tumour cords. J. Theor. Biol. **244**, 388–399 (2007)
14. A. Bertuzzi, A. Fasano, A. Gandolfi, C. Sinisgalli, Reoxygenation and split-dose response to radiation in a tumour model with Krogh-type vascular geometry. Bull. Math. Biol. **70**, 992–1012 (2008)
15. A. Bertuzzi, A. Fasano, A. Gandolfi, C. Sinisgalli, Modelling the evolution of a tumoural multicellular spheroid as a two-fluid Bingham-like system. Math. Mod. Meth. Appl. Sci. **23**, 2561–2602 (2013)
16. A. Bertuzzi, C. Bruni, F. Papa, C. Sinisgalli, Optimal solution for a cancer radiotherapy problem. J. Math. Biol. **66**, 311–349 (2013)
17. A. Brú, S. Albertos, J.L. Subiza, J. López García-Asenjo, I. Brú, The universal dynamics of tumor growth. Biophys. J. **85**, 2948–2961 (2003)
18. H.M. Byrne, J.R. King, D.L.S. McElwain, L. Preziosi, A two-phase model of solid tumour growth. Appl. Math. Lett. **16**, 567–573 (2003)
19. H.M. Byrne, L. Preziosi, Modelling solid tumour growth using the theory of mixtures. Math. Med. Biol. **20**, 341–366 (2003)
20. M.A. Chaplain, S.R. McDougall, A.R. Anderson, Mathematical modeling of tumor-induced angiogenesis. Annu. Rev. Biomed. Eng. **8**, 233–257 (2006)
21. D. Chen, J.M. Roda, C.B. Marsh, T.D. Eubank, A. Friedman, Hypoxia inducible factors-mediated inhibition of cancer by GM-CSF: a mathematical model. Bull. Math. Biol. **74**, 2752–2777 (2012)
22. A. d'Onofrio, A. Gandolfi, Chemotherapy of vascularised tumours: role of vessel density and the effect of vascular "pruning". J. Theor. Biol. **264**, 253–265 (2010)
23. T. Eubank, R.D. Roberts, M. Khan, J. Curry, G.J. Nuovo, P. Kuppusamyl, C. Marsh, Granulocyte macrophage Colony-Stimulating factor inhibits breast cancer growth and metastasis by invoking an anti-angiogenic program in tumor-educated macrophages. Cancer Res. **69**, 2133–2140 (2009)
24. A. Fasano, Glucose metabolism in multicellular spheroids, ATP production and effects of acidity, in *New Challenges for Cancer Systems Biomedicine*, ed. by A. d'Onofrio, Z. Agur, P. Cerrai, A. Gandolfi (Springer, to appear)
25. A. Fasano, A. Gandolfi, The steady state of multicellular tumour spheroids: a modelling challenge, in *Mathematical Methods and Models in Biomedicine*, ed. by U. Ledzewicz, H. Schaettler, A. Friedman, E. Kashdan (Springer, New York, 2012), pp. 161–179
26. A. Fasano, A. Bertuzzi, A. Gandolfi, Mathematical modelling of tumour growth and treatment. In: *Complex Systems in Biomedicine*, ed. by A. Quarteroni, L. Formaggia, A. Veneziani (Springer, Italia, Milano, 2006), pp. 71–108
27. A. Fasano, M.A. Herrero, M. Rocha Rodrigo, Slow and fast invasion waves in a model of acid-mediated tumour growth. Math. Biosci. **220**, 45–56 (2009)

28. A. Fasano, M. Gabrielli, A. Gandolfi, The energy balance in stationary multicellular spheroids. Far East J. Math. Sci. **39**, 105–128 (2010)
29. A. Fasano, M. Gabrielli, A. Gandolfi, Investigating the steady state of multicellular spheroids by revisiting the two-fluid model. Math. Biosci. Eng. **8**, 239–252 (2011)
30. A. Fasano, M. Gabrielli, A. Gandolfi, Erratum to: investigating the steady state of multicellular spheroids by revisiting the two-fluid model. Math. Biosci. Eng. **9**, 697 (2012)
31. J. Folkman, M. Hochberg, Cell-regulation of growth in three dimensions. J. Exp. Med. **138**, 745–753 (1973)
32. J. Folkman, E. Merler, C. Abernathy, G. Williams, Isolation of a tumor fraction responsible for angiogenesis. J. Exp. Med. **133**, 275–288 (1971)
33. J.P. Freyer, R.M. Sutherland, Regulation of growth saturation and development of necrosis in EMT6/Ro multicellular spheroids by the glucose and oxygen supply. Cancer Res. **46**, 3504–3512 (1986)
34. A. Friedman, A hierarchy of cancer models and their mathematical challenges. Discrete Contin. Dyn. Syst. B **4**, 147–159 (2004)
35. R.A. Gatenby, E.T. Gawlinski, A reaction-diffusion model for cancer invasion. Cancer Res. **56**, 5745–5753 (1996)
36. J.B. Gillen, E.A. Gaffney, N.K. Martin, P.K. Maini, A general reaction-diffusion model of acidity in cancer invasion. J. Math. Biol. (2013)
37. S. Goel, D.G. Duda, L. Xu, L.L. Munn, Y. Boucher, D. Fukumura, R.K. Jain, Normalization of the vasculature for treatment of cancer and other diseases. Physiol. Rev. **91**, 1071–1121 (2011)
38. P. Greenspan, Models for the growth of a solid tumour by diffusion. Stud. Appl. Math. **51**, 317–340 (1972)
39. P. Hahnfeldt, D. Panigrahy, J. Folkman, L. Hlatky, Tumor development under angiogenic signaling: a dynamical theory of tumor growth, treatment response, and postvascular dormancy. Cancer Res. **59**, 4770–4775 (1999)
40. M.A. Herrero, Reaction-diffusion systems: a mathematical biology approach, in *Cancer Modelling and Simulation*, ed. by L. Preziosi (Chapman and Hall, Boca Raton, 2003), pp. 367–420
41. T. Hillen, K.J. Painter, A user's guide to pde models for chemotaxis. J. Math. Biol. **58**, 183–217 (2009)
42. T. Hillen, H. Enderling, P. Hahnfeld, The tumor growth paradox and immune system-mediated selection for cancer stem cells. Bull. Math. Biol. **75**, 161–184 (2013)
43. P. Hinow, P. Gerlee, L.J. McCawley, V. Quaranta, M. Ciobanu, J.M. Graham, B.P. Ayati, J. Claridge, K.R. Swanson, M. Loveless, A.R.A. Anderson, A spatial model of tumor-host interaction: application of chemotherapy. Math. Biosci. Eng. **6**, 521–546 (2009)
44. D.G. Hirst, J. Denekamp, Tumour cell proliferation in relation to the vasculature. Cell Tissue Kinet. **12**, 31–42 (1979)
45. A. Iordan, A. Duperray, C. Verdier, A fractal approach to the rheology of concentrated cell suspensions. Phys. Rev. E **77**, 011911 (2008)
46. R.K. Jain, Normalizing tumor vasculature with anti-angiogenic therapy: a new paradigm for combination therapy. Nat. Med. **7**, 987–989 (2001)
47. H.V. Jain, A. Friedman, Modeling prostate cancer response to continuous versus intermittent androgen ablation therapy. Discrete Contin. Dyn. Syst. B **18**, 945–967 (2013)
48. K.A. Landman, C.P. Please, Tumour dynamics and necrosis: surface tension and stability. IMA J. Math. Appl. Med. Biol. **18**, 131–158 (2001)
49. U. Ledzewicz, M. Naghnaeian, H. Schättler, Optimal response to chemotherapy for a mathematical model of tumor-immune dynamics. J. Math. Biol. **64**, 557–77 (2012)
50. H.A. Levine, B.D. Sleeman, M. Nilsen-Hamilton, Mathematical modeling of the onset of capillary formation initiating angiogenesis. J. Math. Biol. **42**, 195–238 (2001)
51. J.S. Lowengrub, H.B. Frieboes, F. Jin, Y.L. Chuang, X. Li, P. Macklin, S.M. Wise, V. Cristini, Nonlinear modelling of cancer: bridging the gap between cells and tumours. Nonlinearity **23**, 1–91 (2010)

52. J.V. Moore, H.A. Hopkins, W.B. Looney, Dynamic histology of a rat hepatoma and the response to 5-fluorouracil. Cell Tissue Kinet. **13**, 53–63 (1980)
53. J.V. Moore, P.S. Hasleton, C.H. Buckley, Tumour cords in 52 human bronchial and cervical squamous cell carcinomas: inferences for their cellular kinetics and radiobiology. Br. J. Cancer **51**, 407–413 (1985)
54. M. Neeman, K.A. Jarrett, L.O. Sillerud, J.P. Freyer, Self-diffusion of water in multicellular spheroids measured by magnetic resonance microimaging. Cancer Res. **51**, 4072–4079 (1991)
55. P. Panorchan, M.S. Thompson, K.J. Davis, Y. Tseng, K. Konstantopoulos, D. Wirtz, Single-molecule analysis of cadherin-mediated cell–cell adhesion. J. Cell Sci. **119**, 66–74 (2006)
56. V.M. Perez-Garcia, G.F. Calvo, J. Belmonte-Beitia, D. Diego, L. Perez-Romasanta, Bright solitary waves in malignant gliomas. Phys. Rev. E **84**, 1–6 (2011)
57. L. Preziosi, G. Vitale, A multiphase model of tumor and tissue growth including cell adhesion and plastic reorganization. Math. Mod. Meth. Appl. Sci. **21**, 1901–1932 (2011)
58. K.R. Rajagopal, L. Tao, *Mechanics of Mixtures* (World Scientific, Singapore, 1995)
59. D. Ribatti, A. Vacca, M. Presta, The discovery of angiogenic factors: a historical review. General Pharmacol. **35**, 227–231 (2002)
60. J.M. Roda, L.A. Summer, R. Evans, G.S. Philips, C.B. Marsh, T.D. Eubank, Hypoxia-inducible factor-2α regulates GM-CSF-derived soluble vascular endothelial growth factor receptor 1 production from macrophages and inhibits tumor growth and angiogenesis. J. Immunol. **187**, 1970–1976 (2011)
61. J.M. Roda, Y. Wang, L. Sumner, G. Phillips, T.D. Eubank, C. Marsh, Stabilization of HIF-2α induces SVEGFR-1 production from Tumor-associated macrophages and enhances the Anti-tumor effects of GM-CSF in murine melanoma model. J. Immunol. **189**, 3168–3177 (2012)
62. A. Stephanou, S.R. McDougall, A.R.A. Anderson, M.A.J. Chaplain, Mathematical modelling of flow in 2d and 3d vascular networks: applications to anti-angiogenic and chemotherapeutic drug strategies. Math. Comput. Model. **41**, 1137–1156 (2005)
63. K.R. Swanson, C. Bridge, J.D. Murray, E.C. Alvord Jr., Virtual and real brain tumors: using mathematical modeling to quantify glioma growth and invasion. J. Neurol. Sci. **216**, 1–10 (2003)
64. K.R. Swanson, R. Rockne, J. Claridge, M.A. Chaplain, E.C. Alvord Jr., A.R.A. Anderson, Quantifying the role of angiogenesis in malignant progression of gliomas: In silico modeling integrates imaging and histology. Cancer Res. **71**, 7366–7375 (2011)
65. I.F. Tannock, The relation between cell proliferation and the vascular system in a transplanted mouse mammary tumour. Br. J. Cancer **22**, 258–273 (1968)
66. Y. Yang, L. Xing, Optimization of radiotherapy dose-time fractionation with consideration of tumor specific biology. Med. Phys. **32**, 3666–3677 (2005)

Avascular Tumor Growth Modelling: Physical Insights to Skin Cancer

Martina Ben Amar

Abstract In this chapter I present the state-of-the-art theoretical models for avascular tumor growth which are well established nowadays. I focus on models able to treat morphologic instabilities and phase segregation, two typical features of skin cancer for example melanoma. Contrary to experiments made in vitro on growing colonies, I show that the geometry of melanoma confined in the epidermis in the early stages of tumor growth suppresses the necrotic core and is responsible of inhomogeneities due to aggregation of cancerous cells. A relatively simple model consisting in the adaptation of the two-phase mixture model is enough to explain the morphologies of the tumor not only qualitatively but also quantitatively. Despite the complexity of the nonlinear partial differential equations that results from this model, I also present analytical treatments based on the techniques of nonlinear physics and W.K.B approximation to explain the observed structures in dermatology.

Keywords Tumor multiphase modeling • Contour instability-Phase segregation • Skin cancer morphology • Clinical dermatology

1 Introduction

Avascular cancer modelling is a well-developed topic nowadays [97], but few of the theoretical studies have been orientated towards skin tumors, even less towards melanomas. By opposition, there is an increase interest for melanoma in molecular genetics and pathology [110]. From a mathematical viewpoint, it is astonishing that experts in modelling neglect this really important disease for the world population and for which many morphological data are available nowadays.

M. Ben Amar (✉)
Laboratoire de Physique Statistique, Ecole Normale Supérieure, UPMC Univ Paris 06, Université Paris Diderot, CNRS, 24 rue Lhomond, 75005 Paris, France

Institut Universitaire de Cancérologie, Faculté de médecine, Université Pierre et Marie Curie-Paris 6, 91 Bd de l'Hôpital, 75013 Paris, France
e-mail: benamar@lps.ens.fr

© Springer Science+Business Media New York 2014
A. d'Onofrio, A. Gandolfi (eds.), *Mathematical Oncology 2013*,
Modeling and Simulation in Science, Engineering and Technology,
DOI 10.1007/978-1-4939-0458-7_3

Skin cancers are observable to the naked eyes, or by dermoscopic inspections in the office of the dermatologist, or by more sophisticated techniques in hospital. In addition biopsy is possible, giving the structure of the tumor inside the first layers of the skin. Sophisticated modelling approaches now available combined with powerful computational methods should test their predictive potentials on melanomas in situ (by opposition to melanocytic cell colonies growing in vitro [16]). The goal of this chapter is to present specific features of melanomas that the mixture model is able to recover. But let us first give a brief reminder of skin biology.

Human skin can be divided into three layers, epidermis, dermis and hypodermis. The epidermis is the superficial layer of the skin and is mainly composed of keratinocytes and to a lesser extent melanocytes. These cells proliferate in a basal monolayer attached to the basement membrane forming the dermal-epidermal junction and migrate towards the skin surface during their differentiation. The stratum corneum is the outermost part of the epidermis and is made of fully differentiated keratinocytes and of non-living corneocytes, in a lipid-rich matrix regulating skin permeability. In healthy skin, each melanocyte remains connected to neighbor keratinocytes and to the basement membrane. Its main role consists in producing the melanin, which is the skin pigment. Melanin is enclosed into vesicles and is then transported by neighbor keratinocytes via endocytosis and exocytosis. Melanocytic lesions such as nevi and melanoma originate from a dysregulation of melanocytes leading to the invasion to the surrounding tissue.

Melanomas are the most deadly skin cancer, being responsible for 75 % of the mortality in this kind of tumors, according to the Skin Cancer Foundation. Unlike cancers affecting other organs, these tumors are directly observable since the primary tumor appears as a pigmented lesion at the surface of the skin. Early detection is therefore made possible by simple skin examination, possibly performed by the patient himself. When a melanoma is detected at an early stage, it can be treated by simple excision and the 10-year survival rate is higher than 99 %. However, the survival rate drops to less than 50 % when it penetrates deeply into the dermis. In the last decades many efforts have been made to improve the methods of differential diagnosis in order to classify malignant and benign melanocytic lesions based on the morphological criteria. Empirical studies from collections of clinical cases have led to the identification of shapes and microstructures, characteristics of melanoma, but the underlying mechanisms generating these structures as well as the morphological differences between malignant and benign tumors remain largely unknown. We recently proposed physical mechanisms controlling the contour regularity of melanocytic tumors [25, 42, 44] and explaining the appearance of microstructures such as pigmented dots and globules [13, 41].

In order to develop methods for more accurate diagnosis, it is essential to better understand the mechanisms which control the evolution of skin tumors. It will allow maximizing the information detected from the structures observed by various imaging techniques. The molecular and cellular biology has made strong progress in oncology very recently thanks to the accumulation of data on the molecular and cellular components involved in the genesis and progression of cancer [43, 76, 77]. More than 200 genes involved in these processes have been identified [64] and the treatment of metastatic melanoma shows significant improvements thanks to the development of targeted therapies and immunotherapies. In particular, the Vemurafenib targets a mutation in the RAS protein present in 50 % of melanoma [89] and the Ipilimumab stimulates the action of T cells interacting with CTLA-4, a molecule that plays a central role in regulating the immune system [99]. A search with PubMed shows that the number of papers in oncology referring to a mathematical modelling increased from 2.4 % to 8.6 % between 1991 and 2011,[1] reflecting an awareness of the importance of these tools for the prevention, diagnosis and treatment of cancer [36, 48]. Depending on the organ affected, on the phase of growth and the issues discussed, a variety of models were used [10, 35, 97]. In this chapter, I briefly describe different models: continuous, stochastic, discrete and hybrid, developed in the framework of avascular solid tumors and I will focus more on multiphase models adapted to skin tumors. I explain the adaptation of this model to the geometry of melanomas. Finally I focus on the two main aspects of their morphology: contour instability and existence of microstructures at the origin of the heterogeneity of the tumor.

2 Historical Overview

2.1 *Continuous Models*

Global Mass Balance and Concept of Dormant State

The first models developed for solid tumors were mainly interested by growth dynamics [29, 54, 70, 128, 132, 137]. Mayneord recognized as early as 1932 [100] that for a long time the diameter of a tumor seems to increase linearly with time, in apparent contradiction with the idea of an exponential proliferation of cancer cells. To explain this phenomenon, he proposed the first model of cell growth confined to the tumor periphery. This model is completed in 1955 by Thomlinson and Gray [137] who explain the presence of a necrotic core in bronchial carcinoma with a gradient of oxygen concentration between the periphery and the center of the cylindrical tumors. In the absence of necrotic core (rate sufficient to ensure cell survival oxygen) and assuming a rate of oxygen consumption δ_n constant, their model gives the oxygen concentration $n(r)$ at a distance r from the center:

$$n(r) = n_0 - \frac{\delta_n}{4D_n}(R^2 - r^2) \tag{1}$$

[1]http://www.ncbi.nlm.nih.gov/ Pubmed. The number of items in oncology has been determined using the keywords “cancer” and “tumor.” The number of articles referring to a mathematical modelling was determined by adding the keywords “modelling” and “mathematical model.” Search on 29th June 2012.

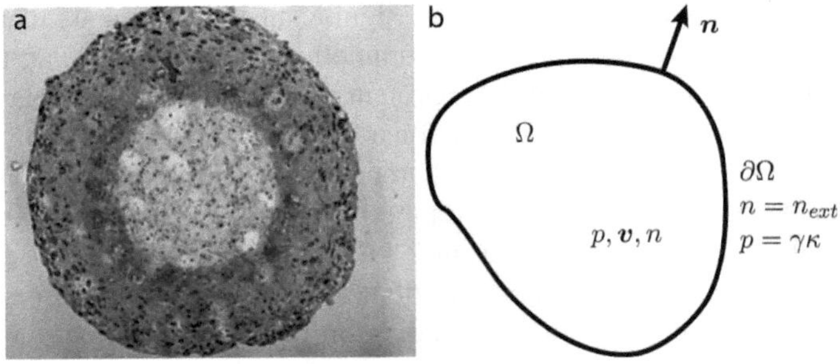

Fig. 1 (**a**) Section of a spheroid of tumor cells cultured in vitro having reached a diameter of 1 mm in 20 days. One can observe the presence of a necrotic core. Image from [60]. (**b**) Hydrodynamic model of growth

with n_0 the concentration at the surface, D_n the diffusion coefficient of oxygen and R the radius of the tumor. For a radius $R > R_{crit} = \sqrt{4D_n n_0/\delta_n} \approx 200\,\mu\text{m}$, the concentration falls to zero in the middle of the tumor and a necrotic core appears (see Fig. 1). This pattern of growth limited by diffusion was further developed by Burton [29] and then by Greenspan in 1972 [70]. By introducing the existence of a layer of quiescent cells (vanishing rate of proliferation), caused by the presence of growth inhibitors and cell wastes, Greenspan succeeds to explain the saturation of the growth for most of the tumors when the tumor size reached the millimeter (dormant). With R being the radius of the tumor, n the nutrient concentration and m the inhibitor concentration, the model of Greenspan reads [70]

$$3R^2\frac{dR}{dt} = \Gamma(R^3 - R_g^3) - \delta R_i^3,$$
$$\frac{1}{r^2}\frac{\partial}{\partial r}\left(r^2\frac{\partial n}{\partial r}\right) = \frac{\delta_n}{D_n}H(r - R_i)H(R - r),$$
$$\frac{1}{r^2}\frac{\partial}{\partial r}\left(r^2\frac{\partial m}{\partial r}\right) = \frac{\gamma_i}{D_i}H(R_i - r),$$

with Γ the rate of cell proliferation, R_i the radius of the necrotic core, R_g the outer radius of the quiescent area, δ the loss rate of dead cells in the necrotic core, γ_i the production rate of inhibitors by the cells, D_i the diffusion coefficient for the inhibitors and H the Heaviside function. The boundary conditions are given by $m(R) = 0$, $m(R_g) = \beta_i$ (the inhibition threshold for proliferation), $n(R) = n_0$ and $n(R_i) = n_i$ (threshold for cell death). The first equation expresses the change in volume of the tumor, with a layer of proliferative growth and with contraction of the necrotic core, and the last two equations mean the equilibrium concentrations of oxygen and inhibitor in the tumor. The cancellation of the right-hand side of the first equation gives the stationary radius of the tumor. Note that this steady state is dynamic, the growth of cells in the outer layer and loss in the necrotic core causing a cell flow to the center of the spheroid. The dormant state concept was experimentally confirmed by Folkman and Hochberg in 1973 [60]. Cultivating spheroids of melanoma cells in an agar gel saturated with nutrients, they discovered that the colonies reach a diameter of 3–4 mm and that in this steady state the cells at the surface migrate towards the necrotic center, thus validating the model of Greenspan (Fig. 1). In 1971 Sutherland et al. also found that cell proliferation decreases gradually inwards the spheroids [135].

Homogeneous Hydrodynamic Models

Based on these observations and the pioneering work of Greenspan, many hydrodynamic models of avascular tumor growth have been developed [18, 30, 46, 71] to understand the internal dynamics of the tumor. According to Lowengrub et al. [97], we give here a typical theoretical framework of these models. Let Ω be the volume of the tumor (possibly divided into several regions Ω_i [70, 71]) and $\partial\Omega$ the border with the surrounding tissue (Fig. 1). Assuming a uniform cell density and diffusion equilibrium in Ω the evolution of the tumor is given by

$$\boldsymbol{v} = -K\nabla p, \quad p = \gamma\kappa \quad \text{on } \partial\Omega, \tag{2}$$

$$\nabla \cdot \boldsymbol{v} = \Gamma, \quad V = \boldsymbol{n} \cdot \boldsymbol{v} \quad \text{on } \partial\Omega, \tag{3}$$

$$D_n \Delta n + S = 0, \quad n = n_{ext} \quad \text{on } \partial\Omega, \tag{4}$$

with $\boldsymbol{v}$ being the velocity of tumor cells, K the cell mobility, p the hydrostatic pressure, γ the surface tension of the interface due to the difference in adhesion between tumor cells with their counterpart and with the exterior environment, κ the mean curvature of the interface $\partial\Omega$, Γ the local variation of cell volume due to proliferation and cell death, V the normal velocity of the interface, $\boldsymbol{n}$ the outer normal of the interface, n the nutrient concentration, S the consumption of nutrients by tumor cells (S being negative) [2,3] and n_{ext} the nutrient concentration at the tumor border. The cell velocity, here given by Darcy's law [Eq. (2)], is represented in some models also by a Stokes' law [62] ($-\nu\delta\boldsymbol{v} + \nabla p - (\nu/3)\nabla\Gamma = 0$ with ν the viscosity of the medium) or a Darcy-Stokes law [152] ($\boldsymbol{v} - \epsilon\Delta\boldsymbol{v} = -K\nabla p$ with ϵ a constant associated with the medium viscosity [19]). These models account for different regimes of spheroid growth and the emergence of a steady state with spherical

[2] In the case of a diffusion-limited growth one often chooses $\Gamma = an - \delta$ [97]. This non-uniform growth including a uniform apoptosis term δ in the tumor was introduced by McElwain and Morris in 1978 [101] according to experiments of Sutherland and Durand [134] showing that a dormant state was reached without the appearance of necrotic cell loss in the tumor center.

[3] For avascular tumors, one often chooses $S = -\delta_n n$, that is, an oxygen consumption proportional to the oxygen concentration as introduced by Deakin [49], according to the experiments of Sutherland and Durand [134] that show an inconsistency with the model of Greenspan. Such a relationship is well justified when the oxygen concentration is low.

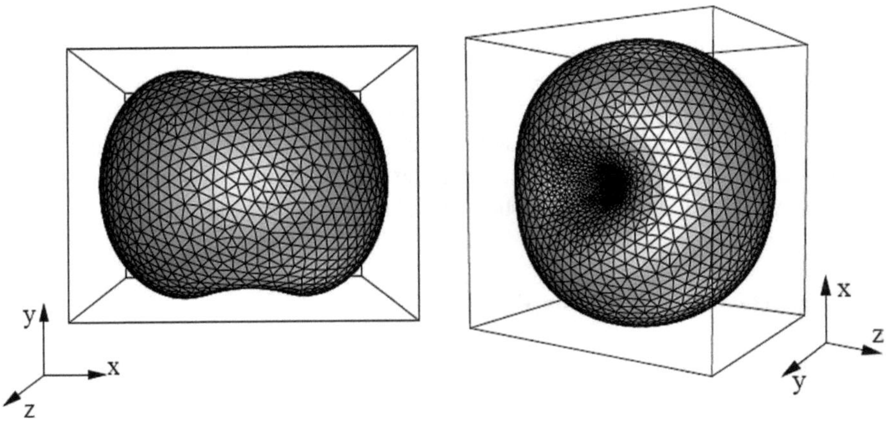

Fig. 2 Destabilization of a spheroid in the model of Li et al. [95]

symmetry when $\int_{\Omega} \Gamma d\Omega = 0$ (indicating a compensation between proliferation and cell death) combined with the flux of cells from the periphery to the center [71, 104, 133].

Hydrodynamic Instabilities

The stability of the spherical shapes was considered as early as 1976 by Greenspan [71]. He shows that when the spheroid exceeds a critical radius the stabilizing effects of surface tension (cell adhesion) no longer compensate the destabilizing effects of cell death in the tumor center leading to destabilization. It may explain the fragmentation of spheroids observed in some experiments by Sutherland et al. [135] and a possible mechanism of invasion. In the framework of these hydrodynamic models [17, 63], this instability has been found in several geometries by linear stability analysis [1, 31] and developed recently numerically by Lowengrub et al. [46, 95] (Fig. 2).

Influence of Mechanical Stresses and Elastic Models

In 1997, Jain and collaborators [78] studied the growth of tumor cell spheroids in agar gels of different stiffness. The spheroids placed in the most rigid environments reach dormant states of smaller diameter, demonstrating the inhibitory effect of strong constraints on the growth. Also from several studies shown later, there is a significant pressure at the center of solid tumors [72, 130], which particularly affect the penetration of therapeutic agents. This phenomenon was predicted by Shannon and Rubinsky [125]. Using a linear elasticity model, they show that any process of inhomogeneous growth of a tumor mass with spherical symmetry led the development of residual mechanical stresses. Existence of residual stresses due to non homogeneous growth is well explained now [23, 24, 68]. Moreover, fibre networks forming the cell skeletons and extracellular matrices confer elastic behavior in living tissues and may be at the origin of instabilities. The mechanical stresses generated by such growth processes play a significant role in their homeostasis and final shapes. These forces are felt by the cells through adhesion molecules, especially influencing the rate of cell proliferation [117] or polarization during cell divisions [59]. These observations are accompanied by the development of models taking into account the effects of mechanical stresses on the tumor progression [97, 142]. Some of these are based on rheology such as the poroelastic [122] and the viscoelastic model [98] or on nonlinear elasticity [4, 5, 40, 52, 53]. The theory of elastic growth developed by Rodriguez et al. [121] provides a fruitful framework for these models [24, 51, 82], the strain tensor $\boldsymbol{F}$[4] being decomposed into a growth tensor $\boldsymbol{G}$ (variation of volume due to growth and cell death) and an elastic tensor $\boldsymbol{F_n}$ (mechanical constraints during growth)

$$\boldsymbol{F} = \boldsymbol{F}_n \boldsymbol{G}, \tag{5}$$

as shown in Fig. 3. Growth may possibly be taken isotropic ($\boldsymbol{G} = g\boldsymbol{I}$) and the solid is often assumed incompressible in the case of living tissues (composed mainly of

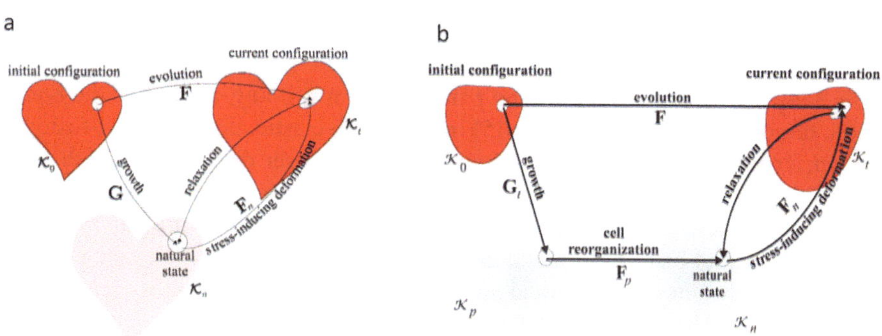

Fig. 3 (**a**) In the model of elastic growth of Rodriguez et al. [121] the strain tensor $\boldsymbol{F}$ is decomposed into a tensor representing the growth $\boldsymbol{G}$, bringing the system to a non-stressed state with a configuration possibly incompatible with the solid integrity (due to breakdown or superposition) or with the boundary conditions, and an elastic deformation tensor $\boldsymbol{F_n}$, bringing the system to the final current configuration. Figure from [69]. (**b**) Ambrosi and Preziosi [7] add an additional step in their model to take into account the cellular reorganization

[4]By definition $F_{ij} = \partial x_i / \partial X_j$ with $\boldsymbol{X}$ identifying points of the solid in the initial state Ω_0 and $\boldsymbol{x}$ in the current state Ω_t [51, 53].

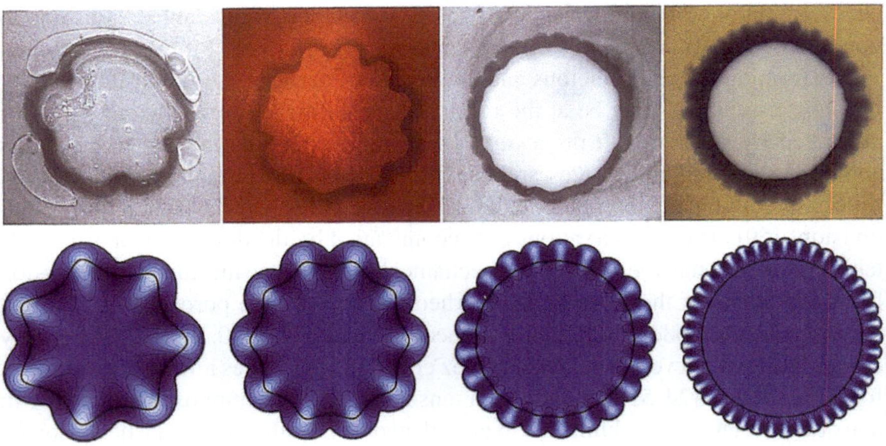

Fig. 4 Instability established by Dervaux and Ben Amar [52, 53] when the growth of an elastic mass is located in the outer part. Swelling gel experiment (*top*) and theoretical predictions in an elastic model (*bottom*). Figures from [51]

water) ($\det(\boldsymbol{F}_n) = 1$). Assuming negligible inertia of the system, the mechanical balance of the tissue imposes

$$\nabla \cdot \boldsymbol{T} = 0, \qquad \boldsymbol{T}\boldsymbol{n} = \boldsymbol{t} \tag{6}$$

with $\boldsymbol{T}$ the Cauchy stress tensor and $\boldsymbol{t}$ the constraints, if any, applied to the border of the tumor. Ambrosi and Mollica [5] introduce an elastic model with diffusion of the nutrient limiting the growth and found qualitatively the dynamical growth of the spheroids. An elastic model of melanoma growth in the epidermis is also developed by Dervaux and Ben Amar [52] (see Fig. 4). Considering the circular symmetry for the tumor and localized growth in an outer layer they show the existence of an elastic instability beyond a critical amount of growth. This instability leads to the breaking of the circular symmetry and to the appearance of ripples on the contour of the tumor itself. It provides a possible mechanism for the appearance of contour irregularities and asymmetrical appearance observed in melanomas.

The existence of residual mechanical stresses is well established in the case of arteries and plays an important role for their correct functioning [38, 141]. In the case of tumors that notion is less clear; tumor cells proliferate in an uncontrolled manner with the extracellular matrix constantly being deteriorated and re-synthesized. Thus the definition of a reference state in purely elastic models is problematic. To take into account the reorganization of bonds within the tissue, Ambrosi and Preziosi [7] add to a previous theoretical framework a plastic deformation given by $\boldsymbol{F}_p$ (Fig. 3).

$$\boldsymbol{F} = \boldsymbol{F}_n \boldsymbol{F}_p \boldsymbol{G}. \tag{7}$$

Assuming a fast reorganization with respect to the growth process, they found a fluid behavior of the tumor. A similar result is obtained by Ranft et al. [118] in a continuous model including a stochastic process of division and apoptosis. In this model a multicellular mass has an effective behavior such as a viscoelastic fluid with a relaxation time determined by the cell proliferation rate. Modelling the cell component as a Bingham fluid has been proposed in [26].

In the case of melanocytic lesions confined to the epidermis, the growth time is long (months or years) compared to cell division (hours or days), tumor proliferation is disorganized and the tissue (the epidermis) does not have extracellular matrix. Although we keep in mind the possible existence of elastic effects, the model that I will discuss hereafter is therefore based on a fluid description of tumors.

Stochastic Models

In biological systems random fluctuations are fundamental to many processes (genetic heterogeneity of a cellular population, signal transduction, dynamics of small systems). Wette et al. [145, 146] in 1974 suggest a model of solid tumor growth based on a Fokker-Planck equation with a stochastic process of mitosis and apoptosis (see also [36, 37]). This approach is developed by Ranft et al. [118] to evaluate the velocity and density fluctuations of the cells and the diffusion coefficient of a single cell within the tumor. In analogy with the nucleation theory, Basan et al. [18] taking into account surface tension between healthy cells and cancer cells show that the diameter for the dormant state is thermodynamically unstable and fluctuations in the number of cells may allow the tumor to continue its expansion. Note however that the latter model includes only an inhibition of proliferation by contact (proportional to the local cell concentration) and does not take into account the influence of the concentration of diffusing molecules in the tissue.

2.2 *Discrete Models*

It is now well established that tumors are far from being monoclonal sets originating from a single renegade cell [143]. Indeed they consist of a heterogeneous population with genetic and epigenetic changes [124]. Testing transplantation of cells from breast cancer [2], brain [127] and colon [120] on immunodeficient mice shows that only a small portion of this population has the potential to regenerate a tumor. Similar to stem cells from healthy tissue, these cancer stem cells can generate a variety of differentiated cells [27]. To improve methods of treatment and diagnosis a better understanding of the influence of this heterogeneous population is also necessary [124].

Discrete models describe the dynamics of the tumor at the level of individual cells, including possibly continuous fields such as nutrient concentration. They allow studying the effect of this heterogeneity by taking into account precisely the interactions and cell cycles. These models are reviewed in detail by Lowengrub et al. [97], Drasdo [56] and Roose et al. [123] and can be grouped between models on grid (cellular automata [8, 9], generalized Potts models [66, 113], Boltzmann methods on grid [3]) and models without grid, where (similar to those of molecular dynamics models) particles represent cells [79, 119]. In the model proposed by Anderson et al. [8,9] cells perform a random walk on a square network, biased by the concentration of extracellular matrix (haptotaxis). They follow an internal program (mitosis, aging, apoptosis, migration) influenced by the oxygen concentration and the number of neighboring cells and can acquire mutations randomly. The authors show that an unfavorable microenvironment (hypoxia, heterogeneous extracellular matrix) exerts a strong selection pressure favoring the growth of more aggressive phenotypes (high proliferation and poor adhesion between cells) and leads to more invasive morphologies. In the model of Graner and Glazier [66, 113] each cell is an extended object represented by an area having the same spin σ corresponding to the index of the cell ($\sigma = 1 \dots N$). The evolution of the system is given by a Monte Carlo method using the Hamiltonian

$$H = \sum_{(i,j)} H_{ij} + \sum_{\sigma} h_{\sigma} \tag{8}$$

with

$$H_{ij} = J(\tau_{\sigma_i}, \tau_{\sigma_j})(1 - \delta_{\sigma_i \sigma_j})$$

(i, j) being indices of neighboring points on the network and

$$h_{\sigma} = \lambda_V(\tau_{\sigma})(V(\sigma) - V_t(\sigma))^2 + \lambda_S(\tau_{\sigma})(S(\sigma) - S_t(\sigma))^2,$$

τ identifying the cell type σ, $J(\tau_1, \tau_2)$ the adhesion energy between cell types τ_1 and τ_2, λ_V the energy associated to changes in cell volume, λ_S the energy associated to changes in the surface of the cell, V_t the average volume of the cell at time t which may vary according to the local nutrient concentration and the cell dividing when it reaches twice its original volume and S_t the average surface cell for imposing a shape close to a sphere. Graner and Glazier [66] have used this model to study the phase separation between cells of different types by the differential adhesion mechanism identified by Townes and Holtfreter [138] (Fig. 5). Based on the similarities of their model with models of directional solidification, they also develop a phase diagram of the growth morphologies, the control parameters being the adhesion between cells (surface tension) and the ratio between the cell proliferation and diffusion of nutrients [113]. The results obtained by the discrete description [57] and the continuous models of growing spheroids have been compared by Byrne and Drasdo [32]. The model of Drasdo and Hoehme [57] has been implemented in the CellSys software [79]. It leads to the same

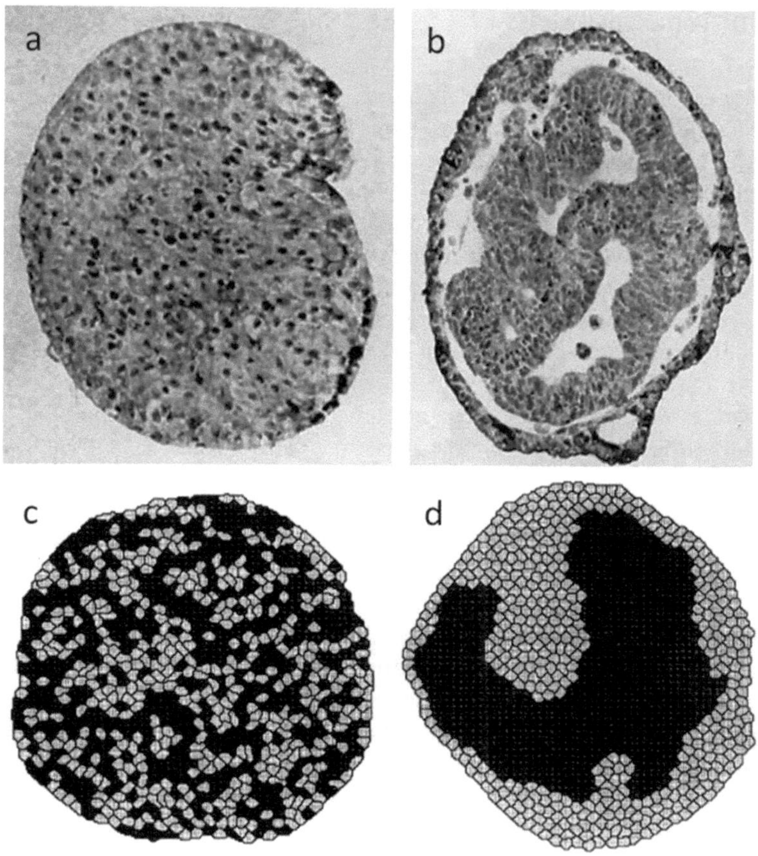

Fig. 5 (**a**, **b**) Experiments of Townes and Holtfreter [138]. (**a**) Amphibian embryo cells are dissociated. (**b**) Cell adhesion causes them to re-aggregate and differences in compliance between cell types reorganize the tissue (here epidermal cells in *black* and cells of the neural plate in *white*). (**c**, **d**) Simulations of Graner and Glazier [66] who study this phase separation with a generalized Potts model

dynamical growth that continuous models such as the work of Greenspan [Eqs. (2) and (3)].

Discrete models, however, are limited by the large number of unknown coupled variables and require almost systematic use of numerical simulations, making it difficult to understand the dynamics of the modelled systems. The large number of variables is limiting also for simulations of large systems (such as melanomas of 6 mm in diameter). Note, however, the recent development of hybrid models [20] adopting a description with both discrete cells (intracellular mechanisms, heterogeneity of the population) and continuous quantities (cell density, continuum mechanics) to reach the scale of molecules and tissues simultaneously.

3 Multiphase Models

In homogeneous continuous models, terms representing source or sink of matter linked to the cellular proliferation [term Γ in the equation Eq. (3)] imply the existence of another phase providing or absorbing this matter that is not modelled explicitly. Living tissue includes local effect of the constituents of different types (different types of cells, extracellular fluid, extracellular matrix, vasculature, etc.). These elements can locally occupy different portions of the space and have different displacements. Also between the different phases, matter exchanges (growth or cell death, synthesis or degradation of extracellular matrix) and mechanical interactions (as viscous drag, cell–cell adhesion, cell adhesion to the extracellular matrix) exist. At a scale larger than the spatial heterogeneity, the mixture theory allows a continuous description of the complete system (including all possible phases constituting the tissue) without the need to define interfaces between different phases.

The mixture theory has been developed from the pioneering work of Truesdell on the mechanics of fluid mixtures in the 1960 [139, 140]. The theory is completed a few years later in particular by Mueller [103] to give the modern thermodynamic framework of multiphase models. It knows important success for the description of granular media [15], mixed dielectric [136] or even meteorological systems [50]. In 1998 Please et al. [111, 112] introduce exchange terms of specific mass in biological growth process and apply the mixture theory to tumor growth particularly for understanding the dynamics of cellular mass lost in the necrotic center of spheroids. This model was subsequently developed to take into account the full balance of interphase forces [34, 92], the inhibition of growth by mechanical constraints [33] and the presence of solid phases and residual stresses [11,115,116].

Concepts and the formalism of multiphase models used in the remainder of this work will now be introduced (Fig. 6).

3.1 Conservation of Mass

All fields of the model are described in a reference Eulerian configuration and explicitly depend on a space variable $\boldsymbol{x}$ in Ω_t and a time variable t. The domain definition Ω_t of the tumor may be time-dependent (invasion of adjacent tissues).

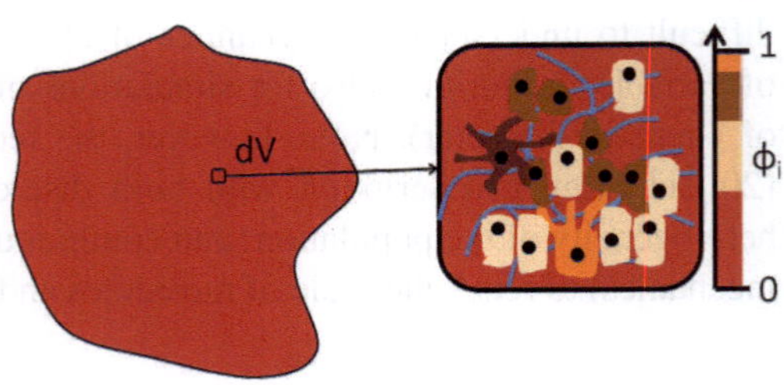

Fig. 6 In the theory of mixtures, the various components of the tissue are described by their local volume fraction $\phi_i(\boldsymbol{x}) = dV_i(\boldsymbol{x})/dV(\boldsymbol{x})$ and their local velocity $\boldsymbol{v}_i$

We consider around each point $\boldsymbol{x}$ a volume $dV(\boldsymbol{x})$ to be large enough compared to the size of the components, but small compared to the size of the whole system. We denote $\phi_i(\boldsymbol{x}) = dV_i(\boldsymbol{x})/dV(\boldsymbol{x})$ the volume fraction occupied by the constituents i and $\boldsymbol{v}_i$ their average velocity over $dV(\boldsymbol{x})$. All phases $i = 1 \dots N$ fill the entire space, requiring a saturation relationship for the mixture

$$\sum_{i=1}^{N} \phi_i = 1. \tag{9}$$

Living tissues being mainly composed of water, we assume the density of all the components to be equal to the water density ρ and the incompressibility of the mixture becomes

$$\nabla \cdot \left(\sum_{i=1}^{N} \phi_i \boldsymbol{v}_i \right) = 0. \tag{10}$$

Assuming $V \subset \Omega_t$ a fixed volume, the change in mass $M_i = \int_V \rho\phi_i d\Omega$ of the constituents of the phase i contained in V is given by

$$\frac{dM_i}{dt} = -\oint_{\partial V} \rho\phi_i \boldsymbol{v}_i \cdot \boldsymbol{n} dS + \int_V \rho\Gamma_i dV \tag{11}$$

with $\boldsymbol{n}$ the normal to the surface ∂V of volume V and $\rho_i \Gamma_i$ a mass exchange term due to biological processes of growth, death, synthesis or degradation. In some models a diffusion term is added to account for cell motility [97]. Applying the Green-Ostrogradski theorem, Eq. (11) can be rewritten:

$$\int_V \frac{\partial \rho\phi_i}{\partial t} dV = \int_V [-\nabla \cdot (\rho\phi_i \boldsymbol{v}_i) + \rho\Gamma_i] dV. \tag{12}$$

The volume V being arbitrary this relationship implies

$$\frac{\partial \phi_i}{\partial t} + \nabla \cdot (\phi_i \boldsymbol{v}_i) = \Gamma_i, \qquad i = 1..N. \tag{13}$$

Summing these equations for $i = 1 \dots N$, the assumption of saturation and incompressibility [Eqs. (9) and (10)] then give the conservation of mass in the system [6]

$$\sum_{i=1}^{N} \Gamma_i = 0. \tag{14}$$

The tissue also contains chemical factors of concentration c_j (nutrients, growth inhibitors) dissolved in the interstitial fluid and satisfying an equation of advection-diffusion-reaction

$$\frac{\partial c_j}{\partial t} + \nabla \cdot (c_j \boldsymbol{v}_l) = \nabla \cdot (D_j \nabla c_j) + S_j, \qquad j = 1..N', \tag{15}$$

with $\boldsymbol{v}_l$ the interstitial fluid velocity, D_j the diffusion coefficient of the factor j and S_j an algebraic term of consumption or degradation (in this case $S_j < 0$), or production ($S_j > 0$). When the evolution of the system is slow we can assume the diffusion at equilibrium and neglect the left side of Eq. (15).

3.2 *Mechanical Balance*

We consider the situation where the growth is slow compared to the time needed for the tissue reorganization, each phase being treated as a fluid (see discussion in the penultimate paragraph of Sect. 2.1). The interactions between the components of the tissue may be assumed weakly nonlocal and the free energy of each phase can be developed at first order [149] as shown below:

$$F_i = \int_{\Omega} \left(\psi(\phi_1, ..\phi_N) + \sum_{j=1}^{N} \frac{\epsilon_{ij}^2}{2} |\nabla \phi_j|^2 \right) d\Omega, \tag{16}$$

with ψ the free energy density for a uniform tissue and the term ϵ_{ij} can be interpreted as a surface tension between phases penalizing strong composition gradients. To take into account the effects of chemotaxis [22], some models also add a term dependent of concentrations c_j [22, 47] not considered here.

Variational Principle

To determine the dynamical state of the system here we follow the approach developed by Doi and Onuki [55] based on a variational formulation. To introduce this concept first we consider a system described by thermodynamic variables $\{x_i\}_{i=1}^N$ and a free energy $\mathcal{F}$. In equilibrium thermodynamics, we have

$$\frac{d\mathcal{F}}{dt} = 0 = \sum_{i=1}^{N} \frac{\partial \mathcal{F}}{\partial x_i} \cdot \frac{dx_i}{dt}. \tag{17}$$

Close to equilibrium, when the thermodynamic forces $\partial \mathcal{F}/\partial x_i$ are slowly varying, a quasi-stationary approximation is used to show that there is a linear relationship [91]

$$\frac{dx_i}{dt} = -\sum_{j=1}^{N} \mathcal{L}_{ij} \frac{\partial \mathcal{F}}{\partial x_i} \tag{18}$$

between velocities and forces [91]. Defining $(\mathcal{M}_{ij})$ the inverse matrix of $(\mathcal{L}_{ij})$, this relation can be inverted and gives

$$-\frac{d\mathcal{F}_i}{\partial x_i} = \sum_{j=1}^{N} \mathcal{M}_{ij} \frac{dx_j}{dt}. \tag{19}$$

This last relation can be recovered by using a variational principle concerning the "Rayleighian" (equivalent to the Lagrangian for Hamiltonian mechanics) defined by

$$W = \sum_{i=1}^{N} \sum_{j=1}^{N} \frac{dx_i}{dt} \mathcal{M}_{ij} \frac{dx_j}{dt}. \tag{20}$$

Defining

$$\mathcal{R} = \frac{1}{2} W + \frac{d\mathcal{F}}{dt}, \tag{21}$$

we get

$$\mathcal{R} = \frac{1}{2} \sum_{i=1}^{N} \sum_{j=1}^{N} \frac{dx_i}{dt} \mathcal{M}_{ij} \frac{dx_j}{dt} + \sum_{i=1}^{N} \frac{\partial \mathcal{F}}{\partial x_i} \frac{dx_i}{dt} \tag{22}$$

and Eq. (19) is obtained by minimization of $\mathcal{R}$ with respect to the velocities dx_i/dt. W means the dissipated energy by the system [55]. This variational principle defines an equilibrium between forces of viscous dissipation obtained from W and the elastic forces obtained from $d\mathcal{F}/dt$. Doi and Onuki [55] apply this principle to the multiphase model, the velocities dx_i/dt being the velocity fields $\boldsymbol{v}_i$, and the dissipation can be written as

$$W = \sum_{i=1}^{N} \int_{\Omega} \left[\sum_{j=1}^{N} \frac{M_{i,j}}{2} (\boldsymbol{v}_i - \boldsymbol{v}_j)^2 + \nu_i (\nabla \cdot \boldsymbol{v}_i)^2 \right] d\Omega. \tag{23}$$

The first term corresponds to the energy dissipation driven by viscous friction between the different constituents where M_{ij} are the friction constants. The second term means the energy dissipation due to the velocity gradients with ν_i the viscosity of phase i. For the system we consider, we will see that it is negligible compared to the first one. Using the variational principle of Rayleigh, one can show that the overdamped dynamics of the system can be determined by minimizing $\mathcal{R}$ with respect to the variables $\boldsymbol{v}_i$, the boundary conditions being vanishing at infinity. The incompressibility constraint [Eq. (10)] is imposed by adding a Lagrange multiplier p which will be identified to the interstitial pressure. Using Eq. (13), the "Rayleighian" can then be written as

$$\mathcal{R} = \sum_{i=1}^{N} \int_{\Omega} \Bigg[\underbrace{\sum_{j=1}^{N} \left(\frac{M_{i,j}}{4} (\boldsymbol{v}_i - \boldsymbol{v}_j)^2 \right) + \frac{\nu_i}{2} (\nabla \cdot \boldsymbol{v}_i)^2}_{\text{dissipation}}$$

$$+ \underbrace{\sum_{j=1}^{N} \left(\frac{\partial \psi_i}{\partial \phi_j} - \epsilon_{ij}^2 \Delta \phi_j \right) \left(-\nabla \cdot (\phi_j \boldsymbol{v}_j) + \Gamma_j \right)}_{\text{variation of the free energy}} - \underbrace{p \nabla \cdot (\phi_i \boldsymbol{v}_i)}_{\text{incompressibility}} \Bigg] d\Omega. \quad (24)$$

The equations corresponding to the balance of forces in each phase are derived by minimization of this functional with respect to $\boldsymbol{v}_i$

$$0 = \underbrace{\sum_{j=1}^{N} \phi_j \nabla \left(\frac{\partial \psi_i}{\partial \phi_j} - \epsilon_{ij}^2 \Delta \phi_j \right)}_{\text{mechanical stresses}} + \underbrace{\sum_{j=1}^{N} M_{i,j} (\boldsymbol{v}_i - \boldsymbol{v}_j) + \nu_i \Delta \boldsymbol{v}_i}_{\text{viscous forces}} + \underbrace{\phi_i \nabla p}_{\text{hydrostatic pressure}} \quad (25)$$

The interstitial liquid can be considered as incompressible [47, 114] and the force equilibrium for this liquid phase l simplifies

$$\sum_{j=1}^{N} M_{l,j} (\boldsymbol{v}_l - \boldsymbol{v}_j) + \phi_l \nabla p = 0. \quad (26)$$

The incompressibility [Eq. (10)] is not sufficient to identify all the variables of the system and one additional assumption on the velocities is necessary. One possibility is to cancel the total velocity of the mixture [42,97], meaning that the center of mass does not move so we get

$$\boldsymbol{v} = \sum_{i=1}^{N} \phi_i \boldsymbol{v}_i = \boldsymbol{0}. \quad (27)$$

This solution satisfies the incompressibility constraint and is physically justified for a very viscous mixture when the external forces on the volume and on the surface of the system are neglected. The interstitial fluid velocity is given by

$$\boldsymbol{v}_l = \frac{-1}{\phi_l} \sum_{i \neq l} \phi_i \boldsymbol{v}_i . \quad (28)$$

3.3 Two-Phase Mixture

For simplicity and in order to adapt this model to tumor growth in the epidermis we consider the case of a mixture with two phases: the cellular phase, labelled by the index c corresponding to melanocytes, and a liquid phase, labelled by the index l, corresponding to the interstitial fluid. The liquid phase contains possibly other cellular phases that are not explicitly distinguished here (e.g.,degrading dead cells, keratinocytes). Also we consider two types of chemical factors: nutrient concentration n and growth inhibitors of concentration m. The liquid phase contains possibly other cellular phases that are not explicitly distinguished here. A more detailed model including also keratinocytes and the basement membrane phase was developed in [44]. We will see, however, that most of the mechanisms of morphogenesis may be found with the simpler model considered here. Thanks to the incompressibility constraint $\phi_c + \phi_l = 1$, the free energy of the cellular phase can be written as a function only of ϕ_c without loss of generality. Choosing a friction rate proportional to the cell number $M_{c,l} = M_{l,c} = \tau\phi_c$, mechanical equilibrium of the two phases reads

$$\tau\phi_c(\boldsymbol{v_c} - \boldsymbol{v_l}) + \phi_c\nabla\left(\frac{\partial\psi}{\partial\phi_c} - \epsilon^2\Delta\phi_c\right)$$

$$-\mu\phi_c\Delta\boldsymbol{v_c} + \phi_c\nabla p = 0, \tag{29}$$

$$\tau\phi_c(\boldsymbol{v_l} - \boldsymbol{v_c}) + (1 - \phi_c)\nabla p = 0. \tag{30}$$

Using Eq. (28) and eliminating the pressure p between Eqs. (29) and (30), the set of equations giving the evolution of the model is then

$$\frac{\partial\phi_c}{\partial t} + \nabla\cdot(\phi_c\boldsymbol{v_c}) = \Gamma_c, \tag{31}$$

$$\boldsymbol{v_c} = K(\phi_c)(\nabla(-\Sigma(\phi_c) + \epsilon^2\Delta\phi_c) + \mu\Delta\boldsymbol{v_c}) \tag{32}$$

$$\frac{\partial n}{\partial t} + \nabla\cdot(n\boldsymbol{v_l}) = \nabla\cdot(D_n\nabla n) + S_n, \tag{33}$$

$$\frac{\partial m}{\partial t} + \nabla\cdot(m\boldsymbol{v_l}) = \nabla\cdot(D_m\nabla n) + S_m, \tag{34}$$

with $\Sigma(\phi_c) = \partial\psi/\partial\phi_c$ the cell–cell adhesion energy and $K(\phi_c) = (1-\phi_c)^2/\tau$. The volume fraction of the interstitial fluid ϕ_l is given by the saturation constrain of the mixture, $\phi_l = 1 - \phi_c$, and the velocity $\boldsymbol{v_l}$ by the Eq. (28), $\boldsymbol{v_l} = -(\phi_c/\phi_l)\boldsymbol{v_c}$. Expressions for the pressure term Σ and the exchange terms Γ_c, S_n and S_m are discussed in Sect. 3.4. In the case of skin tumors, growth rates are low, typically 0.3 mm per month, and the left-hand side of Eqs. (33) and (34) will be taken equal to zero in the following (diffusion equilibrium).

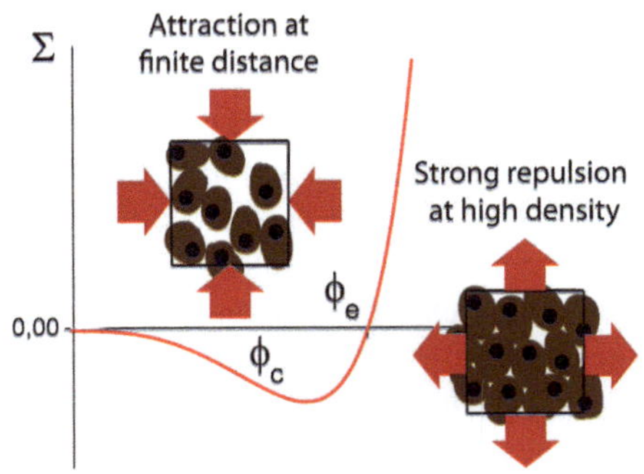

Fig. 7 Qualitative representation of the pressure Σ imposed by the cells according to their volume fraction ϕ_c. The physical consistency of the model requires $\Sigma(0) = 0$, $\Sigma < 0$ at low density (adhesion between cells) and $\Sigma > 0$ for high density (steric repulsion). We denote ϕ_e the equilibrium volume fraction ($\Sigma(\phi_e) = 0$) and ϕ^* the minimum of the pressure Σ

3.4 *Mechanical Interactions*

These interactions must satisfy elementary physical requirements. Low-density adhesion between cells tends to form aggregates leading to an effective negative cell pressure ($\Sigma < 0$), the pressure vanishing obviously in the absence of cells ($\Sigma(0) = 0$). At high density, entropic repulsion dominates ($\Sigma > 0$) to prevent the full space occupation by the cells ($\Sigma \to +\infty$ for $\phi_c \to 1$). These assumptions require the existence of a volume fraction ϕ_e for which the pressure exerted by the cells is zero ($\Sigma(\phi_e) = 0$) which can be identified as the homeostatic equilibrium pressure. The cell–cell attraction at low cell density must be also imposed, giving a domain where the pressure derivative (denoted by $\Sigma' = d\Sigma/d\phi_c$) becomes negative. This zone will be considered unique and the minimum of Σ will be noted by ϕ^*. The cell-pressure Σ satisfying these assumptions is shown qualitatively in Fig. 7. Equation (32) giving the displacement of any point of the cell phase reads

$$\mathbf{v}_c = -K(\phi_c)(\Sigma'(\phi_c)\nabla\phi_c - \nabla(\epsilon^2\Delta\phi_c) - \mu\Delta\mathbf{v}_c). \tag{35}$$

The first term on the right may be assimilated to a diffusion term. By taking $\epsilon = 0$, $\mu = 0$ and $\Gamma_c = 0$ Eqs. (31) and (32) can be rewritten in the form of a nonlinear diffusion equation:

$$\frac{\partial \phi_c}{\partial t} = \nabla \cdot (K(\phi_c)\Sigma'(\phi_c)\nabla\phi_c).$$

In the region where $\Sigma'(\phi_c) < 0$, the diffusion coefficient is negative, potentially leading to a phase separation between domains rich in cancer cells and domains containing only interstitial fluid (this segregation is called spinodal decomposition).

However, the term ϵ prevents the appearance of a discontinuity in the solution fixing a nonzero thickness of the transition zone for these domains [107]. These effects are discussed in detail in the following.

3.5 Exchange Between Components

The fields ϕ_c, n and m are coupled by the terms of mass exchange $\Gamma_c(\phi_c, n, m)$ and consumption rates $S_n(\phi_c, n, m)$, $S_m(\phi_c, n, m)$. Based on experimental observations on melanoma we now give expressions for the possible coupling terms, discussing separately the growth by nutrients, the inhibiting factors and the contact inhibition.

Influence of Nutrients on Proliferation

The maintenance of the cell metabolism requires a number of vital nutrients, including oxygen, without which the cells cannot stay alive. The growth inhibition by lack of nutrients was the first proposed mechanism to describe the growth dynamics of avascular tumors [137]. Greenspan's model [70] assumes the existence of a concentration threshold $n_{\text{threshold}}$, such that the cells proliferate with a constant rate $\Gamma_0 > 0$ for a nutrient concentration $n \geq n_{\text{threshold}}$ and die with a constant rate δ for $n < n_{\text{threshold}}$. However, the experiments of Sutherland and Durand [134] show a non-zero rate of apoptosis in nutrient-rich areas and that the arrest of the expansion of a tumor is not always accompanied by the existence of a necrotic core. Culture assays in vitro of melanoma cells showed that the growth rate increases more or less linearly with the concentration of serum [58]. A simple choice consistent with these observations is given by

$$\Gamma_c = (\gamma_c \frac{n}{n_e} - \delta_c)\phi_c, \tag{36}$$

with $\gamma_c = 0.2$–$0.67\ \text{day}^{-1}$ [39, 45, 58] and $\delta_c = 0.05$–$1.65\ \text{day}^{-1}$ [39, 45], γ_c being the proliferation rate, δ_c the apoptosis rate in the cellular phase and n_e a typical nutrient concentration in the organ considered. In the epidermis for example, the partial pressure of oxygen is like $n_e = 25$–$78\,\text{mmHg}$ [131]. The mass creation is naturally proportional to the number of cells in the absence of contact inhibition Γ_c and is thus proportional to ϕ_c.

Consumption of Nutrients

Greenspan initially introduced a nutrient consumption simply proportional to the amount of cells $S_n = -\delta_n \phi_c$. However, this form is inconsistent with the experimental results of Sutherland and Durand [134]. Studying the xenografts in various tissues, including melanomas, Kallinowski et al. [84] have shown that the consumption of nutrients actually varies linearly with the concentration, for low concentrations, and then saturates to a plateau. Since the environment of the epidermis is very hypoxic [21, 131], we assume the linear regime for melanomas and choose

$$S_n = -\delta_n \phi_c n, \tag{37}$$

with $\delta_n = 1{,}190\text{–}3{,}030\,\text{day}^{-1}$ [80, 131].

Influence of Mechanical Stress on the Proliferation

The initial exponential growth of melanoma cells placed in culture (in a medium full of in nutrients) saturates when the cells reach confluence. Ellem and Kay [58] thus show that the growth rate peaked around a certain density and then decreases rapidly. Experiments by Helmlinger et al. [78] and Puliafito et al. [117] show that this phenomenon is found in avascular tumors and that inhibition of cell proliferation is largely mediated by the mechanical stresses experienced by cells. Then a limitation of the growth occurs by contact inhibition so we have

$$\Gamma_c = \gamma_c\big(\Sigma(\phi_c) - \Sigma(\phi_{inhib})\big)\phi_c. \tag{38}$$

This proliferation vanishes for a certain volume fraction $\phi_c = \phi_{inhib}$, choosing a function γ_c such that $\gamma_c(0) = 0$ (apoptosis compensating exactly the number of cell divisions). Thus ϕ_{inhib} represents the homeostatic volume fraction.

Growth Inhibitors

Finally the cell proliferation may also depend on the concentration of certain kinds of growth inhibitors, cytokines or cellular toxic waste. As in Greenspan's model [70] we assume that these inhibitors are produced by the cells themselves with a constant rate δ_m. Thereby

$$S_m = \delta_m \phi_c. \tag{39}$$

In the case of a growth process limited by the presence of inhibitors, we will take a linear dependence of cellular proliferation as for the nutrients:

$$\Gamma_c = \gamma_c(1 - \frac{m}{m_e})\phi_c, \tag{40}$$

with m_e the inhibitor concentration such that the net cell proliferation vanishes. Typical values of the physical constants for skin tumors can be found in Table 1.

Table 1 Biophysical parameters of the skin and of solid tumors derived from the literature

Parameters	Values	References
Epidermis thickness h in non-glabrous area	100 μm	[65, 85]
Epidermis thickness h in glabrous area	1 mm	[65, 85]
Migration time of keratinocytes from the basal	30 days	[85]
Volumetric cellular fraction ϕ_e in the healthy epidermis	0.57–0.87	[86, 147]
Volumetric cellular fraction ϕ_e in the basocellular carcinoma	0.23	[102]
Proliferation rate γ of melanoma cells	0.2–0.67 day^{-1}	[39, 45, 58]
Oxygen consumption rate δ_n of the healthy skin	1,190–3,030 day^{-1}	[80, 131]
Oxygen diffusion coefficient $D_n^{//}$ parallel to the epidermis layers	39.7 $\text{mm}^2 \cdot \text{day}^{-1}$	[83]
Oxygen diffusion coefficient D_n^z perpendicular to the layers	18.5–26.6 $\text{mm}^2 \cdot \text{day}^{-1}$	[131]
Partial pressure of oxygen in the corneal layer n_{at}	10^4 Pa	[131]
Partial pressure of oxygen in the basal layer n_{derm}	3.33×10^3 Pa	[131]
Typical overpressure χ generated by cell–cell interactions	130–3,700 Pa	[81, 88, 147]
Viscosity of the cellular phase μ	$300 - 10^3$ Pa · s	[90]
Cell/liquid friction coefficient τ	$10^3 - 1.2 \times 10^4\ \text{mm}^{-2}$ Pa · day	[86]
Apoptosis rate of melanoma cell in absence of nutriments δ	0.05–1.65 day^{-1}	[39, 45]

Notice that the measure conditions of some of these parameters are very far from in vivo skin conditions. These values must be considered as order of magnitude and cannot be considered as precise values

After these generalities on tumor modelling and the justification of the two-phase mixture model, we can consider the specific case of melanoma. As mentioned in the introduction, among its specificity, melanomas do no exhibit a necrotic core, characteristic of spheroids, in most cases. This is mostly due to the geometry inside the epidermis closer to the disc than to the sphere. Hereafter, we present some numerical and analytical results.

4 Contour Irregularity

Regularity and symmetry of the contour of a lesion is an important morphological criterion for the differential diagnosis of melanomas and nevi, used for example in the ABCDE rule [144]. Melanoma may appear de novo or from a pre-existing nevus; in both cases there exists a mechanism induced by a genetic mutation to break down the circular geometry during its evolution. Now, we aim to connect the evolution and morphology of an avascular tumor confined in the epidermis to the cell properties and to the microenvironment that make up the tissue, and we focus more precisely on the development of instabilities on the initially circular contour of a tumor. These contour instabilities are characteristic of an inhomogeneous growth process in soft tissues [23, 68] but are also considered as tumor agressiveness. In these last two sections of this review our scope is not to elaborate the most sophisticated approach of avascular tumor growth but simply to explain by the simplest possible model the presence or the absence of a necrotic core, the contour instability and finally the cell segregation. The two-fluid mixture model described above is sufficient for our purpose. More sophisticated and complex models have been treated previously (not for melanomas) and can be found in [44, 97]. However the main features of melanoma growth in the avascular phase is due to the skin geometry and to the cell–cell adhesion property. Up to now, melanoma—which is the subject of a large literature in dermatology and cancer biology—has poorly been considered in cancer modelling. Taking into account the geometry of the thin skin, we first derive a bidimensional model at the lubrication limit. We present numerical and analytical results indicating that the lesion radius grows with constant velocity, which explains eventually the presence of a necrotic core followed by a contour instability, and we finish by the process of segregation for the cell repartition as commonly observed in melanocytic lesions. The results are compared with clinical observations at the end of this section.

4.1 *Mixture Model in Thin Geometry*

The two main components for tumors confined in the epidermis are the proliferative cancer cells (index c) and a phase made by the other constituents of the tumor such as healthy and dead cells and the interstitial liquid. The keratinocytes are supposed dispersed in the interstitial fluid phase and their influence is neglected. The epidermis is represented by a thin layer between the basement membrane ($z = 0$) and the stratum corneum ($z = h$). This layer is wavy but for simplification we take it horizontal. The typical scale of the undulations is about $\sim$ 100 μm which is small compared to the size of the overall pattern of the lesion ($\sim$ 5 mm) and the geometry of the relatively flat epidermis in the hairless areas justifies this assumption. The cell proliferation rate Γ_c is regulated by the cell volume fraction ϕ_c, by the nutrient concentration n and possibly by the concentration of inhibitors m, by-products of the growth, that we neglect here. Cells consume nutrients at a constant rate δ_n.

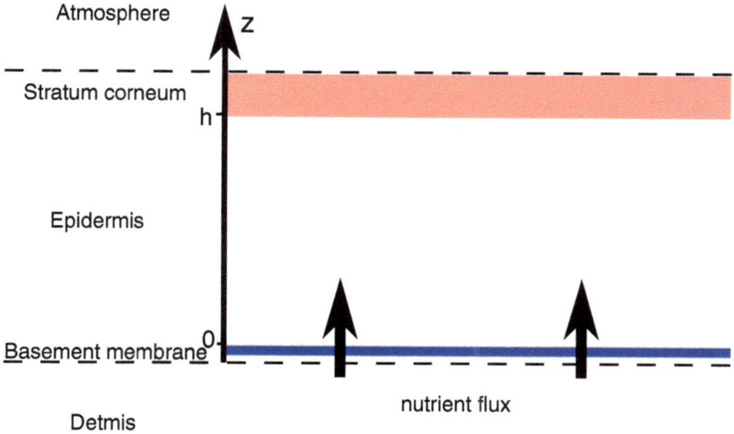

Fig. 8 The epidermis is represented by a layer of thickness $h \sim 0.1$–1 mm [28, 126] confined between a basement membrane ($z = 0$) and a stratum corneum ($z = h$). The concentrations of nutrients in the dermis ($z < 0$) and at the skin surface ($z > h$) are assumed to be fixed. The flux of nutrients J^n_{at} and J^n_{derm} depends on the concentration at $z = h$ and $z = 0$ [Eqs. (41) and (42)]

Nutrients and possibly inhibitors of negligible mass diffuse into the interstitial fluid, through the basement membrane and through the stratum corneum. The equilibrium distribution is considered in the following, due to the low growth rate of the system ($\sim$ 0.3 mm per month typically). Concentrations in the dermis (n_{derm}) and at the surface of the skin (n_{at}) are assumed to be constant, and fluxes through the basement membrane (J^n_{derm}) and through the stratum corneum (J^n_{at}) are given by a balanced distribution (Fig. 8):

$$D^z_n \frac{\partial n}{\partial z} = J^n_{at} = \alpha^n_{at}(n_{at} - n), \quad \text{at } z = h \tag{41}$$

$$D^z_n \frac{\partial n}{\partial z} = J^n_{derm} = \alpha^n_{derm}(n - n_{derm}), \quad \text{at } z = 0 \tag{42}$$

with α^n_{at} and α^n_{derm} being the permeability coefficients of the horny layer and the basement membrane, respectively. The evolution of this two-phase model is given by Eqs. (31)–(33) and (34). Notice that the in-plane diffusion is rather different from the transverse one in the epidermis, and it is represented by a diffusion coefficient $D^{//}_n$ different from the corresponding transverse one, D^z_n, due to the layered structure of the tissue (see Table 1). Boundary conditions at the tumor border, defined by $\mathbf{x} = \mathbf{x_b}$, are simply given by

$$\phi_c = \phi_e, \tag{43}$$

$$n = n_e + a(z - h/2), \tag{44}$$

$$V_f = \boldsymbol{v}_c \cdot \boldsymbol{n}. \tag{45}$$

Equation (43) is the mechanical equilibrium condition at the tumor edge, and Eqs. (44) and (45) mean the continuity of nutrient concentration and velocity at the border. The effects of compression on the tumor by the surrounding environment may be taken into account by imposing $\phi_c = \phi_{e,compress} > \phi_e$ at the border.

4.2 Lubrication Method

The epidermis thickness $h \sim 0.1$–1 mm [28, 126] being small compared to the typical tumor diameter $L \sim 5$ mm suggests that the evolution can be described by a 2D model through a lubrication approximation, a technical method without difficulty, usual in thin film hydrodynamics. Considering averaged quantities given by $\langle(\cdot)\rangle = \frac{1}{h}\int_0^h(\cdot)dz$, one derives effective 2D equations for these quantities to leading orders, the only difference coming from the diffusion equation having now a source term due to the third dimension. We finally focus on the dimensionless 2D set of equations that we plan to analyze in the following and we define: $\bar{x} = \sqrt{\tau\gamma/\chi}x$, $\bar{t} = \gamma_c t$, $\bar{n} = n_e^{-1}n$, $\bar{\Sigma} = \chi^{-1}\Sigma$, $\bar{K}(\phi_c) = \tau K(\phi_c)$, $\bar{\Gamma} = (\gamma_c h)^{-1}\langle\Gamma_c\rangle$. To simplify the notation, hereafter we suppress the bars. The quantities are averaged in the meaning defined above, ϕ is the concentration of cancerous cells, and diffusion concerns only in plane diffusion, perpendicular diffusion being responsible of a source. Then we finally get

$$\frac{\partial\phi}{\partial t} + \nabla\cdot(\phi\boldsymbol{v}) = \Gamma(\phi, n), \;\; \phi(\boldsymbol{x}_b) = \phi_e, \tag{46}$$

$$\boldsymbol{v} = -K(\phi)\nabla\Sigma, \;\; \boldsymbol{n}\cdot d\boldsymbol{x}_b/dt = \boldsymbol{n}\cdot\boldsymbol{v}(\boldsymbol{x}_b) = V_f, \tag{47}$$

$$0 = \Delta n - \beta_1\phi n + \beta_2(1-n), \;\; n(\boldsymbol{x}_b) = 1, \tag{48}$$

with $\boldsymbol{n}$ the external normal to the lateral edge of the tumor at the point $\boldsymbol{x}_b$. All quantities are dimensionless and depend only on the variables x, y and t. Parameters which control the system evolution are functions of Γ, Σ and the constants $\beta_1 = \chi\delta_n/(\tau\gamma D_n)$ and $\beta_2 = \chi D_n^z/(\tau\gamma\delta_n)$ with $D_{b,n}$ the average coefficient representing the flux of nutrient coming either from the upper or the lower horizontal surface. In case of oxygen that we consider as the main nutrient in the avascular phase, it can come either from the atmosphere or from the blood below the basal membrane. This question remains debated, the stratum corneum of the upper skin being a potential barrier. In any case, the origin of the transverse flux does not change the equations.

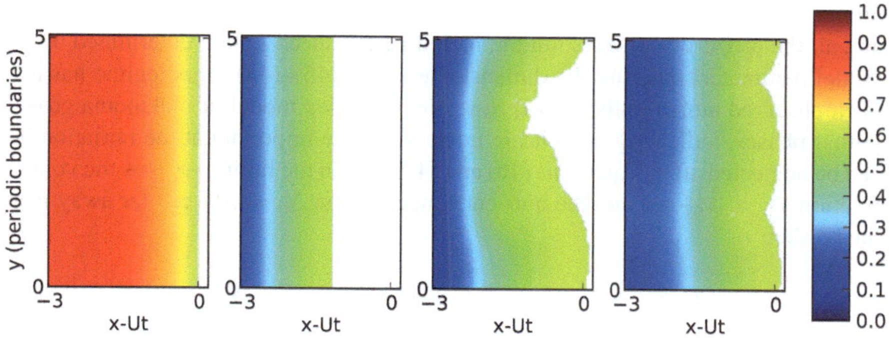

Fig. 9 Profile of cell volume fraction ϕ for a growing contour instability in a plane front propagating along x and initially invariant in y. The ripples on the contour associated to this instability has a typical wavelength Λ and the finite amplitude saturates at long times. The simulation parameters are $\beta_1 = 4$, $\beta_2 = 0$, $\phi_e = 0.6$, $\delta = 0.4$ and $t = 0, 10, 20, 80$ from *left to right*

5 Numerical Results

Let us consider the growth of a tumor with initially a radius R_0. To understand the emergence of asymmetry and contour irregularities observed in melanomas during the horizontal growth phase, we seek here the conditions for a contour instability breaking the circular symmetry.

For numerical simulations we choose Γ as $\Gamma = \phi(n - \delta)$ and Σ is given by

$$\Sigma(\phi) = \frac{(\phi - \phi_e)\phi^p}{1 - \phi}. \tag{49}$$

This form, similar to the choice made in [33], satisfies the phenomenological observations mentioned above: (i) the pressure Σ vanishes in the absence of cells, (ii) cells attract and adhere when they are close causing a negative pressure at low densities ($\phi < \phi_e$) and (iii) when cells are too close they experience a repulsive force causing a positive pressure for higher densities ($\phi > \phi_e$) which diverges when the cells occupy all the available space ($\phi \to 1$). Note that if the literal expressions of Σ and Γ are required for simulations, most results are valid for general functions satisfying the constraints described above. Here we have taken $p = 3$.

First, let us consider a semi-infinite planar front, with a scale invariance along y, the axis perpendicular to the tumor progression. The 1D numerical simulations (see Fig. 9) show a steady growth along the x-axis, the tumor edge being at $x = L(t)$ moving at constant velocity U such as $L(t) = Ut + L_0$. In the referential moving with the edge velocity, the cancer and nutrient concentrations $\phi(z = x - Ut, t)$, $n(z, t)$ become time-independent indicating a travelling wave behavior. Appearing after a short transient, the asymptotic regime is independent of the *initial conditions* and all the properties of the solution, including the speed U, are determined by the model parameters entering the dimensionless Eqs. (46)–(48). This regime has also been observed numerically in [44] in a more complex model of melanoma growth with 4 phases, indicating that elastic interactions are important at the initiation but can be neglected after. Equations (46) and (48) give in the limit $z \to -\infty$ the volume fraction $\Phi_0 = \phi(-\infty)$ and the nutrient concentration $N_0 = n(-\infty)$ far away from the border:

$$\Gamma(\Phi_0, N_0) = 0, \qquad \beta_1 \Phi_0 N_0 + \beta_2 (N_0 - 1) = 0. \tag{50}$$

Recall that the net proliferation $\Gamma(\Phi_0, N_0)$ vanishes for a critical nutrient concentration $N_0 = \delta$. Then Eq. (50) gives $\Phi_0 = (\beta_2/\beta_1)(\delta^{-1} - 1)$. We can conclude that $\Phi_0(z) = \phi(x - Ut, t)$ is a monotonic function if $\Phi_0 > \phi_e$ and has a maximum for $z = z_m < 0$, near the edge of the tumor if $\Phi_0 < \phi_e$. The penetration length for nutrients can be found by WKB techniques:

$$l_p \sim \frac{\sqrt{\beta_1 \phi_e + \beta_2}}{\beta_1 \phi_e}. \tag{51}$$

An important fact in the dynamics is the creation of a necrotic zone after some time and then an instability of the nutrient zone which finally gives rise to a front, roughly steady, evolving with a nonlinear instability at its border. Very few analyses are concerned with stability of steady travelling waves and the interested reader can find a complete analysis in [42]. In the circular regime, numerical simulations also

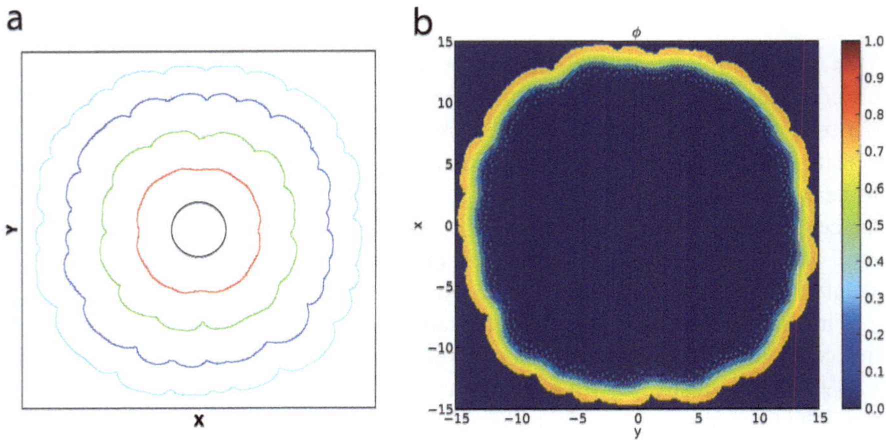

Fig. 10 Border instability developing on a tumor initially circular in radial growth. (**a**) Outline of the tumor border at time $t = 0$, $t = 10$, $t = 20$, $t = 30$ and $t = 40$, from the smallest to the largest diameter. Ripples appear with a typical wavelength. (**b**) Cell volume fraction ϕ at $t = 40$. $\beta_1 = 4$, $\beta_2 = 0$, $\phi_e = 0.6$ and $\delta = 0.4$

indicate a steady growth evolution of the tumor radius at constant velocity after a short transitory regime and the net creation of a necrotic core (see Fig. 10). The fact that the radius grows quite linearly indicates that the growth occurs in a zone of constant thickness close to the border that we identify to the nutrient zone. Above a critical radius one can observe a nascent instability on the simulations which also grows and saturates. Here again mathematical proofs have been given by Ben Amar et al. [25, 42] but let us simply make estimates.

The growth rate of the tumor G is given by the growth rate $\Gamma(\phi_e)$ times the surface of the proliferating ring which has a constant thickness l_p so $G = 2\pi R \Gamma(\phi_e) l_p$ which is also equal to $\phi_e 2\pi R \dot{R} = 2\phi_e \pi R U$ if the circular geometry is preserved as the steadiness of the growth. Then we get $\phi_e U \sim \Gamma l_p$ where $\Gamma = \phi_e(1-\delta)$. Notice that l_p, estimated before by Eq. (51), must be small in comparison to the radius for this estimation to be valid. Moreover knowing that the tumor velocity is the cancer cell velocity at the border fixed by the Darcy law, we also have: $U \sim K(\phi_e)\Sigma' \nabla \phi_e$ (Σ' being the derivative of Σ with respect to ϕ at the border). We finally derive a critical dimensionless parameter for the growth of instabilities at the tumor border

$$T = \frac{2\Gamma K(\phi_e)\Sigma'(\phi_e)}{V_f^2}. \tag{52}$$

The factor 2 is added for convenience. For $T > 1$ the destabilization contour occurs when the radius of the tumor exceeds a critical radius R_e given in physical units by

$$R_e = \frac{2K(\phi)\Sigma'(\phi_e)}{V_f(T-1)} \sim \frac{2l_p}{T-1} \quad \text{being } T > 1. \tag{53}$$

The diagram represented in Fig. 11 shows the evolution of the stability of the circular tumor as a function of the control parameter T, also found in a one phase model [109] . The physical origin of this "buckling" criteria is simply the fact that localized

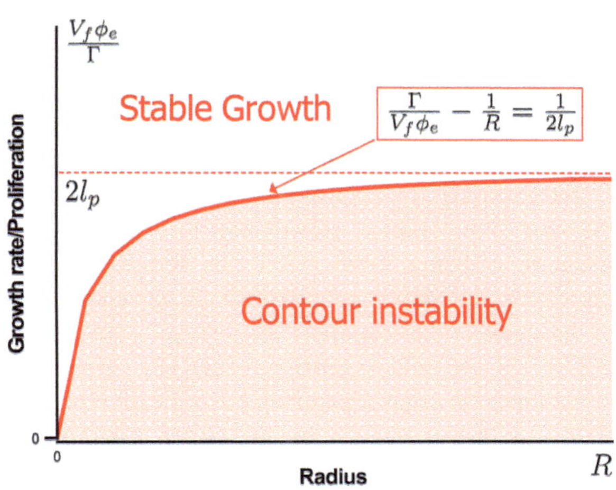

Fig. 11 Stability diagram of the circular tumor growing as a function of its radius R, the growth rate V_f and the cell proliferation rate at the edge of the tumor Γ/ϕ_e. The border of tumor in regression ($V_f < 0$) is always unstable

Table 2 Stability of the circular tumor growth in our model with two phases

Morphology	Stable	Instable contour (undulation wavelength)
T	< 1	> 1
R	$< R_e$	$> R_e$ (Λ decreases with R)
Σ	Weak	Strong ($\Lambda \sim l_p$)
Γ	Weak	Strong ($\Lambda \sim l_p$)
V_f	Fast	Slow ($\Lambda \sim l_p$) regression ($\Lambda \ll l_p$)

The phenotype of cells, mechanical interactions (Σ) and proliferation rate (Γ) are connected to the macroscopic properties of the tumor growth rate V_f, to the radius R, to the size of the proliferative area l_p and to the wavelength of the corrugations (Λ) on the contour of the tumor

Table 3 Abbreviations: superficial spreading melanoma (SSM), acral-lentiginous melanoma (LMM) (ALM), nodular melanoma (NM) and desmoplastic melanoma (DM), med. is for median

Growth velocity [mm day^{-1}]	T	R_e [mm]
0.004 (med. SSM)	11–18	0.021–0.022
0.011 (SSM)	4–6.5	0.065–0.073
0.0043 (med. LMM)	10–17	0.023–0.024
0.013 (med. ALM)	3.4–5.5	0.079–0.092
0.016 (med. NM)	2.75–4.5	0.1–0.13
0.049 (NM)	0.9–1.5	> 0.77
0.062 (DM)	0.71–1.2	> 2.2

The parameter T is estimated with Eq. (52) and R_e from Eq. (53). The growth rates reported here are from [12, 96] and parameter values used are $\gamma = 0.2\text{days}^{-1}$ and $l_p = 0.11$–0.18 mm (Table 1)

proliferative zone cannot remain circular if they are too much proliferative. The tumor changes its shape to be able to accommodate the new cancer cells which are produced. However, even if the tumor velocity is linked to the cell proliferation, because of the mechanics inside the tumor, fast-growing tumors such as nodular melanomas are stable [as shown by Eq. (52)]. Near the threshold of instability, the perturbations developing on the contour have a wavelength given by $\Lambda \sim 2\pi l_p$, compatible with the size of the proliferative zone. A summary of our analytical predictions is presented in Table 2.

Interestingly, the control parameter T predicts a significant correlation between the rate of melanoma growth and the stability of its contour, the fast-growing melanomas being more stable. In addition this model shows that tumors in regression also show ripples with small wavelengths allowing a better understanding of the shape of the scar areas. These results are consistent with clinical observations of Liu et al. [96] and Argenziano et al. [12]. Clinical data reported in Table 1 are used to estimate the control parameter T for different growth rates. One can notice that $T < 1$ for fast-growing melanoma ($V_f > 1.5$ mm per month for nodular melanoma) and $T > 1$ for slow-growing melanomas ($V_f < 0.4$ mm per month for superficial

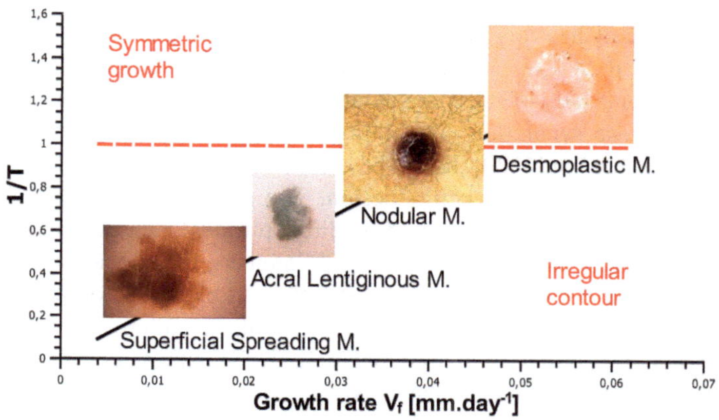

Fig. 12 Estimation of the control parameter T for slow-growing melanomas (superficial spreading melanoma and acral-lentiginous melanoma) and melanomas with rapid growth such as the nodular and desmoplastic melanoma. The median growth rates are taken from [96] and the model parameters from the literature and reported in Table 3. Clinically fast-growing melanomas with $T < 1$ are characterized by symmetrical growth and slow-growing melanomas with $T > 1$ are often asymmetrical. The images are from the Skin Cancer Foundation and from the work of Dr. P. Guitera

spreading melanoma), in agreement with the observed morphology for different types of melanomas (Table 3 and Fig. 12). Notice that nodular and desmoplastic melanomas grow mainly in the dermis, but our model manages to capture their behavior. The experimental data give an estimate of $\Lambda = 0.43$–2.2 mm, compatible with the assumptions of a continuous model, the typical size of a melanoma cells being 6–20 μm. The border instability of melanomas is really a consequence of the small diffusion length of the nutrients as for the spheroids, but another important feature of melanomas is the existence of structures inside the tumors called "nest" and "theque" and for some of them the absence of necrotic core. To explain these structures we go back to phase segregation.

6 Phase Separation in Biology

Phase separation is the spontaneous organization of a cell mixture (with different cellular species) in different part containing the same type of cells. This phenomenon is strikingly illustrated by the experiments of Townes and Holtfreter [138] on embryonic amphibian cells: a mixing of cells from the neural plate and from the epidermis, blended and aggregated in vitro, forms in 20 h a homogeneous mass of central medullary cells surrounded by an epidermal tissue (Fig. 5). The hypothesis called differential adhesion explains this phase separation by analogy with the conventional phase separation in immiscible fluids: the cells reorganize themselves in order to maximize their homotypic (i.e., between cells of the same type) but also heterotypic adhesion energy (between different cell types), given by the number and nature of adhesion molecules expressed at the cell surface. In the non-miscible case, the energy of heterotypic adhesions is lower than that of homotypic bonds and the cells of the same kind have tendency to aggregate. Experiments in vitro show that the dynamics of these aggregates is then similar to observations in the conventional phase separation: smoothing of irregular mass and coalescence of clusters of the same kind [129]. It is to be noted that it is possible to define a surface tension due to differences in cell adhesion. The differential adhesion hypothesis is now widely accepted and supported by direct physical measurements of surface tension. This will include work by Foty and Steinberg [61] demonstrating in vitro the proportionality between the number of cadherins expressed on the cell surface and the surface tension of the cell clusters. Also these authors show that a phase separation may be caused by a slight decrease in the expression of cadherins in a subpopulation of otherwise identical cells.

The importance of phase separation by differential adhesion has been highlighted in various biological contexts. In embryology it has been shown in vivo that the expression of cadherins varies depending on the cell type considered, allowing segregation of tissues and the formation of a well-defined frontier between them [75,106]. Also in developed organisms the differences in adhesion play an important role in the maintenance of tissue organization and development of pathologies [73]. In the context of prostate cancer, the transition to an invasive state has recently been associated to a decrease of the cohesion between tumor cells allowing the tumor the possibility to become miscible in surrounding tissues [108] (Fig. 13). This hypothesis is supported by the experiments of Winters et al. [148] showing in vitro a strong inverse correlation between the surface tension of a spheroid and its potential for invasion. Also let us mention the use of differential adhesion as manufacturer principle in tissue engineering. Neagu et al. [105] elegantly illustrate this technique using a bioprinter to manufacture tubular structures composed of two living cell phases. Thus, they manage to obtain different geometries by only controlling the interactions between cells and consequently their self-organization into two separate phases.

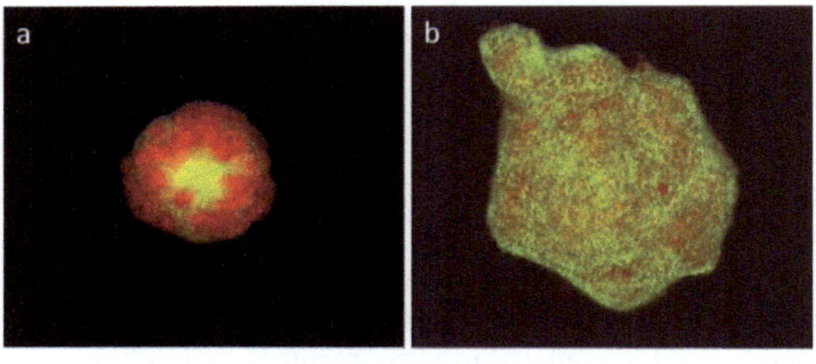

Fig. 13 Joined culture of fibroblasts (*green*) and prostate cancer cell (*red*), in the noninvasive case (**a**) and in the very invasive case (**b**). After 24 h the noninvasive cells get separated from fibroblasts (**a**) while the invasive cells remain mixed (**b**). Figure from [129]

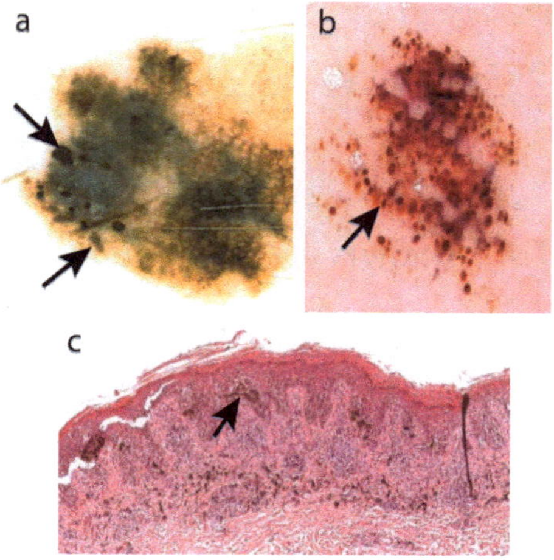

Fig. 14 (**a**) Melanoma presenting globules of different size and shape, irregularly distributed. (**b**) Naevus presenting globules regularly distributed. (**c**) Histopathology cut of a naevus showing melanocytic aggregates. Photos (a) and (b) from [150]; photo (c) is nicely provided by Prof. Giovanni Pellacani

Of course, melanoma is different from the prostate cancer mentioned above. In healthy skin melanocytes are naturally isolated, interacting preferably with neighbor keratinocytes rather than with their counterparts. As described in the introduction, melanoma progression is associated with a decrease of the affinity of keratinocytes for melanocytes joint to an increased affinity with their counterparts. It corresponds to changes in the expression of adhesion molecules for the two types of interactions [74]. Thus it is expected that melanocytes and keratinocytes move from a miscible to an immiscible state. Focussing on the appearance of melanocytic lesions reported in the medical literature, it is often observed both in nevi and melanomas, microstructures described as globules or pigmented dots (Fig. 14). Pigmentation, size, shape and distribution of these cancerous cells are important parameters to identify melanoma, usually associated to strongly pigmented, convex globules varying significantly in size and shape and unevenly distributed in the lesion [150]. Associated to these globules, in histological sections of melanomas we can frequently observe aggregates of melanocytes in the lower epidermis and the dermis (Fig. 14c), while in the upper epidermis wider aggregates are observed.

In order to understand the mechanism of formation of these aggregates of melanocytes, we study the possibility of phase separation in two phases with the same model introduced previously.

7 Spinodal Decomposition in Multiphase Models

As previously, we consider the mixture of two phases consisting of tumor cells and the interstitial fluid (containing other cells such as keratinocytes not explicitly represented here). We limit ourselves to the study of proliferation regulated by the concentration of nutrients diffusing in the interstitial fluid. We consider the set of equations Eqs. (46)–(48) with the same proliferation rate Γ_c while the diffusion equation for nutrients is given by Eq. (48). We only add a surface tension term preventing large ϕ gradient into Eq. (46) and now get

$$\frac{\partial \phi}{\partial t} = D\nabla \cdot \left(\phi K(\phi)\nabla\left[\Sigma(\phi) - \epsilon^2 \Delta\phi\right]\right) + \phi(n-\delta), \tag{54}$$

7.1 Analysis of the Equations

Such system of equations Eqs. (48) and (54) accepts a static solution given by

$$\phi_0 = \frac{\beta(1-\delta)}{\delta} \text{ and } n_0 = \delta \tag{55}$$

with $\beta = \beta_2/\beta_1$. It is the homeostatic state where divisions (respectively the consumption of nutrients) balance exactly the cell death (respectively the nutrient supply). An infinitesimal perturbation of wave vector k around this homogeneous solution can be written as

$$\phi = \phi_0 + \delta\phi \exp(\lambda t)\cos(kx), \tag{56}$$

$$n = n_0 + \delta n \exp(\lambda t)\cos(kx), \tag{57}$$

with $\beta = \beta_2/\beta_1$ and we find the perturbation growth rate λ as

$$\lambda = -\phi_0[K(\phi_0)(f'(\phi_0)k^2 + \epsilon^2 k^4) + \frac{\delta^2}{k^2\delta + \beta}]. \tag{58}$$

The coupling with nutrients is responsible for the last term of this equation, which prevents the growth of disturbances of large wavelength $\lambda(k)$ as shown in Fig. 15.

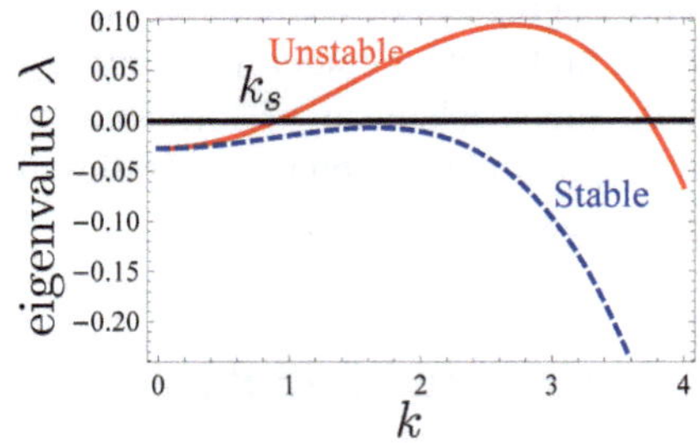

Fig. 15 Growth rate λ of a disturbance with a wave vector k around the stationary homogeneous solution given in Eq. (55). Coupling with nutrients stabilizes the wavelengths, in the range typically larger than the penetration depth of nutrients $k < 1$. Here the mixture is stable for $\Sigma'(\Phi_0) = -0.1$ (*solid line*) and becomes stable for $\Sigma'(\Phi_0) = -0.03 < 0$ (*dashed line*). The model parameters are $\epsilon = 0.084$, $\beta_1 = 2$, $\beta_2 = 0.4$, $\phi_0 = 0.3$ and $\phi_e = 0.6$

Note that this term introduces a second characteristic length in the system, in addition to the interface thickness between domains. This second length is connected to the penetration length of nutrients $\sqrt{D_n/\delta_n}$. The introduction of a second characteristic length in phase separation is generally associated with a selection mechanism. Let us mention the example of the well-known block copolymers consisting of two immiscible parts covalently bonded. The finite length of these polymers creates the phenomenon of micro-separation where the reorganization of the domains stops before reaching macroscopic sizes and generally leads to the appearance of periodic structures [94, 151]. A similar morphogenetic mechanism is also found in the Glozter model [67] where a chemical equilibrium introduces exchange terms between the phases and leads to a saturation of the domain size with the appearance of stripes. In a more immediate context, the dispersion relation shown in Fig. 15 is similar to the one found by Klein et al. [87] with a Cahn-Hilliard model to describe the aggregation of stem cells of the epidermis. In their case a second length is given by the contact inhibition of cell proliferation and differentiation of stem cells [93]. Given the similarities of the dispersion relations, one expects here similar behavior. Assuming $\epsilon \ll 1$ in Eqs. (54), (58) the most unstable length of our model is given by

$$l_s = 2\pi \sqrt{\frac{1 - \frac{\phi_0 K(\phi_0) f'(\phi_0)(\phi_0+\beta)^2}{\phi_0 n_0}}{\phi_0 + \beta}} = \frac{2\pi}{k_s}. \tag{59}$$

The maximum growth rate $\lambda_{max} = \lambda(k_{max})$ is obtained from Eq. (58) and its sign gives the stability of the mixture. Omitting the pre-factors, the positive sign is given by the function:

$$\begin{aligned} g(\beta, \delta, f', \epsilon) = -2(\sqrt{(\beta\epsilon^2 - \delta f')^2 + \beta\epsilon^2\delta f'} - (\epsilon^2\beta + \delta f'))\,((\beta\epsilon^2 - \delta f')^2 \\ +\beta\epsilon^2\delta f') \quad - 3\beta\epsilon^2\delta f'(\beta\epsilon^2 + \delta f') + \frac{(3\epsilon^2\delta^2)^3}{\epsilon^2(\delta(1+\beta) - \beta)^2}, \end{aligned} \tag{60}$$

$g > 0$ indicating the instability of the mixture and the onset of phase separation. The sign of g is shown in Fig. 16 and the stability of the mixture seems to depend mainly on the initial volume fraction ϕ_0 since the limits of unstable areas roughly correspond to the iso-surfaces $\phi_0(\beta, \delta, f', \epsilon)$ being constant.

8 Simulations

We first consider simulations on a square lattice. We can show typical examples of patterning with dots or labyrinths in Figs. 17 and 18. A coarsening effect has been studied in details in [41], and examples are given in Fig. 18. These simulations allow studying the two-phase mixture model from the viewpoint of statistical physics which is interesting in itself. One important feature is the fact that the

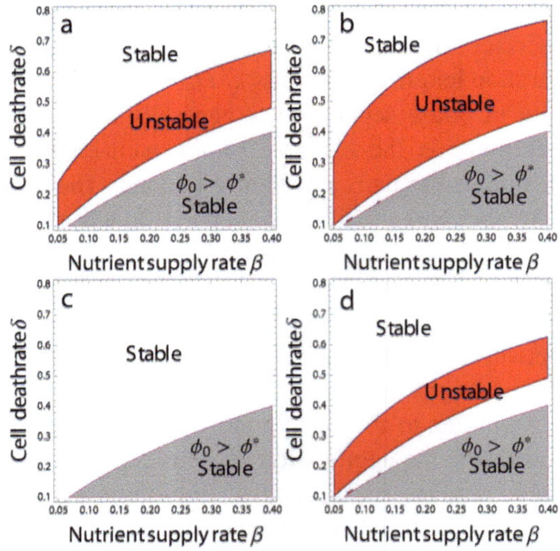

Fig. 16 Phase diagram for a mixing initially homogeneous with a volumetric fraction ϕ_0 and a nutrient concentration n_0, given by the dimensionless parameter δ (proportional to the cell death rate) and β (ratio between β_2 and β_1 proportional to the rate in nutriments). The parameters of the model are $\epsilon = 0.084$ (**a**, **c**), $\epsilon = 0.2$ (**b**, **d**) and $\beta_1 = 2$ (**a**, **b**), $\beta_1 = 20$ (**c**, **d**). The *grey areas* correspond to $\phi_0 > \phi^*$, the mixing being always stable. The coupling with the nutrients tends to stabilize the mixing and the unstable zone is reduced to the area in red. One can notice that the limit between domains seems to correspond to the iso-surface ϕ_0 constant

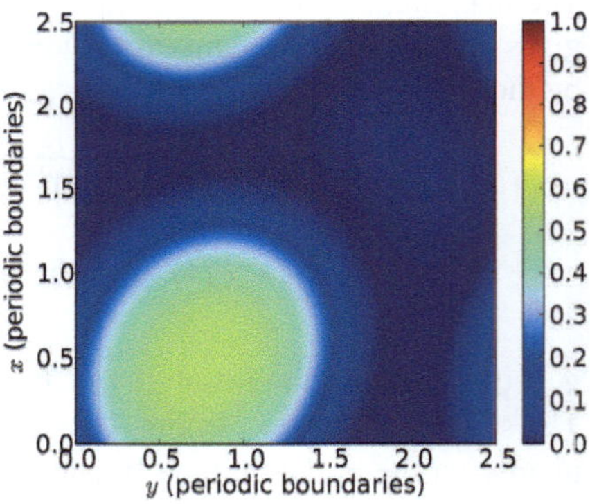

Fig. 17 Aggregation formation of tumor cells surrounded by a domain without cancerous cells (*blue*). Initially $\phi_0 = 0.5$ and there is no mass exchange. Parameter of the model is for $\phi_e = 1$

exponents for self-similarity of the factor of structure and for domain sizes are different from that of the Cahn-Hilliard model proving that asymptotic behavior is dependent of the nonlinearities. However, it turns out that these structures appearing in simulations are related to clinical observations. The presence of aggregates of melanocytes appearing as points in dermoscopy (Fig. 14) is often observed in melanocytic lesions, nevi and melanomas. Simulations of this model exhibit several behaviors that are also seen in vivo. Microstructural patterns appear as growth proceeds giving dots observed in dermoscopy (see Fig. 14a, b) and to the nests seen

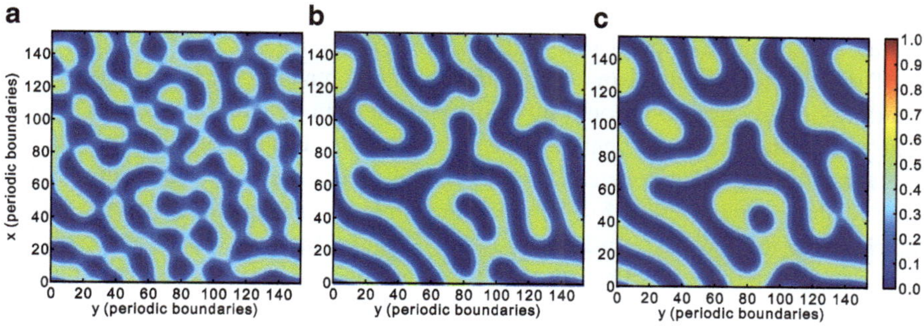

Fig. 18 Time evolution of a pattern with labyrinths obtained for $\phi_0 = 0.25$ and without mass exchange. The simulations is represented at several times $t = 1,000, 18,000, 28,000$. Parameter of the model is for $\phi_e = 0.6$

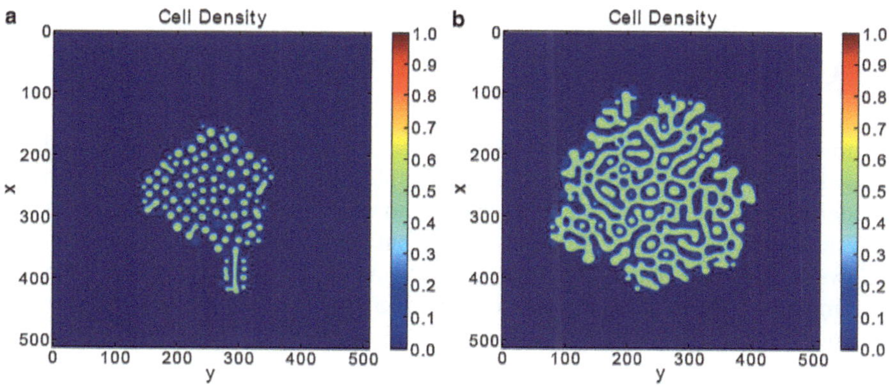

Fig. 19 Two characteristic patterns obtained by simulations on a 512×512 grid. The step-sizes, dx, are taken equal to 0.12 and the step-times to 10^{-4}. The same form of Σ is chosen for all pictures, $p = 2$ and $\phi_e = 0.6$ [see Eq. (49)]. Different values of ϕ_0 are taken: 0.33 for (**a**), 0.25 for (**b**)

in histopathology (see Fig. 14c). Since the concentration inside microstructures is close to the mechanical equilibrium concentration ϕ_0, the simulations show that with this separation process, cells insure their mechanical equilibrium at small scales. In the same time, they insure their equilibrium at large scales, because the long-scale concentration is given by the interaction with nutrients and the difference between proliferation and death as predicted previously.

Moreover, this phase separation leads to a growth asymmetry as shown in Fig. 19. Note that Fig. 19a contains mainly dots while Fig. 19b exhibits a labyrinth pattern in growth. This model explains differences between nevi and melanomas, nevi being homogeneous lesions. However, the asymmetry and irregularity displayed can be attributed to either a melanoma or a dysplastic nevus that actually show the same aspect at the tissue level. While the micro-environment can be responsible for the different characteristics between a nevus, a dysplastic nevus and a melanoma, the difference between dysplastic nevus and melanoma is the existence of growth arrest for the nevus which stops its final size and prevents metastasis.

The formation mechanism of structures in melanoma is well explained by a two-fluid model taking into account a realistic mechanism of adhesion between cells. The appearance of these aggregates is explained by a phase separation between melanocytes and keratinocytes. The theoretical model predicts a saturation of the size of these structures with a length determined by the length of penetration of nutrients, here $l_s \approx 2l_n = 2\sqrt{D_n/\delta_n} \approx 0.2\,\text{mm}$ (Table 1), consistent with the size of aggregation seen clinically. Phase separation in this system is similar to the one involved in a mixture of two immiscible fluids below a critical temperature. This result corresponds to the change in the expression level of cadherins observed clinically in melanoma progression, when melanocyte interactions increase with their peers and decrease with keratinocytes [74].

9 Conclusion

In this chapter, I give an overview of cancer modelling in the avascular phase joined to pioneering experiments on cell colonies. Then I focus on a simple model which contains the main ingredients to describe melanomas. I present the physical hypothesis, some numerical results but also a mathematical treatment that allows predicting quantitatively instabilities knowing the value of the biomechanics parameters of the skin. Not presented here is a special study on acral melanoma, a disease concerning mostly the non-eurasian population. These melanomas are located in glabrous area on the body, where fingerprints are strongly marked and their topology depends on the geometry of the basal layer [14]. The aim of this chapter is to show that we are able to understand from physical insights the avascular tumor phase and that the comparison with clinical data, possible for melanomas, gives very rich and fruitful opportunities.

Acknowledgements This chapter is a summary of results about tumor modelling obtained in Ecole Normale Supérieure during these last years. I strongly acknowledge three Ph.D students, Julien Dervaux, Clément Chatelain and Thibaut Balois, for their intensive collaboration on this very rich topic between mathematics, nonlinear and soft matter physics and biology. This work is supported in part by AAP INSERM 2012.

References

1. J. Adam, A mathematical model of tumor growth. ii. Effects of geometry and spatial nonuniformity on stability. Math. Biosci. **86**, 183–211 (1987)
2. M. Al-Hajj, M.S. Wicha, A. Benito-Hernandez, S.J. Morrison, M.F. Clarke, Prospective identification of tumorigenic breast cancer cells. Proc. Natl. Acad. Sci. USA **100**(7), 3983–3988 (2003)

3. D. Alemani, F. Pappalardo, S. Pennisi, M. Motta, V. Brusic, Combining cellular automata and lattice boltzmann method to model multiscale avascular tumor growth coupled with nutrient diffusion and immune competition. J. Immunol. Methods **376**(1–2), 55–68 (2012)
4. D. Ambrosi, F. Mollica, On the mechanics of a growing tumor. Int. J. Eng. Sci. **40**, 1297–1316 (2002)
5. D. Ambrosi, F. Mollica, The role of stress in the growth of a multicell spheroid. J. Math. Biol. **48**(5), 477–499 (2004)
6. D. Ambrosi, P. Preziosi, On the closure of mass balance models for tumour growth. Math. Model Methods Appl. Sci. **12**, 737–754 (2002)
7. D. Ambrosi, L. Preziosi, Cell adhesion mechanisms and stress relaxation in the mechanics of tumours. Biomech. Model Mechanobiol. **8**(5), 397–413 (2009)
8. A.R. Anderson, A hybrid mathematical model of solid tumour invasion: the importance of cell adhesion. Math. Med. Biol. **22**(2), 163–186 (2005)
9. A.R. Anderson, A.M. Weaver, P.T. Cummings, V. Quaranta, Tumor morphology and phenotypic evolution driven by selective pressure from the microenvironment. Cell **127**(5), 905–915 (2006)
10. R.P. Araujo, D.L. McElwain, A history of the study of solid tumour growth: the contribution of mathematical modelling. Bull. Math. Biol. **66**, 1039–1091 (2004)
11. R.P. Araujo, D.L.S. McElwain, A mixture theory for the genesis of residual stresses in growing tissues. SIAM J. Appl. Math. **65**, 1261–1284 (2005)
12. G. Argenziano, H. Kittler, G. Ferrara, P. Rubegni, J. Malvehy, S. Puig, L. Cowell, I. Stanganelli, V. De Giorgi, L. Thomas, P. Bahadoran, S.W. Menzies, D. Piccolo, A.A. Marghoob, I. Zalaudek, Slow-growing melanoma: a dermoscopy follow-up study. Br. J. Dermatol. **162**(2), 267–273 (2010)
13. T. Balois, M. Ben Amar, Morphology of melanocytic lesions *in situ*, Nat. Sci. Rep. **4**, 3622 (2014). doi:10.1038/srep03622
14. T. Balois, C. Chatelain, M. Ben Amar, Patterns in melanocytic lesions: impact of the geometry on growth and transport inside the epidermis, Journ. Roy. Soc. Interface, **11**, 20140339 (2014)
15. M.R. Baer, J.W. Nunziato, A two-phase mixture theory for the deflagration-to-detonation transition (ddt) in reactive granular materials. Int. J. Multiphase Flow **12**, 861–889 (1986)
16. M.B. Baraldi, A.A. Alemi, J.P. Sethna, S. Caracciolo, C.A.M. La Porta, S. Zapperi, Growth and form of melanoma cell colonies. J. Stat. Mech. **02**, P02032 (2013)
17. B. Basaliy, A. Friedman, Global existence and asymptotic stability for an elliptic-parabolic free boundary problem: an application to a model of tumor growth. Indiana Univ. Math. J. **52**, 1265–1303 (2003)
18. M. Basan, T. Risler, J.F. Joanny, X. Sastre-Garau, J. Prost, Homeostatic competition drives tumor growth and metastasis nucleation. HFSP J. **3**(4), 265–272 (2003)
19. G.K. Batchelor, *An Introduction to Fluid Dynamics* (Cambridge University Press, Cambridge, 1967), pp. 131–173
20. E.L. Bearer, J.S. Lowengrub, H.B. Frieboes, Y.-L. Chuang, F. Jin, S.M. Wise, M. Ferrari, D.B. Agus, V. Cristini, Multiparameter computational modelling of tumor invasion. Cancer Res. **69**(10), 4493–4501 (2009)
21. B. Bedogni, S.M. Welford, D.S. Cassarino, B.J. Nickoloff, A.J. Giaccia, M.B. Powell, The hypoxic microenvironment of the skin contributes to Akt-mediated melanocyte transformation. Cancer Cell **8**, 443–454 (2005)
22. M. Ben Amar, Chemotaxis migration and morphogenesis of living colonies. Eur. Phys. J. E **36**, 64–76 (2013). doi:10.1140/epje/i2013-13064-5
23. M. Ben Amar, A. Goriely, Growth and instability in elastic tissue. J. Mech. Phys. Solids **53**(20), 2284–2319 (2005)
24. M. Ben Amar, F. Jia, Anisotropic growth shapes intestinal tissues during embryogenesis. Proc. Natl. Acad. Sci. USA **110**(26), 10525–10530 (2013)
25. M. Ben Amar, C. Chatelain, P. Ciarletta, Contour instabilities in early tumor growth models. Phys. Rev. Lett. **106**, 148101–148104 (2011)

26. A. Bertuzzi, A. Fasano, A. Gandolfi, C. Sinisgalli, Modelling the evolution of a tumoral multicellular spheroid as a two-fluid Bingham-like system. Math. Mod. Meth. Appl. Sci. **23**, 2561–2602 (2013)
27. T. Borovski, L. Vermeulen, M.R. Sprick, J.P. Medema, One renegade cancer stem cell? Cell Cycle **8**(6), 803–808 (2009)
28. A. Breslow, Thickness, cross-sectional areas and depth of invasion in the prognosis of cutaneous melanoma. Ann. Surg. **172**(5), 902–908 (1970)
29. A.C. Burton, Rate of growth of solid tumours as a problem of diffusion. Growth **30**(2), 157–176 (1966)
30. H.M. Byrne, M.A. Chaplain, Growth of nonnecrotic tumors in the presence and absence of inhibitors. Math. Biosci. **130**(2), 151–181 (1995)
31. H.M. Byrne, M.A. Chaplain, Free boundary value problems associated with the growth and development of multicellular spheroids. Eur. J. Appl. Math. **8**, 639–658 (1997)
32. H.M. Byrne, D. Drasdo Individual-based and continuum models of growing cell populations: a comparison. J. Math. Biol. **58**(4–5), 657–687 (2009)
33. H.M. Byrne, L. Preziosi, Modelling solid tumour growth using the theory of mixtures. Math. Med. Biol. **20**(4), 341–366 (2003)
34. H.M. Byrne, J.R. King, D.L.S. McElwain, L. Preziosi, A two-phase model of solid tumour growth. Appl. Math. Lett. **16**, 567–573 (2003)
35. H.M. Byrne, T. Alarcon, M.R. Owen, S.D. Webb, P.K. Maini, Modelling aspects of cancer dynamics: a review. Philos. Transact. A Math. Phys. Eng. Sci. **364**(1843), 1563–1578 (2006)
36. G. Caravagna, R. Barbuti, A. d'Onofrio, Fine-tuning anti-tumor immunotherapies via stochastic simulations. BMC Bioinform.**12**, S8 (2012)
37. G. Caravagna, G. Mauri, A. d'Onofrio, The interplay between intrinsic and extrinsic noise in biomolecular networks. PLoS ONE **8**(2), e51174 (2013)
38. L. Cardamone, A. Valentín, J.F. Eberth, J.D. Humphrey, Origin of axial prestretch and residual stress in arteries. Biomech. Model Mechanobiol. **8**(6), 431–446 (2009)
39. L.I. Cardenas-Navia, D. Yu, R.D. Braun, D.M. Brizel, T.W. Secomb, M.W. Dewhirst, Tumor-dependent kinetics of partial pressure of oxygen fluctuations during air and oxygen breathing. Cancer Res. **64**(17), 6 (2004)
40. M.A. Chaplain, B.D. Sleeman, Modelling the growth of solid tumours and incorporating a method for their classification using nonlinear elasticity theory. J. Math. Biol. **31**(5), 431–473 (1993)
41. C. Chatelain, T. Balois, P. Ciarletta, M. Ben Amar, Emergence of microstructural patterns in skin cancer: a phase separation analysis in a binary mixture. New J. Phys. **13**, 115013 (2011)
42. C. Chatelain, P. Ciarletta, M. Ben Amar, Morphological changes in early melanoma development: influence of nutrients, growth inhibitors and cell-adhesion mechanisms. J. Theor. Biol. **290**, 46–59 (2011)
43. L. Chin, L.A. Garraway, D.E. Fisher, Malignant melanoma: genetics and therapeutics in the genomic era. Genes Dev. **20**(16), 2149–2182 (2006)
44. P. Ciarletta, L. Foret, M. Ben Amar, The radial growth phase of malignant melanoma: multi-phase modelling, numerical simulations and linear stability analysis. J. R. Soc. Interface **8**(56), 345–368 (2011)
45. A.A. Creasey, H.S. Smith, A.J. Hackett, K. Fukuyama, W.L. Epstein, S.H. Madin, Biological properties of human melanoma cells in culture. In Vitro **15**(5), 342–350 (1979)
46. V. Cristini, J. Lowengrub, Q. Nie, Nonlinear simulation of tumor growth. J. Math. Biol. **46**(3), 191–224 (2003)
47. V. Cristini, X. Li, J.S. Lowengrub, S.M. Wise, Nonlinear simulations of solid tumor growth using a mixture model: invasion and branching. J. Math. Biol. **58**(4–5), 723–763 (2009)
48. A. d'Onofrio, Rapidly acting antitumoral antiangiogenic therapies. Phys. Rev. E **76**, 031920–031927 (2007)
49. A.S. Deakin, Model for the growth of a solid in vitro tumor. Growth **39**(1), 159–165 (1975)
50. R. Decker, R.L. Brown, A turbulent mixture theory for the atmospheric mixture of snow and air. Ann. Glaciol. **4**, 37–41 (1983)

51. J. Dervaux, Morphogenèse et Elasticité en géométrie mince. Ph.D. thesis, Université Paris Diderot, 2010
52. J. Dervaux, M. Ben Amar, Buckling condensation in constrained growth. J. Mech. Phys. Solids **59**, 538–560 (2011)
53. J. Dervaux, M. Ben Amar, Mechanical instabilities of gels. Ann. Rev. Condensed Matter Phys. **3**, 311–332 (2012)
54. L.A. Dethlefsen, J.M. Prewitt, M.L. Mendelsohn, Analysis of tumor growth curves. J. Natl. Cancer Inst. **40**(2), 389–405 (1968)
55. M. Doi, A. Onuki, Dynamic coupling between stress and composition in polymer solutions and blends. J. Phys. II **2**, 1631–1656 (1992)
56. D. Drasdo, On selected individual-based approaches to the dynamics of multicellular systems, in *Multiscale Modelling*, ed. by W. Alt, M. Chaplain, M. Griebel, J. Lenz (Birkhäuser, Basel, 2003), pp. 169–205
57. D. Drasdo, S. Hoehme, A single-cell-based model of tumor growth in vitro: monolayers and spheroids. Phys. Biol. **2**(3), 133–147 (2005)
58. K.A. Ellem, G.F. Kay, The nature of conditioning nutrients for human malignant melanoma cultures. J. Cell Sci. **62**, 249–266 (1983)
59. J. Fink, N. Carpi, T. Betz, A. Bétard, M. Chebah, A. Azioune, M. Bornens, C. Sykes, L. Fetler, D. Cuvelier, M. Piel, External forces control mitotic spindle positioning. Nat. Cell Biol. **13**(7), 771–778 (2011)
60. J. Folkman, M. Hochberg, Self-regulation of growth in three dimensions. J. Exp. Med. **138**(4), 745–753 (1973)
61. R.A. Foty, M.S. Steinberg, The differential adhesion hypothesis: a direct evaluation. Dev. Biol. **278**(1), 255–263 (2005)
62. A. Friedman, B. Hu, Bifurcation from stability to instability for a free boundary problem modelling tumor growth by Stokes equation. J. Math. Anal. Appl. **327**, 643–664 (2007)
63. A. Friedman, F. Reitich, Symmetry-breaking bifurcation of analytic solutions to free boundary problems: an application to a model of tumor growth. Trans. Am. Math. Soc. **353**, 1587–1634 (2001)
64. R.A. Gatenby, P.K. Maini, Mathematical oncology: cancer summed up. Nature **421**(6921), 321–321 (2003)
65. E. Gentilhomme, Y. Neveux, *Physiologie de la peau et explorations fonctionnelles cutanées*, ed. by P. Agache (Editions Médicales Internationales, Cachan, 2000), pp. 165–172
66. J.A. Glazier, F. Graner, Simulation of the differential adhesion driven rearrangement of biological cells. Phys. Rev. E Stat. Phys. Plasmas Fluids Relat. Interdiscip. Topics **47**(3), 2128–2154 (1993)
67. S.C. Glotzer, E.A. Di Marzio, M. Muthukumar, Reaction-controlled morphology of phase-separating mixtures. Phys. Rev. Lett. **74**(11), 2034–2037 (1995)
68. A. Goriely, M. Ben Amar, Differential growth and instability of spherical shells. Phys. Rev. Lett. **94**, 198103–198106 (2005)
69. L. Graziano, L. Preziosi, Modelling of biological materials, in *Mechanics in Tumor Growth* (Birkhauser, Basel, 2007), pp. 267–328
70. H.P. Greenspan, Models for the growth of a solid tumor by diffusion. Stud. Appl. Math. **LI**, 317–340 (1972)
71. H.P. Greenspan, On the growth and stability of cell cultures and solid tumors. J. Theor. Biol. **56**, 229–242 (1976)
72. G. Griffon-Etienne, Y. Boucher, C. Brekken, H.D. Suit, R.K. Jain, Taxane-induced apoptosis decompresses blood vessels and lowers interstitial fluid pressure in solid tumors: clinical implications. Cancer Res. **59**(15), 3776–3782 (1999)
73. B.M. Gumbiner, Cell adhesion: the molecular basis of tissue architecture and morphogenesis. Cell **84**(3), 345–357 (1996)
74. N.K. Haass, K.S.M. Smalley, M. Herlyn, The role of altered cell–cell communication in melanoma progression. J. Mol. Histol. **35**(3), 309–318 (2004)

75. J.M. Halbleib, W.J. Nelson, Cadherins in development: cell adhesion, sorting, and tissue morphogenesis. Genes Dev. **20**(23), 3199–3214 (2006)
76. F.G. Haluska, H. Tsao, H. Wu, F.S. Haluska, V. Lazar, A. Goel, Genetic alterations in signaling pathways in melanoma. Clin. Cancer Res. **12**, 2301–2307 (2006)
77. D. Hanahan, R.A. Weinberg, Hallmarks of cancer: the next generation. Cell **144**(5), 646–674 (2011)
78. G. Helmlinger, P.A. Netti, H.C. Lichtenbeld, R.J. Melder, R.K. Jain, Solid stress inhibits the growth of multicellular tumor spheroids. Nat. Biotechnol. **15**(8), 778–783 (1997)
79. S. Hoehme, D. Drasdo, A cell-based simulation software for multi-cellular systems. Bioinformatics **26**(20), 2641–2642 (2010)
80. M.E. Hystad, E.K. Rofstad, Oxygen consumption rate and mitochondrial density in human melanoma monolayer cultures and multicellular spheroids. Int. J. Cancer **57**(4), 532–537 (1994)
81. R.K. Jain, Transport of molecules in the tumor interstitium: a review. Cancer Res. **47**(12), 3039–3051 (1987)
82. F. Jia, M. Ben Amar, Theoretical analysis of growth or swelling wrinkles on constrained soft slabs. Soft Matter **9**, 8216–8226 (2013)
83. M.E. Johnson, D. Blankschtein, R. Langer, Evaluation of solute permeation through the stratum corneum: lateral bilayer diffusion as the primary transport mechanism. J. Pharm. Sci. **86**(10), 1162–1172 (1997)
84. F. Kallinowski, K.H. Schlenger, S. Runkel, M. Kloes, M. Stohrer, P. Okunieff, P. Vaupel, Blood flow, metabolism, cellular microenvironment, and growth rate of human tumor xenografts. Cancer Res. **49**(14), 3759–3764 (1989)
85. J. Kanitis, *Biologie de la peau humaine*, ed. by D. Schmitt (Les éditions INSERM, Paris, 1997), pp. 1–20
86. P.M. van Kemenade, J.M. Huyghe, L.F.A. Douven, Triphasic fe modelling of skin water barrier. Transport Porous Media **50**, 93–109 (2003)
87. A.M. Klein, V. Nikolaidou-Neokosmidou, D.P. Doupè, P.H. Jones, B.D. Simons, Patterning as a signature of human epidermal stem cell regulation. J. R. Soc. Interface **8**(65), 1815–1824 (2011)
88. C.A. Kristensen, M. Nozue, Y. Boucher, R.K. Jain, Reduction of interstitial fluid pressure after tnf-alpha treatment of three human melanoma xenografts. Br. J. Cancer **74**(4), 533–536 (1996)
89. S. Kuphal, A. Bosserhoff, Recent progress in understanding the pathology of malignant melanoma. J. Pathol. **219**(4), 400–409 (2009)
90. Y. Kuwahara, Y. Shima, D. Shirayama, M. Kawai, K. Hagihara, T. Hirano, J. Arimitsu, A. Ogata, T. Tanaka, I. Kawase, Quantification of hardness, elasticity and viscosity of the skin of patients with systemic sclerosis using a novel sensing device (vesmeter): a proposal for a new outcome measurement procedure. Rheumatology (Oxford) **47**(7), 1018–1024 (2008)
91. L.D. Landau, E.M. Lifshitz, *Statistical Physics*, 1st edn. (Pergamon Press, Oxford, 1980)
92. K.A. Landman, C.P. Please, Tumour dynamics and necrosis: surface tension and stability. IMA J. Math. Appl. Med. Biol. **18**(2), 131–158 (2001)
93. C.A.M. La Porta, S. Zapperi, J.P. Sethna, Senescent cells in growing tumors: population dynamics and cancer stem cells. PLoS Comput. Biol. **8**, e1002316 (2012)
94. L. Leibler, Theory of microphase separation in block copolymers. Macromolecules **13**, 1602–1617 (1980)
95. X. Li, V. Cristini, Q. Nie, J. Lowengrub, Nonlinear three-dimensional simulation of solid tumor growth. Discrete Dyn. Contin. Dyn. Syst. B **7**, 581–604 (2007)
96. W. Liu, J.P. Dowling, W.K. Murray, G.A. McArthur, J.F. Thompson, R. Wolfe, J.W. Kelly, Rate of growth in melanomas: characteristics and associations of rapidly growing melanomas. Arch. Dermatol. **142**(12), 1551–1558 (2006)
97. J.S. Lowengrub, H.B. Frieboes, F. Jin, Y.L. Chuang, X. Li, P. Macklin, S.M. Wise, V. Cristini, Nonlinear modelling of cancer: bridging the gap between cells and tumours. Nonlinearity **23**(1), R1–R9 (2010)

98. B.D. MacArthur, C.P. Please, Residual stress generation and necrosis formation in multi-cell tumour spheroids. J. Math. Biol. **49**(6), 537–552 (2004)
99. C. Mateus, C. Robert, Major therapeutic advances in the treatment of metastatic melanoma. Bull. Cancer **99**(6), 619–625 (2012)
100. W.V. Mayneord, On the law of growth of Jensen's rat sarcoma. Am. J. Cancer **16**, 841–846 (1932)
101. D.L.S. McElwain, L.E. Morris, Apoptosis as a volume loss mechanism in mathematical models of solid tumor growth. Math. Biosci. **39**, 147–157 (1978)
102. S.J. Miller, Biology of basal cell carcinoma (Part II). J. Am. Acad. Dermatol. **24**, 161–175 (1991)
103. T. Mueller, A thermodynamic theory of mixtures of fluids. Arch. Ration. Mech. Anal. **28**, 1–39 (1968)
104. W. Mueller-Klieser, Multicellular spheroids. A review on cellular aggregates in cancer research. J. Cancer Res. Clin. Oncol. **113**(2), 101–122 (1987)
105. A. Neagu, K. Jakab, R. Jamison, G. Forgacs, Role of physical mechanisms in biological self-organization. Phys. Rev. Lett. **95**(17), 178104–178107 (2005)
106. A. Nose, A. Nagafuchi, M. Takeichi, Expressed recombinant cadherins mediate cell sorting in model systems. Cell **54**(7), 993–1001 (1988)
107. A. Onuki, *Phase Transition Dynamics* (Cambridge University Press, Cambridge, 2002)
108. J.M. Pérez-Pomares, R.A. Foty, Tissue fusion and cell sorting in embryonic development and disease: biomedical implications. Bioessays **28**(8), 809–821 (2006)
109. K. Pham, H.B. Frieboes, V. Cristini, J.S. Lowengrub, Predictions of tumour morphological stability and evaluation against experimental observations. J. R. Soc. Interface **8**(54), 16–29 (2011)
110. J.M. Pimiento, E.M. Larkin, K.S.M. Smalley, G.L. Wiersma, N.R. Monks, I.V. Fedorenko, C.A. Peterson, B.J. Nickoloff, Melanoma genotypes and phenotypes get personal. Lab. Invest. **93**, 858–867 (2013)
111. C.P. Please, G.J. Pettet, D.L.S. McElwain, A new approach to modelling the formation of necrotic regions in tumours. Appl. Math. Lett. **11**, 89–94 (1998)
112. C.P. Please, G.J. Pettet, D.L.S. McElwain, Avascular tumour dynamics and necrosis. Math. Model Methods Appl. Sci. **9**, 569–579 (1999)
113. N.J. Poplawskia, U. Agero, J.S. Gens, M. Swat, J.A. Glazier, A.R. Anderson, Front instabilities and invasiveness of simulated avascular tumors. Bull. Math. Biol. **71**, 1189–1227 (2009)
114. L. Preziosi, H.M. Byrne, modelling solid tumor growth using the theory of mixtures. Math. Med. Biol. **20**, 341–366 (2003)
115. L. Preziosi, A. Farina, On Darcy law for growing porous media. Int. J. Nonlinear Mech. **37**, 485–491 (2002)
116. L. Preziosi, G. Vitale, A multiphase model of tumor and tissue growth including cell adhesion and plastic reorganization. Math. Model Meth. Appl. S **21**, 1901–1932 (2011)
117. A. Puliafito, L. Hufnagel, P. Neveu, S. Streichan, A. Sigal, R.M. Fygenson, B.I. Shraiman, Collective and single cell behavior in epithelial contact inhibition. Proc. Natl. Acad. Sci. USA **109**(3), 739–744 (2012)
118. J. Ranft, M. Basan, J. Elgeti, J.F. Joanny, J. Prost, F. Juelicher, Fluidization of tissues by cell division and apoptosis. Proc. Natl. Acad. Sci. USA **107**(49), 20863–20868 (2010)
119. K.A. Rejniak, An immersed boundary framework for modelling the growth of individual cells: an application to the early tumour development. J. Theor. Biol. **247**, 186–204 (2007)
120. L. Ricci-Vitiani, D.G. Lombardi, E. Pilozzi, M. Biffoni, M. Todaro, C. Peschle, R. De Maria, Identification and expansion of human colon-cancer-initiating cells. Nature **445**(7123), 111–115 (2007)
121. E.K. Rodriguez, A. Hoger, A.D. McCulloch, Stress-dependent finite growth in soft elastic tissues. J. Biomech. **27**(4), 455–467 (1994)

122. T. Roose, P.A. Netti, L.L. Munn, Y. Boucher, R.K. Jain, Solid stress generated by spheroid growth estimated using a linear poroelasticity model small star, filled. Microvasc. Res. **66**(3), 204–212 (2003)
123. T. Roose, S.J. Chapman, P.K. Maini, Mathematical models of avascular tumor growth. SIAM Rev. **49**, 179–208 (2007)
124. M. Shackleton, E. Quintana, E.R. Fearon, S.J. Morrison, Heterogeneity in cancer: cancer stem cells versus clonal evolution. Cell **138**(5), 822–829 (2009)
125. M.A. Shannon, B. Rubinsky, The effect of tumor growth on the stress distribution in tissue. Adv. Biol. Mass Transfer HTD ASME **231**, 35–39 (1992)
126. H.M. Shaw, W.H. McCarthy, Small-diameter malignant melanoma: a common diagnosis in new South Wales, Australia. J. Am. A Dermatol. **27**, 679–682 (1992)
127. S.K. Singh, C. Hawkins, I.D. Clarke, J.A. Squire, T. Bayani, J. Hide, R. Henkelman, M.D. Cusimano, P.B. Dirks, Identification of human brain tumour initiating cells. Nature **432**(7015), 396–340 (2004)
128. G.G. Steel, L.F. Lamerton, The growth rate of human tumours. Br. J. Cancer **20**(1), 74–86 (1966)
129. M.S. Steinberg, Differential adhesion in morphogenesis: a modern view. Curr. Opin. Genet. Dev. **17**(4), 281–286 (2007)
130. M. Stohrer, Y. Boucher, M. Stangassinger, R.K. Jain, Oncotic pressure in solid tumors is elevated. Cancer Res. **60**(15), 4251–4255 (2000)
131. M. Stücker, A. Struk, P. Altmeyer, M. Herde, H. Baumgärtl, D.W. Lübbers, The cutaneous uptake of atmospheric oxygen contributes significantly to the oxygen supply of human dermis and epidermis. J. Physiol. **538**, 985–994 (2002)
132. W.C. Summers, Dynamics of tumor growth: a mathematical model. Growth **30**(3), 333–338 (1966)
133. R.M. Sutherland, Cell and environment interactions in tumor microregions: the multicell spheroid model. Science **240**(4849), 177–184 (1988)
134. R.M. Sutherland, R.E. Durand, Hypoxic cells in an in vitro tumour model. Int. J. Radiat. Biol. Relat. Stud. Phys. Chem. Med. **23**(3), 235–246 (1973)
135. R.M. Sutherland, J.A. McCredie, W.R. Inch, Growth of multicell spheroids in tissue culture as a model of nodular carcinomas. J. Natl. Cancer Inst. **46**(1), 113–120 (1971)
136. W.R. Tinga, A.G. Voss, D.F. Blossey, Generalized approach to multiphase dielectric mixture theory. J. Appl. Phys. **44**, 3897–3902 (1973)
137. R.H. Thomlison, L.H. Gray, The histological structure of some human lung cancers and the possible implications for radiotherapy. Br. J. Cancer **9**(4), 539–549 (1955)
138. P.L. Townes, J. Holfreter, Directed movements and selective adhesion of embryonic amphibian cells. J. Exp. Zool. **128**, 53–120 (1995)
139. C. Truesdell, *Rational Thermodynamics* (Springer, New York, 1984)
140. C. Truesdell, R. Toupin, The classical field theory, in *Handbuch der Physik* (Springer, Berlin, 1960)
141. R. Vandiver, A. Goriely, Differential growth and residual stress in cylindrical elastic structures. Philos. Transact. A Math. Phys. Eng. Sci. **367**(1902), 3607–3630 (2009)
142. K.Y. Volokh, Stresses in growing soft tissues. Acta Biomater. **2**(5), 493–504 (2006)
143. R.A. Weinberg, *One Renegade Cell: How Cancer Begins* (Phoenix, London, 2009)
144. M.A. Weinstock, Cutaneous melanoma: public health approach to early detection Dermatol. Ther. **19**, 26–31 (2006)
145. R. Wette, E.Y. Rodin, B. Katz, Stochastic processes for solid tumor kinetics i. Surface-regulated growth. Math. Biosci. **19**, 231–255 (1974)
146. R. Wette, B. Katz, E.Y. Rodin, Stochastic processes for solid tumor kinetics ii. Diffusion-regulated growth. Math. Biosci. **21**, 311–338 (1974)
147. H. Wiig, R.K. Reed, O. Tenstad, Interstitial fluid pressure, composition of interstitium, and interstitial exclusion of albumin in hypothyroid rats. Am. J. Physiol. Heart Circ. Physiol. **278**(5), H1627–H1639 (2000)

148. B.S. Winters, S.R. Shepard, R.A. Foty, Biophysical measurement of brain tumor cohesion. Int. J. Cancer **114**(3), 371–379 (2005)
149. S.M. Wise, J.S. Lowengrub, H.B. Frieboes, V. Cristini, Three-dimensional multispecies nonlinear tumor growth:model and numerical method. J. Theor. Biol. **253**(3), 524–543 (2008)
150. J. Xu, K. Gupta, W.V. Stoecker, Y. Krishnamurthy, H.S. Rabinovitz, A. Bangert, D. Calcara, M. Oliviero, J.M. Malters, R. Drugge, R.J. Stanley, R.H. Moss, M.E. Celebi, Analysis of globule types in malignant melanoma. Arch. Dermatol. **145**(11), 1245–1251 (2009)
151. H. Yabu, T. Higuchi, M. Shimomura, Unique phase-separation structures of block-copolymer nanoparticles. Adv. Mater. **17**, 2062–2065 (2005)
152. X. Zheng, S.M. Wise, V. Cristini, Nonlinear simulation of tumor necrosis, neo-vascularization and tissue invasion via an adaptive finite-element/level-set method. Bull. Math. Biol. **67**(2), 211–259 (2005)

Part II
Tumor and Inter-Cellular Interactions

A Cell Population Model Structured by Cell Age Incorporating Cell–Cell Adhesion

Janet Dyson and Glenn F. Webb

Abstract An analysis is given of a continuum model of a proliferating cell population, which incorporates cell movement in space and cell progression through the cell cycle. The model consists of a nonlinear partial differential equation for the cell density in the spatial position and the cell age coordinates. The equation contains a diffusion term corresponding to random cell movement, a nonlocal dispersion term corresponding to cell–cell adhesion, a cell age-dependent boundary condition corresponding to cell division, and a nonlinear logistic term corresponding to constrained population growth. Basic properties of the solutions are proved, including existence, uniqueness, positivity, and long-term behavior dependent on parametric input. The model is illustrated by simulations applicable to in vitro wound closure experiments, which are widely used for experimental testing of cancer therapies.

Keywords Cell age • Cell adhesion • Non-local • Reaction-diffusion • Analytic semigroup • Fractional power • Existence • Regularity • Positivity

1 Introduction

Mathematical models of cancer progression incorporating both microscale features, such as cell movement and division, and macroscale features, such as tumor growth and metastasis, are necessarily complex in their formulation and analysis. A variety of approaches have been developed, including continuum differential equations models, probabilistic individual based models, and combinations of these two types (see, for example, review articles in [4,7,13,22,23,36]). The advantage of continuum models is their tractability for analysis based on initialization and parameterization.

J. Dyson (✉)
Mansfield College, University of Oxford, Oxford, OX1 3TF, UK
e-mail: janet.dyson@mansfield.ox.ac.uk

G.F. Webb
Department of Mathematics, Vanderbilt University, Nashville, TN 37240, USA
e-mail: glenn.f.webb@vanderbilt.edu

© Springer Science+Business Media New York 2014
A. d'Onofrio, A. Gandolfi (eds.), *Mathematical Oncology 2013*,
Modeling and Simulation in Science, Engineering and Technology,
DOI 10.1007/978-1-4939-0458-7_4

A disadvantage is their continuum idealization of physical processes that are fundamentally discrete. Individual based models are inherently well founded for description of individual cell behavior, but lack the rigorous analysis available for classical continuum models. A major difficulty in resolving the continuum-discrete dichotomy of mathematical models of cancer is the modeling of cell–cell communication processes. These processes are necessarily specific to individual cells, but a new continuum approach has recently been developed to model cell–cell communications as nonlocal phenomena in differential equations models of cell population dynamics.

The model we consider here was originally proposed by Armstrong, Painter, and Sherratt in [2] and developed further in [3, 17, 18, 29, 43, 52]. The equations of these models have nonlocal flux terms corresponding to the component of cell motion attributable to cells bonding to other cells within a specified sensing radius of each individual cell. Rigorous analyses of such models have been carried out in [17, 18].

Our objective here is to extend this work to continuum spatial nonlocal models which also include the cell cycle as determined by continuum cell age. The advantage of cell age structure is that individual cell cycle behavior can be incorporated into a continuum model and, specifically, the variable distribution of age of mitosis as given in (2) below. Continuum models of cell population dynamics combining space and age variables have been developed by many authors and we first review some of those which are relevant to our present problem.

Individual structure in biological populations has many possible interpretations as continuum variables. The most used continuum structure variable is age, which can be interpreted variously as chronological, stage, or phase age. Age as a structure variable has obvious significance, but can be difficult to measure in many biological applications. Another frequently used continuum structure variable is size, which can be interpreted as mass, length, volume, or other readily measurable physical characteristic. Other examples of individual structure variables include biochemical content, maturity stage, or formed pattern levels. In principle, spatial structure can be combined with multiple designations of individual structure variables to provide detailed quantification of complex biological behavior in spatial context [31, 39, 54, 71].

Spatial variables in combination with individual structure variables in biological populations can also be interpreted in many ways, with correspondence to specific locations in biological applications. Examples range from micro-locations of proteins, viruses, bacteria, and cells to macro-locations of tissues, in-hosts settings, and geographical regions. Examples in the literature for combined structure and spatial continuum variables include bacterial pattern formation of *Proteus mirabilis* swarming on agar surfaces [35], tumor growth in in vivo or in vitro settings [3, 14–16, 18, 29, 43, 52, 56, 58], regulatory features of intestinal crypt behavior such as control via morphogens [41, 42], epidermal growth dynamics [27, 28], predator-prey, competitive, and cooperative systems [62–64], and epidemic diseases in geographical settings [8, 9, 11, 12, 20, 21, 33, 34, 48, 68, 69]. Recent treatments of the mathematical theory of continuum spatial and age population models are found in [53, 59–61, 65, 66, 71].

In [71] an analysis is given of problems with age, size, and spatial structure, using the theory of linear and nonlinear semigroups. We will review only the case with age and spatial structure. Let Z be a Banach space of functions on $\Omega \subset \mathbb{R}^N$ and $Y = L^1((0, a_1); Z)$ where $a_1 \leq \infty$ is the maximum age of individuals. Let $T(t) : Z \to Z$ be a strongly continuous semigroup of bounded linear operators in Z with infinitesimal generator A. Let $p(x, a, t)$ be the population density at position $x \in \Omega$, age $a \in (0, a_1)$, and time $t > 0$, so that the total population at time t is $\int_\Omega \int_0^{a_1} p(x, a, t)\text{da}\,\text{dx}$. The density $p(x, a, t) = p(a, t)(x)$ is viewed as a function from $(0, a_1)$ to Z. Equations of the form

$$\frac{\partial p(x,a,t)}{\partial t} + \frac{\partial p(x,a,t)}{\partial a} = Ap(x,a,t) + G(p(x,a,t)),$$

$$a \in (0, a_1), x \in \Omega,\ t > 0 \tag{1}$$

$$p(x,0,t) = 2\int_0^{a_1} \theta(a)p(a,t)\text{da},\ x \in \Omega, t > 0 \tag{2}$$

$$p(x,a,0) = \phi_0(x,a),\ a \in (0,a_1),\ x \in \Omega,\ \phi \in Y.$$

are considered. It is shown that in the case where $G = 0$ the solutions of this problem are given by $p(a, t) = (S(t)\phi)(a)$ where $S(t) : Y \to Y$ is a strongly continuous semigroup of bounded linear operators on Y given by

$$(S(t)\phi)(a) = \begin{cases} T(t)\phi(a-t), & t < a < a_1, \\ T(a)b_\phi(t-a) & 0 \leq a < t \end{cases} \tag{3}$$

where b_ϕ is the unique solution of the linear Volterra equation for b_ϕ in Z

$$b_\phi(t) = 2\int_0^t \theta(a)T(a)b_\phi(t-a)\text{da} + 2\int_t^{a_1} \theta(a)T(t)\phi(a-t)\text{da}.$$

It is then shown that if $G : Y \to Y$ is globally Lipschitz, then Eq. (1) has a unique global mild solution $U(t)\phi$, and $U(t), t \geq 0$ is a strongly continuous semigroup of Lipschitz continuous nonlinear operators in Y. Further, if Z is a Banach lattice and $T(t), t \geq 0$ is a positive semigroup in Z, then $U(t), t \geq 0$ is a positive semigroup in Y.

The results of [71] are exploited in [14, 16], in which a model of tumor growth into surrounding tissue is considered. The model consists of a system of nonlinear partial differential equations for the population densities of tumor cells, $p(x, a, t)$, extracellular matrix macromolecules, $f(x, t)$, oxygen concentration, $w(x, t)$, and extracellular matrix degradative enzyme concentration, $m(x, t)$. The spatial growth of the tumor involves the directed movement of tumor cells toward the extracellular matrix through haptotaxis. The equations of the model are as follows, with $P(x,t) = \int_0^{a_1} p(x,a,t)\mathrm{d}a$ and $\Omega \subset \mathbb{R}^3$ a nonempty open bounded set with smooth boundary:

$$\frac{\partial}{\partial t} f(x,t) = \underbrace{\delta_f \Delta f(x,t)}_{motility} - \underbrace{\omega_f f(x,t)}_{loss} - \underbrace{\lambda_f(x, m(\cdot,t), w(\cdot,t), P(\cdot,t)) f(x,t)}_{enzyme\ induced\ degradation} \quad (4)$$

$$\frac{\partial}{\partial t} m(x,t) = \underbrace{\delta_m \Delta m(x,t)}_{diffusion} - \underbrace{\omega_m m(x,t)}_{decay} + \underbrace{\rho_m(x, P(\cdot,t)) P(x,t)}_{production} \quad (5)$$

$$\frac{\partial}{\partial t} w(x,t) = \underbrace{\delta_w \Delta w(x,t)}_{diffusion} - \underbrace{\omega_w w(x,t)}_{decay} + \underbrace{\beta_w(x, f(\cdot,t)) f(x,t)}_{production} - \underbrace{\mu_w(x, w(\cdot,t), P(\cdot,t)) w(x,t)}_{uptake\ by\ tumor\ cells} \quad (6)$$

$$\frac{\partial}{\partial t} p(x,a,t) = \underbrace{-\frac{\partial}{\partial a} p(x,a,t)}_{cell\ aging} + \underbrace{\delta_p \Delta p(x,a,t)}_{random\ motility} - \underbrace{\omega_p p(x,a,t)}_{loss} - \underbrace{\chi_p \nabla \cdot (p(x,a,t) \nabla f(x,t))}_{haptotaxis} - \underbrace{\mu_p(x, w(\cdot,t), P(\cdot,t)) p(x,a,t)}_{loss\ due\ to\ death\ or\ division} \quad (7)$$

$$p(x,0,t) = \underbrace{2 \int_0^{a_1} \theta(a) p(x,a,t)\, \mathrm{d}a}_{rate\ of\ cell\ division} . \quad (8)$$

Neumann boundary conditions are imposed and initial conditions are given. Equation (7) for the density of tumor cells involves the diffusion term, $A = \delta_p \Delta - \omega_p$, which therefore generates an analytic semigroup of bounded linear operators $T(t)$, $t \geq 0$. The main difficulty for the analysis is that the nonlinear haptotaxis term in (7) is not Lipschitz continuous and the key assumption is a nonlocal spatial dependence in the nonlinear enzyme degradation term in (4). The problem can then be handled using the theory of fractional powers of the generators of analytic semigroups applied to A. The form of $S(t)$, $t \geq 0$ above means that, although $S(t)$ is not itself analytic, it inherits many similar properties to do with the fractional powers of A from $T(t)$. Existence, regularity, positivity, and global bounds for solutions of the model are obtained. In [15] this theory is extended to handle a model of tumor growth in which there is age, size, and spatial structure and which includes proliferating and quiescent compartments of tumor cells indexed by successively mutated cell phenotypes of increasingly proliferative aggressiveness. In [57] Walker treats a similar system to that in [14, 16], except that in the equation for the extracellular matrix macromolecule concentration there is local spatial dependence and no diffusion. A rather different approach is used, starting with the introduction of an age-diffusion semigroup. These results are extended in [56, 58] to allow nonlinear age-boundary conditions.

We present here a rigorous analysis of an age and spatially structured continuum model of cell population dynamics that also includes cell–cell communication via the devise of a sensing radius of individual cells. Cell–cell adhesion is a cell communication process by which cells adhere to each other through the binding of cell adhesion molecules at the cell surface. It is a fundamental mechanism in many areas of biology and, for example, is largely responsible for morphogenesis, stabilization, and degeneration of tissue. Selective adhesion of embryonic cells allows cells to organize themselves into patterns that develop into tissues and organs, and changes in adhesiveness properties are central to the invasive stage of cancer (see, for example, [3, 6, 10, 29, 43, 47, 49, 52, 55, 67]).

It is difficult to model cell adhesion using continuous mathematical models, because in such an approach one uses continuous variables to describe aggregate cell densities. Individual cells are not recognized as such and therefore, there is no representation of cell boundaries. In [18] the following model for cell–cell adhesion in N spatial dimensions was treated: for a cell density $p(x,t)$, with $x \in \mathbb{R}^N$, $t > 0$, and with B_ρ the open ball in $\mathbb{R}^N$, radius ρ, center 0,

$$\frac{\partial p(x,t)}{\partial t} = \underbrace{\delta \Delta p(x,t)}_{\text{random cell motility}} \underbrace{- \nabla \cdot \left(p(x,t) \int_{B_\rho} g(p(x+\xi,t)) \xi h(|\xi|)\, \mathrm{d}\xi \right)}_{\text{cell adhesion}} + \underbrace{F(p(x,t))}_{\text{cell loss and gain}} \tag{9}$$

with initial condition

$$p(x,0) = \phi_0(x). \tag{10}$$

The integral term is a nonlocal flux term which represents the cell–cell contact and the function $F(p)$ models cell loss and cell gain.

Here we consider how age structure may be incorporated into such a model. In our model individual cells are distinguished by spatial position $x \in \mathbb{R}^N$ and by cell cycle age $a \in (0,\infty)$. Let $p(x,a,t)$ be cell density, at position x, age a, and time $t > 0$, and $P(x,t) = \int_0^\infty p(x,a,t)\mathrm{da}$, the total population density of cells at x, with B_ρ as above. We consider the partial differential equation with boundary condition at age 0:

$$\frac{\partial p(x,a,t)}{\partial t} + \underbrace{\frac{\partial p(x,a,t)}{\partial a}}_{\textit{cell aging}} = \underbrace{\delta \Delta p(x,a,t)}_{\textit{random motility}} \underbrace{- \nabla \cdot \left(p(x,a,t) \int_{B_\rho} g(P(x+\xi,t)) \xi h(|\xi|)\mathrm{d}\xi \right)}_{\textit{cell adhesion}} \underbrace{- p(x,a,t) f(P(x,t))}_{\textit{loss}} \tag{11}$$

$$p(x,0,t) = \underbrace{2\int_0^\infty \theta(a)p(x,a,t)\mathrm{da}}_{\textit{rate of cell division}} \tag{12}$$

with initial condition

$$p(x,a,0) = \phi_0(x,a). \tag{13}$$

Here the function $-pf(P)$ models cell loss and now

$$\text{total cell loss plus cell gain } = -Pf(P) + 2\int_0^\infty \theta(a)p(a)\mathrm{d}a.$$

We will require that for all $p(a)$ such that $\int_0^\infty p(a)\mathrm{d}a = P$ and P large enough:

$$-Pf(P) + 2\int_0^\infty \theta(a)p(a)\mathrm{d}a < 0.$$

Thus, at large enough densities, cell loss occurs more rapidly than the generation of new cells via division. Note that if $\beta_0 > 0$ is the intrinsic growth constant (that is the unique β_0 such that $2\int_0^\infty \theta(a)e^{-\beta_0 a}\,da = 1$) and if $P_1 > 0$ is such that $f(P_1) = \beta_0$, then $q(a) = \beta_0 P_1 e^{-\beta_0 a}$ is a stationary solution and we have $-P_1 f(P_1) + 2\int_0^\infty \theta(a)q(a)\mathrm{da} = 0$. So at this point, for the stationary solution, cell loss and cell gain are in balance.

The other two terms in the right-hand side of (11) are flux terms. The equation admits the possibility of two components to the cell motion: One is Fickian diffusion giving rise to the Laplacian term. The second is a cell adhesion term which models the component of the cell motion caused by forces due to adhesive bonds between cells and their neighbors, which can be viewed as an advective flux term. Cells can sense their surroundings over some radius ρ via protrusions and this sensing radius will be considerably larger than the mean cell radius because of the extension of the cell protrusions such as filopodia.

These protrusions cause adhesive bonds to form with other cells, and the resulting forces cause cell movement. In fact, the cell adhesion term in (11) is an advective flux term of the form $-\nabla \cdot (p(x,a,t)\mathbf{U}(x,t))$ where $\mathbf{U}(x,t)$ is the velocity of the cells at $x \in \mathbb{R}^N$ at time t. If we imagine the cells as tiny spheres moving through a viscous fluid, then, by Stokes' law, they are subject to a resistive force which is directly proportional to the velocity. So we are led to suppose that the velocity of a cell is proportional to the net adhesive force on it due to the bonds formed with nearby cells. Absorbing constants such as viscosity into g or h, this force is taken to be of the form

$$\int_{B_\rho} g(P(x+\xi,t))\xi h(|\xi|)\,\mathrm{d}\xi. \tag{14}$$

We are asserting that the total force on a cell at position x is the sum, over all cells within the sensing radius ρ, of the local forces attributable to bonds formed with the cells at the nearby points $x + \xi$, $\xi \in B_\rho$. The function $h : \mathbb{R} \to \mathbb{R}$ describes how the magnitude of the force depends on $|\xi|$ and is a positive function since adhesive forces are always directed toward cell centers. The vector ξ in front of $h(|\xi|)$ gives direction to the force on cells at x due to bonds with cells at $x + \xi$.

The magnitude of the local forces from the nearby location $x + \xi$ depends on the number of adhesive attachments made by cells at $x + \xi$ to a cell at x and hence on the cell density at $x + \xi$. The function g describes how the forces depend on the local cell density. To be realistic biologically $g(P)$ should increase linearly with P if cell density is not too great, then decrease for P above some threshold, and be zero for all P above some critical cell density $P_2 > 0$ which corresponds to close cell packing. For example $g(P) = \max\{\mu P(P_2 - P), 0\}$. For our results it will usually be enough to require only that $g(P) \geq 0$ for $P \in [0, P_2]$. This condition will lead to $P(x,t) \in [0, P_2]$, so that values of g outside this interval are irrelevant to the mathematics. The shift in balance between cell division and cell loss should occur at an achievable density, so before P_2, and therefore in biologically realistic cases we will have $P_2 > P_1$.

In Sect. 2 we obtain existence and stability results for the abstract version of (11)–(13). In Sect. 3 we consider initial data in the space $L^1((0,\infty); BUC^{(k)}(\mathbb{R}^N))$ in the two cases where either the initial data tends to zero as $|x| \to \infty$ or it tends to the stationary solution as $|x| \to \infty$. We obtain the global existence of the unique mild solution. First we prove local existence. We then prove that solutions are positive if the initial data is positive by looking at the equation integrated along the characteristics. Finally, for suitable initial data, we prove boundedness of solutions by looking at the equation integrated with respect to a. This leads to global existence. In Sect. 4 we apply the results of Sect. 2 to obtain conditions for the stability of the stationary solution of the problem in $L^1((0,\infty); L^2(\mathbb{R}^N))$. In Sect. 5 we provide numerical simulations of the model applicable to wound healing experiments.

2 Abstract Results

First we consider the abstract version of (11)–(13) and derive existence and stability results, analogous to those in [32].

Let $\tilde{X}$ be a Banach space and $-\tilde{A} : \mathscr{D}(\tilde{A}) \subset \tilde{X} \to \tilde{X}$ be the generator of an analytic semigroup $T(t) : \tilde{X} \to \tilde{X}$ such that there exist $\tilde{\omega} > 0$ and $M \geq 1$ with $||T(t)|| \leq Me^{\tilde{\omega}t}$. Set $\tilde{Y} = L^1((0,\infty); \tilde{X})$. Let $\omega \in \mathbb{R}$ be such that $A = \tilde{A} + \omega I$ has its spectrum in the open right half plane. For $0 < \alpha < 1$ denote $\tilde{Y}^\alpha = L^1((0,\infty); \mathscr{D}(A^\alpha))$, with $||\phi||_{\tilde{Y}^\alpha} = ||A^\alpha \phi||_{\tilde{Y}}$. Recall that (see [32, 37, 38,

44]) if $\alpha > 0$, then for every $t > 0$ the operator $A^\alpha T(t)$ is a bounded linear operator in $\tilde{X}$ and we can take $\tilde{\omega} > 0$ and $M \geq 1$ such that also

$$||A^\alpha T(t)|| \leq M t^{-\alpha} e^{\tilde{\omega} t},$$

and also for every $U \in \mathscr{D}(A^\alpha)$, $T(t)A^\alpha U = A^\alpha T(t)U$. Furthermore, if $0 \leq \alpha \leq 1$, then $A^{-\alpha} : \tilde{X} \to \tilde{X}$ is a bounded linear operator so there exists a constant $C_1 \geq 1$ such that for all $u \in \tilde{X}$,

$$||A^{-\alpha} u||_{\tilde{X}} \leq C_1 ||u||_{\tilde{X}}. \tag{15}$$

Hence for $u \in \tilde{Y}$

$$||A^{-\alpha} u||_{\tilde{Y}} \leq C_1 ||u||_{\tilde{Y}}. \tag{16}$$

We now let $H : \tilde{Y}^\alpha \to \tilde{Y}$, and suppose that

(H.1) $\theta \in C([0,\infty); \mathbb{R}_+)$ and there exist $0 < \epsilon < \bar{a} < \infty$, $\bar{\theta} > 0$ such that $\theta(a) \equiv 0$ for $a \in [0,\epsilon] \cup [\bar{a},\infty)$ and $\theta(a) \leq \bar{\theta}$ for $a \in (0,\infty)$.

Note that $\beta_0 \leq 2\bar{\theta}$, where β_0 is the intrinsic growth constant. We will assume throughout that $\beta_0 > 0$.

We consider the abstract problem (NA):

$$\begin{aligned} \frac{du}{dt} &= -\frac{\partial u}{\partial a} - \tilde{A}u + H(u), \quad t > 0, \\ u(0,t) &= 2\int_0^\infty \theta(a)u(a,t)\mathrm{d}a \\ u(a,0) &= \phi(a) \in \tilde{Y}. \end{aligned}$$

As in [71] we consider the following linear partial differential equation associated with (NA):

$$\frac{\partial}{\partial t}u(a,t) + \frac{\partial}{\partial a}u(a,t) = -\tilde{A}u(a,t) \tag{17}$$

$$u(0,t) = 2\int_0^\infty \theta(a)u(a,t)\mathrm{d}a \tag{18}$$

$$u(a,0) = \phi(a) \in \tilde{Y}. \tag{19}$$

To obtain the semigroup of the solutions of (17)–(19), we define the family of operators in $\tilde{Y}$, $\{S(t)\}_{t\geq 0}$, as follows:

$$(S(t)\phi)(a) = \begin{cases} T(t)\phi(a-t), & t < a < \infty, \\ T(a)b_\phi(t-a) & 0 \leq a < t \end{cases} \tag{20}$$

where $b_\phi \in C([0,\infty), \tilde{X})$ is the unique solution of the linear Volterra equation in $\tilde{X}$

$$b_\phi(t) = 2\int_0^t \theta(a)T(a)b_\phi(t-a)\mathrm{d}a + 2\int_t^\infty \theta(a)T(t)\phi(a-t)\mathrm{d}a.$$

In [71] Webb proves that $\{S(t)\}_{t\geq 0}$ is a strongly continuous positive semigroup of bounded linear operators in $\tilde{Y}$ such that $||S(t)\phi||_{\tilde{Y}} \leq Me^{\tilde{\omega}_a t}||\phi||_{\tilde{Y}}$, $\phi \in \tilde{Y}$, $t \geq 0$, where $\tilde{\omega}_a := 2M\bar{\theta} + \tilde{\omega}$.

Note that $b_\phi(0) = 2\int_0^\infty \theta(a)\phi(a)\mathrm{d}a$. Also $(S(t)\phi)(a)$ is continuous in a for $t > a$ (as $b_\phi(t)$ is continuous), $(S(t)\phi)(0) = b_\phi(t)$ so that for $t > a$

$$(S(t)\phi)(a) = T(a)((S(t-a)\phi)(0)), \tag{21}$$

and $(S(t)\phi)(0)$ is continuous in t and satisfies for $t > 0$,

$$(S(t)\phi)(0) = 2\int_0^\infty \theta(\hat{a})(S(t)\phi)(\hat{a})\mathrm{d}\hat{a}. \tag{22}$$

Furthermore, for each $0 < \alpha < 1$, we can also choose $M \geq 1$ and $\tilde{\omega}_a > 0$ such that

$$||A^\alpha S(t)|| \leq Mt^{-\alpha}e^{\tilde{\omega}_a t} \text{ for } t > 0, \tag{23}$$

(see [14] Lemma 2). Also, if $\phi \in \tilde{Y}^\alpha$, $A^\alpha b_\phi(t) = b_{A^\alpha\phi}(t)$, so that $A^\alpha S(t)\phi = S(t)A^\alpha\phi$ and

$$||S(t)\phi||_{\tilde{Y}^\alpha} = ||S(t)A^\alpha\phi||_{\tilde{Y}} \leq Me^{\tilde{\omega}_a t}||\phi||_{\tilde{Y}^\alpha}. \tag{24}$$

Note that $S(t) : \tilde{Y} \to \tilde{Y}^\alpha$ so that, as $S(t)$ is strongly continuous, $\tilde{Y}^\alpha$ is dense in $\tilde{Y}$.

We say that $u(t)$ is a local mild solution of (NA) if there exists $T_0 > 0$ such that $u(t) \in C([0, T_0]; \tilde{Y}^\alpha)$ and for $0 \leq t \leq T_0$

$$u(t) = S(t)\phi + \int_0^t S(t-s)H(u(s))\mathrm{d}s. \tag{25}$$

We suppose that

(H.2) $H : \tilde{Y}^\alpha \to \tilde{Y}$, $H(0) = 0$, and for some $\frac{1}{2} < \alpha < 1$, if $\phi_i \in \tilde{Y}^\alpha$, $i = 1, 2$, with $||A^\alpha\phi_i||_{\tilde{Y}} \leq R$, then there exists $K(R)$ such that

$$||H(\phi_1) - H(\phi_2)||_{\tilde{Y}} \leq K(R)||A^\alpha(\phi_1 - \phi_2)||_{\tilde{Y}}.$$

The following existence result is proved similarly to [32] Theorem 3.3.3, Corollary 3.3.5, and Theorem 3.4.1, working in $C([0, T_0]; \tilde{Y}^\alpha)$, with (25) operated on by A^α and using (23).

Theorem 1. *Suppose that $H : \tilde{Y}^{\alpha} \to \tilde{Y}$ satisfies (H.2) and $\phi \in \tilde{Y}^{\alpha}$. Then there exists $T_0 > 0$ such that the problem (NA) has a unique mild solution $u(t)$ with $u(t) \in C([0, T_0]; \tilde{Y}^{\alpha})$.*

There is continuous dependence on the initial data in the following sense. Let $[0, \hat{T}_0)$ be the maximal interval of existence of the solution $u(t)$. Now let $u_n(t)$ be the solution of (25) with initial data $u_n(x, a, 0) = \phi_n(x, a)$ where $\phi_n \in \tilde{Y}^{\alpha}$. Suppose that $||A^{\alpha}(\phi_n - \phi)||_{\tilde{Y}} \to 0$ as $n \to \infty$. Then given any $t_1 \in [0, \hat{T}_0)$, $u_n(t)$ is defined on $[0, t_1]$ for large enough n and $||A^{\alpha}(u_n(t) - u(t))||_{\tilde{Y}} \to 0$ as $n \to \infty$, uniformly on $[0, t_1]$.

The solution is global if $||A^{\alpha}u(t)||_{\tilde{Y}}$ is bounded on bounded intervals of time. In particular this is true if there exists a constant $K > 0$ such that for all $\phi \in \tilde{Y}^{\alpha}$

$$||H(\phi)||_{\tilde{Y}} \leq K \left(||A^{\alpha}\phi||_{\tilde{Y}} + 1\right).$$

Now we look for conditions such that the zero solution of (NA) is asymptotically stable. Suppose that $H(\phi) = B\phi + J(\phi)$ and satisfies

(H.3) For some $\frac{1}{2} < \alpha < 1$, $B : \tilde{Y}^{\alpha} \to \tilde{Y}$ is a bounded linear operator;

$$J : \tilde{Y}^{\alpha} \to \tilde{Y} \quad \text{and} \quad ||J(\phi)||_{\tilde{Y}} = o(||A^{\alpha}\phi||_{\tilde{Y}}) \quad \text{as} \quad ||A^{\alpha}\phi||_{\tilde{Y}} \to 0.$$

We consider first the linearized problem. It follows from Theorem 1 that

Lemma 1. *Suppose that for some $0 < \alpha < 1$, $B : \tilde{Y}^{\alpha} \to \tilde{Y}$ is a bounded linear operator, and consider the linearized problem (LA):*

$$\frac{dr}{dt} = -\frac{\partial r}{\partial a} - \tilde{A}r + Br, \quad t > 0,$$
$$r(0, t) = 2\int_0^{\infty} \theta(a) r(a, t) \mathrm{d}a$$
$$r(a, 0) = \phi(a) \in \tilde{Y}.$$

For $\phi \in \tilde{Y}^{\alpha}$ this problem has a unique mild solution $r(t) = R(t)\phi$ in the sense that for all $\phi \in \tilde{Y}^{\alpha}$, $R(t)\phi \in C([0, \infty); \tilde{Y}^{\alpha})$ is the unique solution, for $t > 0$, of

$$R(t)\phi = S(t)\phi + \int_0^t S(t - s)BR(s)\phi \mathrm{d}s. \tag{26}$$

$\{R(t)\}_{t \geq 0}$ is a strongly continuous semigroup on $\tilde{Y}^{\alpha}$.

Lemma 2. *Suppose now that there exist $\gamma > 0$, M_1 such that for $\phi \in \tilde{Y}^\alpha$*

$$||R(t)\phi||_{\tilde{Y}} \leq M_1 e^{-\gamma t} ||\phi||_{\tilde{Y}}. \tag{27}$$

Thus $R(t)$ can be extended to a strongly continuous semigroup on $\tilde{Y}$. Then for any γ_2 such that $0 < \gamma_2 < \gamma$, there exists $M_2 \geq 1$ such that for $t > 0$ and for all $\phi \in \tilde{Y}^\alpha$

$$||R(t)\phi||_{\tilde{Y}^\alpha} \leq M_2 t^{-\alpha} e^{-\gamma_2 t} ||\phi||_{\tilde{Y}}, \text{ and} \tag{28}$$

$$||R(t)\phi||_{\tilde{Y}^\alpha} \leq M_2 e^{-\gamma_2 t} ||\phi||_{\tilde{Y}^\alpha}. \tag{29}$$

Furthermore $R(t)\phi \in \tilde{Y}^\alpha$ for all $\phi \in \tilde{Y}$, (28) holds for all $\phi \in \tilde{Y}$, and $R(t)\phi$ satisfies (26) for all $\phi \in \tilde{Y}$.

Proof. First we prove (28). Take $\phi \in \tilde{Y}^\alpha$ and suppose we can prove that for $0 < t \leq 1$ there exists a constant K_1 such that

$$||R(t)\phi||_{\tilde{Y}^\alpha} \leq K_1 t^{-\alpha} ||\phi||_{\tilde{Y}}, \tag{30}$$

then for $t > 1$

$$||R(t)\phi||_{\tilde{Y}^\alpha} \leq K_1 ||R(t-1)\phi||_{\tilde{Y}} \leq K_2 e^{-\gamma(t-1)} ||\phi||_{\tilde{Y}} \leq K_3 t^{-\alpha} e^{-\gamma_2 t} ||\phi||_{\tilde{Y}}.$$

Thus it is sufficient to prove (30). But for $\phi \in \tilde{Y}^\alpha$

$$A^\alpha R(t)\phi = A^\alpha S(t)\phi + \int_0^t A^\alpha S(t-s) B R(s)\phi \mathrm{d}s, \tag{31}$$

so, using (23), for $0 < t \leq 1$

$$||A^\alpha R(t)\phi||_{\tilde{Y}} \leq M e^{\tilde{\omega}_a t} t^{-\alpha} ||\phi||_{\tilde{Y}} + \int_0^t M(t-s)^{-\alpha} e^{\tilde{\omega}_a (t-s)} ||B||_{\mathscr{L}(\tilde{Y}^\alpha, \tilde{Y})} ||A^\alpha R(s)\phi||_{\tilde{Y}} \mathrm{d}s,$$

so as $t \leq 1$, (30) follows using the Gronwall-type inequality given in [32] Sect. 1.2.1.

We now note that it follows from (28) and (16) that (29) is true for $t \geq 1$. For $t \leq 1$ we again use (31), but now we apply (24).

Now, given $\phi \in \tilde{Y}$, take $\phi_n \in \tilde{Y}^\alpha$ such that $\phi_n \to \phi$. Then, by (28), for $t > 0$, $A^\alpha R(t)\phi_n$ converges to $h(t) \in \tilde{Y}^\alpha$, say, and again using (16), it is easy to see that $R(t)\phi \in \tilde{Y}^\alpha$ and $A^\alpha R(t)\phi_n$ converges to $A^\alpha R(t)\phi$. So (28) holds for all $\phi \in \tilde{Y}$. To prove the final result put $\phi = \phi_n$ in (26). Then using (28)

$$\left\| \int_0^t S(t-s)BR(s)(\phi_n - \phi)\mathrm{d}s \right\|_{\tilde{Y}}$$

$$\leq \int_0^t Me^{\tilde{\omega}_a(t-s)}||B||_{\mathscr{L}(\tilde{Y}^\alpha,\tilde{Y})}M_2 s^{-\alpha}e^{-\gamma_2 s}||\phi_n - \phi||_{\tilde{Y}}\mathrm{d}s \to 0 \text{ as } n \to \infty$$

and it follows that (26) holds for all $\phi \in \tilde{Y}$. □

Lemma 3. *For $\phi \in \tilde{Y}^\alpha$, $u(t) \in C([0,T_0];\tilde{Y}^\alpha)$ satisfies Eq. (25) if and only if*

$$u(t) = R(t)\phi + \int_0^t R(t-s)J(u(s))\mathrm{d}s. \tag{32}$$

Proof. For $u(t) \in C([0,T_0];\tilde{Y}^\alpha)$ set

$$w_1(t) = u(t) - R(t)\phi - \int_0^t R(t-s)J(u(s))\mathrm{d}s$$

and

$$w_2(t) = u(t) - S(t)\phi - \int_0^t S(t-s)H(u(s))\mathrm{d}s.$$

Then as $R(t)\phi = S(t)\phi + \int_0^t S(t-s)BR(s)\phi\mathrm{d}s$ and $J = H - B$

$$\begin{aligned}
w_1(t) &= u(t) - \Bigg\{ S(t)\phi + \int_0^t S(t-s)BR(s)\phi\mathrm{d}s \\
&\quad + \int_0^t \Bigg\{ S(t-s)J(u(s)) + \int_0^{t-s} S(t-s-\sigma)BR(\sigma)J(u(s)\mathrm{d}\sigma \Bigg\}\, \mathrm{d}s \Bigg\} \\
&= w_2(t) + \int_0^t S(t-s)B(u(s) - R(s)\phi)\mathrm{d}s \\
&\quad - \int_0^t \int_0^{t-\sigma} S(t-s-\sigma)BR(s)J(u(\sigma))\mathrm{d}s\mathrm{d}\sigma \\
&= w_2(t) + \int_0^t S(t-s)B(u(s) - R(s)\phi)\mathrm{d}s \\
&\quad - \int_0^t \int_0^\tau S(t-\tau)BR(\tau-\sigma)J(u(\sigma))\mathrm{d}\sigma\mathrm{d}\tau.
\end{aligned}$$

where in the last line we have made a change of variable $s = \tau - \sigma$ and then interchanged the integrals. Hence

$$w_1(t) = w_2(t) + \int_0^t S(t-s)Bw_1(s)\mathrm{d}s,$$

so that $w_1(t) = 0$ for all $t \in [0, T_0]$ if and only if $w_2(t) = 0$ for all $t \in [0, T_0]$ and the result follows. □

We are now in a position to prove the following stability result for the nonlinear problem.

Theorem 2. *Let $H(\phi) = B\phi + J(\phi)$ satisfy (H.2) and (H.3). Suppose that there exist $\gamma > 0$, M_1 such that (27) holds.*

Then the zero solution of (NA) is uniformly asymptotically stable in $\tilde{Y}^\alpha$ in the sense that for any $\gamma_1 < \gamma$ there exists $\rho > 0$ and $M_3 \geq 1$ such that $||\phi||_{\tilde{Y}^\alpha} \leq \rho$ implies there exists a unique global mild solution u of (NA) such that for $t > 0$

$$||u(t)||_{\tilde{Y}^\alpha} \leq M_3 e^{-\gamma_1 t}||\phi||_{\tilde{Y}^\alpha}.$$

Proof. From Lemma 3 u is the mild solution of (NA) if and only if u satisfies (32). The argument will proceed as in [32] Theorem 5.1.1, using (28) and (29). Take $0 < \gamma_1 < \gamma_2 < \gamma$.

Now analogously to [32] Theorem 5.1.1, choose $\sigma > 0$ such that $M_2\sigma \int_0^\infty s^{-\alpha}e^{-(\gamma_2-\gamma_1)s}\mathrm{d}s < 1/2$ and $\rho_1 > 0$ such that $||J(u)||_{\tilde{Y}} \leq \sigma||u||_{\tilde{Y}^\alpha}$ when $||u||_{\tilde{Y}^\alpha} \leq \rho_1$.

Suppose now that $||\phi||_{\tilde{Y}^\alpha} \leq \frac{\rho_1}{2M_2}$. Then the mild solution, u, exists on some time interval and has $||u(t)||_{\tilde{Y}^\alpha} \leq \rho_1$. So provided it remains true that $||u(t)||_{\tilde{Y}^\alpha} \leq \rho_1$, then from (32),

$$\begin{aligned}||u(t)||_{\tilde{Y}^\alpha} &\leq M_2 e^{-\gamma_1 t}||\phi||_{\tilde{Y}^\alpha} + \int_0^t \sigma M_2(t-s)^{-\alpha}e^{-\gamma_2(t-s)}||u(s)||_{\tilde{Y}^\alpha}\mathrm{d}s\\ &\leq \rho_1/2 + \rho_1\sigma M_2\int_0^\infty s^{-\alpha}e^{-\gamma_2 s}\mathrm{d}s < \rho_1,\end{aligned}$$

so there is global existence. Finally

$$\begin{aligned}||u(t)||_{\tilde{Y}^\alpha}e^{\gamma_1 t} &\leq M_2||\phi||_{\tilde{Y}^\alpha} + \sigma M_2\int_0^t (t-s)^{-\alpha}e^{-(\gamma_2-\gamma_1)(t-s)}||u(s||_{\tilde{Y}^\alpha}e^{\gamma_1 s}\mathrm{d}s\\ &\leq M_2||\phi||_{\tilde{Y}^\alpha} + \frac{1}{2}\sup_{s\in[0,t]}(||u(s||_{\tilde{Y}^\alpha}e^{\gamma_1 s}),\end{aligned}$$

so that $\sup_{s\in[0,t]}(||u(s||_{\tilde{Y}^\alpha}e^{\gamma_1 s}) \leq 2M_2||\phi||_{\tilde{Y}^\alpha}$ as required. □

3 The Problem in $L^1((0,\infty); BUC^{(k)}(\mathbb{R}^N))$

We now consider the global existence of a classical solution of problem (11)–(13) in $L^1((0,\infty); BUC^{(k)}(\mathbb{R}^N))$.

3.1 Preliminaries

Let $X^k = BUC^{(k)}(\mathbb{R}^N) = \{u : \mathbb{R}^N \to \mathbb{R} : u$ has uniformly continuous derivatives up to the kth order $\}$ and $X_0^k = \{u \in X^k : u \to 0$ as $|x| \to \infty\}$. In either case take norm $||u||_k = \sum_{|\lambda|\leq k} ||D^\lambda u||_\infty$, where for $i = 1,\dots,N$, $D_i : \mathscr{D}(D_i) \subset X \to X$,

$$D_i u = \frac{\partial u}{\partial x_i}, \quad \text{with } \mathscr{D}(D_i) = X^1,$$

so D_i is a closed linear operator. Set $D = (D_1,\dots,D_N)$, with domain $\mathscr{D}(D) = \mathscr{D}(D_1) \times \cdots \times \mathscr{D}(D_N)$. If $\lambda = (\lambda_1,\dots,\lambda_N)$, $D^\lambda = D_1^{\lambda_1}\cdots D_N^{\lambda_N}$. We define $|\lambda| = \lambda_1 + \cdots + \lambda_N$ and say $\lambda \leq \mu$ for N-tuples λ and μ if the inequality holds componentwise. Note that from [40] p. 33, if $u \in X_0^k$, then for $|\lambda| \leq k$, $D^\lambda u \to 0$ as $|x| \to \infty$. We write $\hat{X}^k$ when results apply to either X^k or X_0^k and set $X = X^0$.

We define $\tilde{A} : \mathscr{D}(\tilde{A}) \subset X \to X$, by

$$\tilde{A}u = -\delta\Delta u, \quad \mathscr{D}(\tilde{A}) = \{u \in BUC(\mathbb{R}^N) \bigcap_{1\leq q<\infty} W^{2,q}_{loc}(\mathbb{R}^N) : Au \in BUC(\mathbb{R}^N)\}.$$

For $\omega > 0$ set $A = \tilde{A} + \omega I$, $\mathscr{D}(A) = \mathscr{D}(\tilde{A})$, so that the spectrum of A is contained in the open right half plane. We will work in the spaces $Y^k = L^1((0,\infty); X^k)$, with norm $||\cdot||_{Y^k}$ and similarly for Y_0^k and $\hat{Y}^k$. Set $Y = Y^0$.

We rewrite the problem (11)–(13) abstractly as

$$\frac{\mathrm{d}}{dt}p(t) = -\frac{\partial}{\partial a}p(t) - \tilde{A}p(t) - D\cdot\Big(p(t)K(P(t))\Big) - p(t)f(P(t)), \ p(0) = \phi_0.$$

Here $p(t) : [0,\infty) \to Y$, $P(t)(x) = \int_0^\infty p(t)(x,a)\mathrm{d}a$, $t \geq 0$, $f : \mathbb{R} \to \mathbb{R}$, $g : \mathbb{R} \to \mathbb{R}$, and

$$K(\Phi)(x) = \int_{B_\rho} g(\Phi(x+\xi))\xi h(|\xi|)\mathrm{d}\xi, \ \Phi \in X, \ x \in \mathbb{R}^N,$$

with ith component $K_i(\Phi)(x)$.

Let $\{T(t)\}_{t\geq 0}$ be the analytic semigroup in X generated by $-\tilde{A}$, so

$$T(t)\phi(x) = \frac{1}{(4\pi t\delta)^{N/2}} \int_{\mathbb{R}^N} e^{\frac{-|y|^2}{4t\delta}} \phi(x-y)\,\mathrm{d}y,$$
$$||T(t)\phi||_X \leq ||\phi||_X,$$

$T(t) : X_0 \to X_0$ and $T(t) : X^k \to X^k$.

Let $T_k(t)$ be the analytic semigroup which is $T(t)$ restricted to the space $\hat{X}^k$ (k an integer); the corresponding generators $-\tilde{A}_k$ are given by the operator $-\tilde{A}$ but restricted to different domains (see [40]).

It is easy to see that for $\phi \in X^k$, $k > 0$,

$$||T_k(t)\phi||_k \leq ||\phi||_k,$$

and

$$\mathscr{D}(\tilde{A}_k) = \{\phi \in X^k;\ \phi \in \mathscr{D}(\tilde{A}),\ \tilde{A}\phi \in X^k\}.$$

Set $A_k = \tilde{A}_k + \omega I$. We will exploit the result that, for suitable α, the operator $D_i A_k^{-\alpha} : X^k \to X^k$ is bounded; see [40]. Thus $D_i A_k^{-\alpha} : Y^k \to Y^k$ is also bounded. We will write $\mathscr{D}_0^{\alpha,k} = \mathscr{D}(A_k^{\alpha})$ in the space X_0^k and $\hat{\mathscr{D}}^{\alpha,k} = \mathscr{D}(A_k^{\alpha})$ in the space $\hat{X}^k$. We give these spaces the graph norm. Denote $Y^{\alpha,k} = L^1((0,\infty);\mathscr{D}(A_k^{\alpha}))$ and $Y_0^{\alpha,k} = L^1((0,\infty);\mathscr{D}_0^{\alpha,k}))$. Let $S_k(t) : Y^k \to Y^k$ be the semigroup related to $T_k(t)$ as in (20).

Thus the mild form of (11)–(13) in Y^k is

$$p(t) = S_k(t)\phi_0 - \int_0^t S_k(t-s)\Big(D\cdot\Big(p(s)K(P(s))\Big)+p(s)f(P(s))\Big)\,\mathrm{d}s \quad (33)$$

$$= S_k(t)\phi_0 - \int_0^t S_k(t-s)G(A_k^{\alpha}p(s))\,\mathrm{d}s, \quad (34)$$

where, if $\Phi = \int_0^{\infty}\phi(a)\mathrm{d}a$,

$$G(\phi) = D\cdot(A^{-\alpha}\phi\, K\,(A^{-\alpha}\Phi)) + A^{-\alpha}\phi f\,(A^{-\alpha}\Phi)\,.$$

Observe that for $\Phi \in \mathscr{D}(D)$, $g \in C^1(\mathbb{R})$, $h \in L^1(0,\rho)$, we have $K(\Phi) \in \mathscr{D}(D)$ and

$$D\cdot K(\Phi)(x) = \sum_i \int_{B_\rho} g'(\Phi(x+\xi))\, D_i\Phi(x+\xi)\xi_i\, h(|\xi|)\,\mathrm{d}\xi, \qquad \text{for } x \in \mathbb{R}^N,$$

and

$$G(\phi) = \sum \left(D_i A^{-\alpha}\phi \right) \left(K_i \left(A^{-\alpha}\Phi \right) \right) \\ + \left(A^{-\alpha}\phi \right) \left(\sum D_i K_i \left(A^{-\alpha}\Phi \right) \right) + A^{-\alpha}\phi f \left(A^{-\alpha}\Phi \right).$$

For $\phi \in Y$ and $\Psi \in X$ we set

$$F(\phi, \Psi) = D \cdot \left(A^{-\alpha}\phi K(A^{-\alpha}\Psi) \right) + A^{-\alpha}\phi f(A^{-\alpha}\Psi),$$

so that $G(\phi) = F(\phi, \Phi)$, with $\Phi = \int_0^\infty \phi(a)\mathrm{d}a$. For $\Phi, \Psi \in X$ we define $F(\Phi, \Psi)$ analogously and $G(\Phi) = F(\Phi, \Phi)$.

3.2 *Stationary Solution of (11)–(12)*

As has already been noted, if $P_1 > 0$ is such that $f(P_1) = \beta_0$, then a nonzero stationary solution of (11)–(12) is

$$q(a) = \beta_0 P_1 e^{-\beta_0 a}.$$

Write $w(x,a,t) = p(x,a,t) - q(a)$, $W(x,t) = \int_0^\infty w(x,a,t)\,da$ in (11) to get

$$\frac{\partial w(x,a,t)}{\partial t} = -\frac{\partial w(x,a,t)}{\partial a} + \delta \Delta w(x,a,t) \\ -\nabla \cdot \left((w(x,a,t) + q(a)) \int_{B_\rho} \bar{g}(W(x+\xi,t))\xi h(|\xi|)\mathrm{d}\xi \right) \\ -\bar{f}(w(x,a,t))(a) \tag{35}$$

$$w(x,0,t) = 2\int_0^\infty \theta(a) w(x,a,t)\mathrm{da}, \tag{36}$$

$x \in \mathbb{R}^N$, $a \in (0,\infty)$, $t > 0$ with

$$w(x,a,0) = \phi_0(x,a) - q(a), \tag{37}$$

where $\bar{g}(W) = g(W + P_1)$, $\bar{f}(w)(a) = wf(W + P_1) + q(a)(f(W + P_1) - \beta_0)$. If

$$\bar{K}(\Phi)(x) = \int_{B_\rho} \bar{g}(\Phi(x+\xi)\xi h(|\xi|)\,\mathrm{d}\xi,$$

with components $\bar{K}_i(\Phi)$, the abstract form of (35) is

$$\frac{\mathrm{d}}{\mathrm{d}t}w(t) = -\frac{\partial w(t)}{\partial a} - \tilde{A}w(t) - D\cdot\Big((w(t)+q(a))\bar{K}(W(t))\Big) - \bar{f}(w(t)).$$

We also set

$$\bar{G}(\phi)(a) = \sum\left(D_i A_k^{-\alpha}\phi\right)\left(\bar{K}_i\left(A_k^{-\alpha}\Phi\right)\right) + \left(A_k^{-\alpha}\phi\right)\left(\sum D_i\bar{K}_i\left(A_k^{-\alpha}\Phi\right)\right)$$
$$+q(a)\left(\sum D_i\bar{K}_i\left(A_k^{-\alpha}\Phi\right)\right) + \bar{f}\left(A_k^{-\alpha}\phi\right)(a),$$

where $\Phi = \int_0^\infty \phi(a)\mathrm{d}a$.

3.3 Local Existence

We can now prove local existence of solutions, using Theorem 1.

Note that in our proofs of existence we will take $f, g : \mathbb{R} \to \mathbb{R}$ but we will then prove that if $\phi_0 \geq 0$ then $p(t) \geq 0$. Thus provided there is enough smoothness at 0 it is only the properties of f and g on $[0,\infty)$ that are relevant as we can then define $f(P)$ and $g(P)$ appropriately for $P < 0$.

We have

(a) If $\alpha > 0$, then for every $t > 0$ the operator $A_k^\alpha T_k(t)$ is a bounded linear operator in $\hat{X}^k$ and the operator $A_k^\alpha S_k(t)$ is a bounded linear operator in $\hat{Y}^k$ and for any $\beta > 0$ there exists $C_2 > 0$ such that

$$||A_k^\alpha T_k(t)|| \leq C_2 t^{-\alpha} e^{\beta t};\ \text{and} \tag{38}$$

$$||A_k^\alpha S_k(t)|| \leq C_2 t^{-\alpha} e^{\bar{\beta} t},\ \ \text{where}\ \bar{\beta} = 2\bar{\theta} + \beta. \tag{39}$$

(b) If $\frac{1}{2} < \alpha < 1$, then $D_i A_k^{-\alpha} : \hat{X}^k \to \hat{X}^k$ is a bounded linear operator so there exists a constant $C_3 \geq 1$ such that for all $u \in X^k$,

$$||D_i A_k^{-\alpha} u||_k \leq C_3 ||u||_k. \tag{40}$$

Hence for $u \in Y^k$

$$||D_i A_k^{-\alpha} u||_{Y^k} \leq C_3 ||u||_{Y^k} \tag{41}$$

(see [40] Theorem 2.4).

The following lemma is central to the proof of existence of solutions of various problems.

Lemma 4. *Suppose that for some* $k \geq 0$, $f \in C^{k+1}(\mathbb{R})$, $g \in C^{k+2}(\mathbb{R})$ *and* $h \in L^1(0,\rho)$, $\frac{1}{2} < \alpha < 1$ *and* $\beta_0 > 0$. *If* $\phi \in \hat{Y}^k$ *and* $\Psi \in \hat{X}^k$, *then* $F(\phi,\Psi) \in \hat{Y}^k$, *and if* $f(P_1) = \beta_0$, $\bar{G}(\phi) \in \hat{Y}^k$. *If* $\Phi \in \hat{X}^k$ *and* $\Psi \in \hat{X}^k$, *then* $F(\Phi,\Psi) \in \hat{X}^k$.

(a) If $||A_k^{-\alpha}\Psi||_k \leq R_1$, *there exists a constant* $K_1(R_1)$ *such that*

$$||F(\phi,\Psi)||_{Y^k} \leq K_1(R_1)(\|\phi\|_{Y^k} + \|\Psi\|_k \|A_k^{-\alpha}\phi\|_{Y^k}) \text{ and} \tag{42}$$

$$||F(\Phi,\Psi)||_k \leq K_1(R_1)(\|\Phi\|_k + \|\Psi\|_k \|A_k^{-\alpha}\Phi\|_k). \tag{43}$$

(b) If $||\phi_i||_{Y^k} \leq R_2$, $i = 1,2$, *then there exists a constant* $K_2(R_2)$ *such that*

$$||G(\phi_1) - G(\phi_2)||_{Y^k} \leq K_2(R_2)||\phi_1 - \phi_2||_{Y^k}; \tag{44}$$

and if $||\Phi_i||_k \leq R_2$ *and* $||\Psi_i||_k \leq R_2$, $i = 1,2$, *then we can also take* $K_2(R_2)$ *such that*

$$||F(\Phi_1,\Psi_1) - F(\Phi_2,\Psi_2)||_k \leq K_2(R_2)(||\Phi_1 - \Phi_2||_k + ||\Psi_1 - \Psi_2||_k). \tag{45}$$

Condition (44) is also satisfied by $\bar{G}(\phi)$.

(c) If $k \geq 2$, *and* $||\phi||_{Y^{k-2}} \leq R_3$,

$$||G(\phi)||_{Y^k} \leq K_3(R_3)\|\phi\|_{Y^k}. \tag{46}$$

From (45) if $\phi(a) \in X^k$ *is continuous at* a_0, *then also* $F(\phi(a),\Psi) \in X^k$ *and is continuous at* a_0.

Proof. We work first with $k = 0$. Note that as f, g, and g' are all locally Lipschitz continuous and $\Psi \in X$, $f(\Psi(x))$, $g(\Psi(x))$, and $g'(\Psi(x))$ are all in X and hence, for each i, $K_i(A^{-\alpha}\Psi)$ and $D_i K_i(A^{-\alpha}\Psi)$ are in X. Thus $F : Y \times X \to Y$. Let $\Psi \in X$, $||A^{-\alpha}\Psi||_X \leq R_1$, and then it is easy to see that there exist constants $K_4 \cdots K_7$ such that for each $i = 1 \cdots N$,

$$|K_i(A^{-\alpha}\Psi_1)(x)| \leq K_4(R_1), \tag{47}$$

$$|K_i(A^{-\alpha}\Psi_1)(x) - K_i(A^{-\alpha}\Psi_2)(x)| \leq K_5(R_2)||\Psi_1 - \Psi_2||_X, \tag{48}$$

$$|D_i K_i(A^{-\alpha}\Psi)(x)| \leq K_6(R_1)||\Psi||_X, \quad \text{and} \tag{49}$$

$$|D_i K_i(A^{-\alpha}\Psi_1)(x) - D_i K_i(A^{-\alpha}\Psi_2)(x)| \leq K_7(R_2)||\Psi_1 - \Psi_2||_X. \tag{50}$$

Thus (42)–(45) hold for $k = 0$. Similarly for $\bar{G}(\phi)$.

Now consider $k > 0$. Let $\Psi \in X^k$, $||A_k^{-\alpha}\Psi||_{X^k} \le R_1$. We want to estimate $||D^\lambda F(\phi,\Psi)||_Y$, $|\lambda| \le k$. Now

$$D^\lambda F(\phi,\Psi) = \left\{ \sum_{\mu\le\lambda} \binom{\lambda}{\mu} \left(\sum_i (D^\mu D_i A_k^{-\alpha}\phi)(D^{\lambda-\mu} K_i(A_k^{-\alpha}\Psi)) \right.\right.$$
$$\left.\left. +(D^\mu A_k^{-\alpha}\phi)(\sum_i D^{\lambda-\mu} D_i K_i(A_k^{-\alpha}\Psi)\right)\right\} + D^\lambda(A_k^{-\alpha}\phi f(A_k^{-\alpha}\Psi)).$$

But

$$||D^\mu D_i A_k^{-\alpha}\phi||_Y \le ||D_i A_k^{-\alpha}\phi||_{Y^{|\mu|}} \le C_3||\phi||_{Y^{|\mu|}},$$

and

$$||D^\mu D_i K_i(A_k^{-\alpha}\Psi)||_X = \left\| \int_{B_\rho} D^\mu D_i g(A_k^{-\alpha}\Psi(\cdot+\xi))\xi_i h(|\xi|)\,\mathrm{d}\xi \right\|_X .$$

Now

$$\begin{aligned} &D^\mu D_i g(A_k^{-\alpha}\Psi(x+\xi)) \\ &= g'(A_k^{-\alpha}\Psi(x+\xi))D^\mu D_i A_k^{-\alpha}\Psi(x+\xi) + \text{ linear combination of products in} \\ &g^{(r)}(A_k^{-\alpha}\Psi((x+\xi))) \text{ and } D^\sigma A_k^{-\alpha}\Psi(x+\xi), \text{ for } 2\le r\le|\mu|+1 \\ &\text{and } 1\le|\sigma|\le|\mu|+1, \end{aligned} \tag{51}$$

and in each product the powers of D sum to $|\mu|+1$, so that

$$||D^\mu D_i g(A_k^{-\alpha}\Psi)||_X \le K_8(R_1)||\Psi||_{X^{|\mu|}}.$$

Further each term in (51) is uniformly continuous, so we can see that $D^\mu D_i K(A_k^{-\alpha}\Psi) \in X$. We can deal similarly with other terms in the sum and with $D^\lambda(A_k^{-\alpha}\phi f(A_k^{-\alpha}\Psi))$, so that $F(\phi,\Psi) \in Y^k$. Thus (42) and (43) can also be shown to hold for all integers $k > 0$. Equations (44)–(45) are similar.

For (46) look at the norm of each product in $D^\lambda G(\phi)$. Recall that if $\Phi = \int_0^\infty \phi(a)\mathrm{d}a$, then $||\Phi||_{X^k} \le ||\phi||_{Y^k}$. We then note that the norm of the term with highest derivative in each product is bounded by constant times $||\phi||_{Y^k}$, whereas the norm of each term with lower derivative is bounded by constant times $||\phi||_{Y^{k-2}}$ and the result follows.

Further, using (47) and (49), we see that if $\phi(x) \in Y_0$, $\Psi \in X$, then $F(\phi,\Psi)(x) \in Y_0$. Thus $\phi \in Y_0^k$, $\Psi \in X^k$, implies that $F(\phi,\Psi)(x) \in Y_0^k$. As $\phi \in \hat{Y}^k$ implies $\Phi \in \hat{X}^k$ one can immediately deduce equivalent results for G.

□

Remark. We can deduce from (42) and (43) that there exists a constant $\bar{K}_1(R_1)$ such that:

$$\text{If} \quad ||A_k^{-\alpha}\Phi||_k \leq R_1, \text{ then } ||G(\Phi)||_k \leq \bar{K}_1(R_1)||\Phi||_k, \tag{52}$$

$$\text{and if } ||\Phi||_k \leq R_1 \text{ (where } \Phi = \int_0^\infty \phi(a)\text{da}), \text{ then } ||G(\phi)||_{Y^k} \leq \bar{K}_1(R_1)||\phi||_{Y^k}, \tag{53}$$

$$\text{and if } ||\Psi||_k \leq R_1, \text{ then } ||F(\Phi,\Psi)||_k \leq \bar{K}_1(R_1)\|\Phi\|_k. \tag{54}$$

Now from Theorem 1, using (44), and working with (35)–(37) in place of (11)–(13) in the case $Q = P_1$, we get

Theorem 3. *Suppose that for some $k \geq 0$, $f \in C^{k+1}(\mathbb{R})$, $g \in C^{k+2}(\mathbb{R})$, $h \in L^1(0,\rho)$, θ satisfies (H.1), $\beta_0 > 0$, and there exists $P_1 > 0$ such that $f(P_1) = \beta_0$. Let $\phi_0(a) - \beta_0 Q e^{-\beta_0 a} \in Y_0^{\alpha,k}$ where $\frac{1}{2} < \alpha < 1$ and $Q = 0$ or $Q = P_1$.*

Then there exists $T_0 > 0$ such that the problem (11)–(13) has a unique mild solution $p(t)$ such that $p(t) - \beta Q e^{-\beta_0 a} \in C([0,T_0]; Y_0^{\alpha,k})$. Furthermore there is continuous dependence on the initial data, in the sense of Theorem 1.

Note that from (22) and (33)

$$p(t)(0) = 2\int_0^\infty \theta(\hat{a})p(t)(\hat{a})\text{d}\hat{a}. \tag{55}$$

3.4 Equation for P

Our goal now is to prove the global existence of the mild solution of (11)–(13). To achieve this we will need to prove the positivity and boundedness of solutions. Because of the lack of regularity of $p(x,a,t)$, we will work with the integrated equation in $P(x,t)$ and with the equation along the characteristics. This will enable us to exploit the analyticity of the semigroup $T_k(t)$.

First we construct the mild equation in P and then, treating p as a known function, we show that this equation has a unique classical solution.

Write

$$I(\phi) = 2\int_0^\infty \theta(\hat{a})\phi(\hat{a})\text{d}\hat{a}.$$

Note that $\phi \in Y^k$ implies $I(\phi) \in X^k$ and

$$||I(\phi)||_{X^k} \leq 2\bar{\theta}||\phi||_{Y^k}. \tag{56}$$

Set $\Phi_0(x) = \int_0^\infty \phi_0(x,a)\text{da}$.

Theorem 4. *Suppose that for some* $k \geq 0$, $f \in C^{k+1}(\mathbb{R})$, $g \in C^{k+2}(\mathbb{R})$, $h \in L^1(0,\rho)$, θ *satisfies (H.1),* $\beta_0 > 0$, *and there exists* $P_1 > 0$ *such that* $f(P_1) = \beta_0$. *Let* $\phi_0(a) - \beta_0 Q e^{-\beta_0 a} \in Y_0^{\alpha,k}$ *where* $\frac{1}{2} < \alpha < 1$ *and* $Q = 0$ *or* $Q = P_1$. *Suppose also that the problem (11)–(13) has a unique mild solution* $p(t)$ *for* $t \in [0, T_0]$. *Then, for* $0 \leq t \leq T_0$, $P(t) = \int_0^\infty p(t)(a)\mathrm{d}a$ *satisfies*

$$P(t) = T_k(t)\Phi_0 - \int_0^t T_k(t-\hat{\tau})(G(A_k^\alpha P(\hat{\tau})) - I(p(\hat{\tau})))\,\mathrm{d}\hat{\tau}, \tag{57}$$

and this equation with initial condition $P(x,0) = \Phi_0(x)$ *has a unique classical solution* P *in the sense that* $P - Q \in C([0,T_0]; X_0^k) \cap C^1((0,T_0]; X_0^k)$, *and for* $0 < t \leq T_0$, $P(t) \in \mathscr{D}(A_k)$, *and* P *satisfies*

$$\frac{\partial P(x,t)}{\partial t} = 2\int_0^\infty \theta(a)p(x,a,t)\mathrm{d}a + \delta\Delta P(x,t)$$
$$-\nabla\cdot\left(P(x,t)\int_{B_\rho} g(P(x+\xi,t))\xi h(|\xi|)\mathrm{d}\xi\right) - P(x,t)f(P(x,t)). \tag{58}$$

Further, for $k \geq 2$, *we also have* $P(t) - Q \in C^1([0,T_0]; X_0^{k-2})$ *and (58) also holds at* $t = 0$.

Proof. Integrate (34) with respect to a, so for $0 \leq t \leq T_0$

$$\begin{aligned}P(t) &= \int_0^\infty (S_k(t)\phi_0)(a)\mathrm{d}a - \int_0^t\int_0^\infty (S_k(t-s)G(A_k^\alpha p(s)))(a)\mathrm{d}a\mathrm{d}s\\ &= \int_t^\infty T_k(t)\phi_0(a-t)\mathrm{d}a + \int_0^t T_k(a)((S_k(t-a)\phi_0)(0))\mathrm{d}a\\ &\quad -\int_0^t\int_0^{t-s} T_k(a)((S_k(t-s-a)G(A_k^\alpha p(s)))(0))\mathrm{d}a\mathrm{d}s\\ &\quad -\int_0^t\int_{t-s}^\infty T_k(t-s)G(A_k^\alpha p(s))(a-t+s)\mathrm{d}a\mathrm{d}s.\end{aligned}$$

Now, changing the variable in the first and fourth integrals and reversing the order of integration in the third integral we get

$$\begin{aligned}P(t) &= T_k(t)\Phi_0\\ &\quad + \int_0^t T_k(a)\left\{(S_k(t-a)\phi_0)(0) - \int_0^{t-a}(S_k(t-s-a)G(A_k^\alpha p(s)))(0)\,\mathrm{d}s\right\}\mathrm{d}a\\ &\quad - \int_0^t T_k(t-s)G(A_k^\alpha P(s))\mathrm{d}s\end{aligned}$$

$$= T_k(t)\Phi_0 + \int_0^t T_k(a)p(t-a)(0)\mathrm{d}a - \int_0^t T_k(t-s)G(A_k^\alpha P(s))\mathrm{d}s$$

$$= T_k(t)\Phi_0 - \int_0^t T_k(t-s)\Big\{G(A_k^\alpha P(s)) - I(p(s))\Big\}\mathrm{d}s,$$

where in the last line we have changed variable and used (55).

To show that $P(t)$ is a classical solution we note that, from (45) and (56), $G(A_k^\alpha P(t)) - I(p(t)) \in C([0, T_0]; X^k)$ and so, as in [44] Theorem 6.3.1, $A_k^\alpha P(t)$, and hence also $P(t)$ is locally Hölder continuous on $(0, T_0]$. Thus $G(A_k^\alpha P(t))$ is locally Hölder continuous. Furthermore $I(p(t)) \in D(A_k^\alpha)$ and $||A_k^\alpha I(p(t))||_k$ is bounded. The result follows from a combination of [44] Theorem 4.3.2 and Theorem 4.3.6.

If $k \geq 2$, $\frac{\partial P}{\partial t}$ exists for $t \in (0, T_0]$, $P(x,t) \in C(\mathbb{R}^N \times [0, T_0])$. Also the right-hand side is in $C([0, T_0]; X^{k-2})$, so that $\lim_{t\to 0+} \frac{\partial P}{\partial t}$ exists. Thus $\frac{\partial P}{\partial t}(x, 0)$ exists and equals the right-hand side at 0. Thus $P \in C^1([0, T_0]; X^{k-2})$. That $P(x,t) - Q \in X_0^k$ when $p(x,a,t) - \beta_0 Q e^{-\beta_0 a} \in Y_0^k$ follows from Lebesgue's dominated convergence theorem. □

3.5 Method of Characteristics, Leading to Regularity and Positivity

We now set up the characteristic equations.

We know from Theorems 3 and 4 that there exists T_0 such that for $t \in [0, T_0]$ there is a unique mild solution $p(t)$ of (11)–(13) and that $P(t) = \int_0^\infty p(t)(a)\mathrm{d}a$ satisfies Theorem 4 and in particular (57); so we can regard the coefficients $K(P)$, $f(P)$, as known functions of x and t. In setting up the characteristic equations there is a technical problem that the null sets in (20) depend on t. This is overcome by assuming that $\phi_0(a) \in C_b([0,\infty); \mathscr{D}(A^{\alpha,k}))$, for some $\frac{1}{2} < \alpha < 1$, where $C_b([0,\infty); \mathscr{D}(A^{\alpha,k})) = \{u : [0,\infty) \to \hat{\mathscr{D}}^{\alpha,k} : u \text{ is continuous and bounded}\}$. We will see that this implies that $p(t)(a)$ is continuous as a function of a. Using the regularity properties of P we will find that the characteristic equations have a classical solution.

We observe first that $b_\phi(t)$ is continuous and so for all $\phi \in Y^k$, $(S_k(t)\phi)(a)$ is continuous in a for $0 \leq a < t$, and the left-hand limit exists and equals $2T_k(t)\int_0^\infty \theta(a)\phi(a)\,\mathrm{d}a$. Furthermore if $\phi \in C([0,\infty); X^k)$ then $(S_k(t)\phi)(a)$ is also continuous in a for $a > t$ and is continuous at $a = t$ if

$$\phi(0) = 2\int_0^\infty \theta(a)\phi(a)\mathrm{d}a.$$

Exactly as in [16], using (21) and (22), we have for almost all $a < t$

$$\begin{aligned} p(t)(a) &= (S_k(t)\phi_0)(a) - \int_0^{t-a} \left(S_k(t-s)\Big(F(A_k^\alpha p(s), A_k^\alpha P(s))\Big)\right)(a)\mathrm{d}s \\ &\quad - \int_{t-a}^{t} \left(S_k(t-s)\Big(F(A_k^\alpha p(s), A_k^\alpha P(s))\Big)\right)(a)\mathrm{d}s \\ &= T_k(a)2\int_0^\infty \theta(\hat{a})p(\hat{a}, t-a)\mathrm{d}\hat{a} \\ &\quad - \int_{t-a}^{t} \left(S_k(t-s)\Big(F(A_k^\alpha p(s), A_k^\alpha P(s))\Big)\right)(a)\mathrm{d}s, \end{aligned}$$

and is continuous in a. So if $A_k^\alpha p(t)(a)$ (and hence $F(A_k^\alpha p(t)(a), A_k^\alpha P(t)))$ were a continuous function of a, using (20), we would have if $t > a$,

$$\begin{aligned} p(t)(a) &= T_k(a)2\int_0^\infty \theta(\hat{a})p(\hat{a}, t-a)\mathrm{d}\hat{a} \\ &\quad - \int_{t-a}^{t} T_k(t-s)\Big(F(A_k^\alpha p(a-t+s, s), A_k^\alpha P(s))\Big)\mathrm{d}s, \end{aligned} \tag{59}$$

while for $t < a$,

$$\begin{aligned} p(t)(a) &= T_k(t)\phi_0(a-t) \\ &\quad - \int_0^t T_k(t-s)\Big(F(A_k^\alpha p(a-t+s, s), A_k^\alpha P(s))\Big)\mathrm{d}s. \end{aligned} \tag{60}$$

Motivated by this we take the characteristic equations along the characteristic lines $a - t = c$

$$p_c(t) = T_k(t-t_c)p_c(t_c) - \int_{t_c}^{t} T_k(t-s)\Big(F(A_k^\alpha p_c(s), A_k^\alpha P(s))\Big)\mathrm{d}s, \tag{61}$$

where if $c \geq 0$ then $t_c = 0$ and $p_c(t_c) = \phi_0(c)$, and if $c < 0$ then $t_c = -c$ and $p_c(t_c) = 2\int_0^\infty \theta(\hat{a})p(\hat{a}, -c)\mathrm{d}\hat{a}$. We will show that for $c > -T_0$, $p_c(x,t) = p(x, t+c, t)$.

As in [70] Chap. 1 we define

$$\mathrm{d}p(a,t) = \lim_{h \to 0} \frac{p(a+h, t+h) - p(a,t)}{h}.$$

Theorem 5. *Suppose that for some $k \geq 0$, $f \in C^{k+1}(\mathbb{R})$, $g \in C^{k+2}(\mathbb{R})$, $h \in L^1(0,\rho)$, θ satisfies (H.1), $\beta_0 > 0$, and there exists $P_1 > 0$ such that $f(P_1) = \beta_0$. Let $\phi_0(a) - \beta_0 Q e^{-\beta_0 a} \in C_b([0,\infty)); \mathscr{D}_0^{\alpha,k}) \cap Y_0^{\alpha,k}$ where $\frac{1}{2} < \alpha < 1$ and $Q = 0$ or $Q = P_1$.*

Suppose that the problem (11)–(13) has a unique mild solution $p(t)$ for $t \in [0, T_0]$. Then for $c > -T_0$ the problem (61) has a unique mild solution $p_c(t) \in C([t_c, T_0); \mathscr{D}(A_k^\alpha))$. This mild solution is a classical solution in the sense that

$$p_c(t) \in C([t_c, T_0); X^k) \cap C^1((t_c, T_0); X^k), \tag{62}$$

and for $t_c < t < T_0$, $p_c(t) \in \mathscr{D}(A_k)$, and

$$\frac{\mathrm{d}}{\mathrm{d}t} p_c(t) = \delta \Delta p_c(t) - \nabla \cdot \left(p_c(t) \int_{B_\rho} g(P(x+\xi,t))\xi h(|\xi|) d\xi \right) - p_c(t) f(P(t)). \tag{63}$$

$$p_c(t_c) = \phi_0(c), \text{ if } c \geq 0 \tag{64}$$

$$= 2\int_0^\infty \theta(\hat{a}) p(\hat{a}, -c) \mathrm{d}\hat{a}, \text{ if } -T_0 < c < 0. \tag{65}$$

Further, if $k \geq 2$, $p_c(t) \in C^1([t_c, T_0); X^{k-2})$ and (63) holds also at $t = t_c$.

Also, for $t \in [0, T_0)$, $p(t)(a)$ is continuous from $[0,\infty)$ to X^k except possibly at $a = t$ and

$$p(a,t) = p_{a-t}(t), \tag{66}$$

so $p_c(x,t) - Q\beta_0 e^{-\beta_0(t+c)} \in X_0^k$ and

$$\mathrm{d}p = \delta \Delta p(t) - \nabla \cdot \left(p(t) \int_{B_\rho} g(P(x+\xi,t))\xi h(|\xi|) \mathrm{d}\xi \right) - p(t) f(P(t)). \tag{67}$$

Proof. Take $c > -T_0$. Note first that $p_c(t_c) \in \mathscr{D}(A_k^\alpha)$. Then, as $A_k^\alpha P(t)$ is continuous and locally Hölder continuous, it follows from (45) and (54) that $f(t, \Phi) := F(A_k^\alpha \Phi, A_k^\alpha P(t))$ satisfies the conditions of Theorems 6.3.1 and 6.3.3 in [44], suitably adapted. Thus the problem (63)–(65) has a unique mild solution which is also a classical solution.

Then similar to [16] Section 4 we can show that for each $t \in [0, T_0)$, $A_k^\alpha p(a,t)$, and hence also $p(a,t)$ is continuous in a for $a \neq t$ and (66) holds.

Finally, $\phi_0(a) - \beta_0 Q e^{-\beta_0 a} \in X_0^k$ implies that $p(x,a,t) - \beta_0 Q e^{-\beta_0 a} \in X_0^k$, so also $p_c(x,t) - Q\beta_0 e^{-\beta_0(t+c)} = p(x, t+c, t) - Q\beta_0 e^{-\beta_0(t+c)} \in X_0^k$. Equation (67) follows immediately. □

3.6 Positivity and Boundedness

In order to get sufficient regularity in x we work in the space X^2. First we use the characteristic equation to get positivity and then the equation in P to give conditions for boundedness.

Proposition 1. *Suppose that $f \in C^3(\mathbb{R})$, $g \in C^4(\mathbb{R})$, $h \in C^1[0,\rho]$, θ satisfies (H.1), $\beta_0 > 0$, and there exists $P_1 > 0$ such that $f(P_1) = \beta_0$. Let $\phi_0(a) - \beta_0 Q e^{-\beta_0 a} \in Y_0^{\alpha,2}$ where $\frac{1}{2} < \alpha < 1$, and $Q = 0$ or $Q = P_1$. Assume also that $\phi_0(x,a) \geq 0$ for all $x \in \mathbb{R}^N$ and almost all $a \geq 0$. Suppose that the problem (11)–(13) has a unique mild solution $p(t)$ for $t \in [0, T_0]$. Then $p(x,a,t) \geq 0$ for all $0 \leq t \leq T_0$, $x \in \mathbb{R}^N$ and almost all $a \geq 0$.*

Proof. Suppose first that $\phi_0(a) - \beta_0 Q e^{-\beta_0 a} \in C_b([0,\infty)); \mathscr{D}_0^{\alpha,2}) \cap Y_0^{\alpha,2}$, so we can use the characteristic equations (63)–(65). Fix $\bar{T}_0 < T_0$ and first take $c \geq 0$, $t \in [0, \bar{T}_0]$. In this case $t_c = 0$ and $p_c(t_c) = \phi_0(c) \geq 0$. We can now use a maximum principle argument as in [18] Proposition 1 to see that $p_c(t) \geq 0$ for $c \geq 0$. So $p(x,a,t) \geq 0$ for $0 \leq t \leq a$, $x \in R^N$, $t \in [0, \bar{T}_0]$.

We now use the case $-\bar{T}_0 \leq c < 0$ to get $p(x,a,t) \geq 0$ for $0 < a < t \leq \bar{T}_0$, $x \in R^N$. Again we see that $p_c(x,t) \geq 0$ provided $p_c(t_c) \geq 0$. But

$$p_c(t_c) = 2\int_0^\infty \theta(\hat{a}) p(\hat{a}, -c) \mathrm{d}\hat{a} = 2\int_\epsilon^\infty \theta(\hat{a}) p(\hat{a}, -c) \mathrm{d}\hat{a} \geq 0$$

when $-\epsilon \leq c \leq 0$. Hence $p(a,t) = p_{a-t}(t) \geq 0$ for $a < t \leq a + \epsilon$, $0 \leq t \leq \bar{T}_0$. Now continue by induction to show that $p(x,a,t) \geq 0$ for $0 \leq t \leq \bar{T}_0$, $a \geq 0$, $x \in \mathbb{R}^N$. But this is true for all $\bar{T}_0 < T_0$ and the result follows from the continuity of $p(t)$.

Now consider the general case $\psi_0 := \phi_0(a) - \beta_0 Q e^{-\beta_0 a} \in Y_0^{\alpha,2}$. Then, as $C_{+,c}(([0,\infty); \mathscr{D}_0^{\alpha,2})$ is dense in $L^1_+((0,\infty); \mathscr{D}_0^{\alpha,2})$ (where $C_{+,c}$ denotes the continuous, positive functions with compact support), there exist $\psi_n \in C_{+,c}([0,\infty); \mathscr{D}_0^{\alpha,2})$ such that $||A_2^\alpha(\psi_n - \psi_0)||_{Y^2} \to 0$ as $n \to \infty$. So if $p_n(t)$ is the solution of (11)–(12) with initial data $\psi_n + \beta_0 Q e^{-\beta_0 a}$, by continuous dependence on the initial data, for $t \in [0, T_0)$, $||A_2^\alpha(p_n(t) - p(t))||_{Y^2} \to 0$ as $n \to \infty$. Hence there exists a subsequence $p_{n_r}(x,a,t) \to p(x,a,t)$ a.e. and $p(x,a,t) \geq 0$ a.e. for $t \in [0, T_0]$.

□

Let ω_N be the surface area of ∂B_1.

Proposition 2. *Suppose that $f \in C^3(\mathbb{R})$, $g \in C^4(\mathbb{R})$, $h \in C^1[0,\rho]$, $h \geq 0$, θ satisfies (H.1), $\beta_0 > 0$, and there exists $P_1 > 0$ such that $f(P_1) = \beta_0$. Let $\phi_0(a) - \beta_0 Q e^{-\beta_0 a} \in Y_0^{\alpha,2}$ where $Q = 0$ or $Q = P_1$, and $\frac{1}{2} < \alpha < 1$ and $\phi_0(a,x) \geq 0$ for all $x \in \mathbb{R}^N$ and almost all $a \geq 0$. Suppose also that either:*

(a) There exists $P_2 > P_1$ such that $g(P) \geq 0$ for $P \in [0, P_2]$, that $\Phi_0 \leq P_2$ for all $x \in \mathbb{R}^N$, and that for all $p(a) \in L^1(0,\infty)$ such that $p(a) \geq 0$ a.e. and $\int_0^\infty p(a) = 1$,

$$f(P_2) > 2\int_0^\infty \theta(a)p(a)\mathrm{d}a, \tag{68}$$

and

$$\max_{P\in[0,P_2]} g(P)\sum\int_{B_\rho}\left|\frac{\partial}{\partial\xi}(\xi_i h(|\xi|))\right| \mathrm{d}\xi < f(P_2) - 2\int_0^\infty \theta(a)p(a)\mathrm{d}a. \tag{69}$$

Or

(b) There exists P_3 such that for all $P > P_3$ and $p(a) \in L^1(0,\infty)$ such that $p(a) \geq 0$ a.e. and $\int_0^\infty p(a) = 1$,

$$f(P) > 2\int_0^\infty \theta(a)p(a)\mathrm{d}a \tag{70}$$

and

$$\sup_{P\in[0,\infty)} g(P)\left\{h(\rho)\omega_N\rho^N + \sum\int_{B_\rho}\left|\frac{\partial}{\partial\xi}(\xi_i h(|\xi|))\right| \mathrm{d}\xi\right\} < f(P) - 2\int_0^\infty \theta(a)p(a)\mathrm{d}a. \tag{71}$$

Choose P_3 such that also $\Phi_0(x) \leq P_3$ for all $x \in \mathbb{R}^N$.

Suppose that the problem (11)–(13) has a unique mild solution $p(t)$ for $t \in [0, T_0]$. Then $P(t) = \int_0^\infty p(t)(a)\mathrm{d}a$ satisfies $0 \leq P(x,t) \leq P_i$ for all $0 \leq t \leq T_0$, $x \in \mathbb{R}^N$, where $i = 2$ in case (i) and $i = 3$ in case (ii).

Proof. The proof uses a very similar argument to that used in [17] Propositions 2 and 3. For part (a), suppose solutions do not remain bounded by P_2 and let

$$t^* = \inf\{t \in [0, T_0] : \exists\, x \text{ such that } P(x,t) > P_2\}. \tag{72}$$

Then $t^* \geq 0$ and by continuity $P(x,t^*) \leq P_2$, for all $x \in \mathbb{R}^N$. There exists (x_i, t_i) such that $t_i > t^*$ for each i, $t_i \to t^*$ and $P(x_i,t_i) > P_2$, but $P(x,t) \to Q < P_2$ as $|x| \to \infty$, so we can take a subsequence (call it x_i again), such that $x_i \to x^*$, say. Then $P(x_i,t_i) \to P(x^*,t^*)$. So we have

$$P(x^*,t^*) = P_2, \quad \frac{\partial P(x^*,t^*)}{\partial x_i} = 0, \quad \frac{\partial^2 P(x^*,t^*)}{\partial x_i^2} \leq 0, \quad \partial P(x^*,t^*)/\partial t \geq 0 \tag{73}$$

and $0 \leq P(x,t^*) \leq P_2$ for all $x \in \mathbb{R}^N$. We will now show that $\partial P(x^*,t^*)/\partial t < 0$ contradicting (73).

Note that, by Theorem 4 , $\partial P(x,t)/\partial t$ exists and is continuous even at $t = 0$, so that (58) holds for $t \in [0, T_0]$. Evaluating (58) at (x^*, t^*), making use of (73), and integrating by parts we obtain

$$\frac{\partial P(x^*,t^*)}{\partial t} = P_2\Big[2\int_0^\infty \theta(a)\frac{p(x^*,a,t^*)}{P_2}\mathrm{d}a - h(\rho)\sum\int_{\partial B_\rho} g(P(x^*+\xi,t^*))\frac{\xi_i^2}{\rho}\mathrm{d}S$$
$$+\sum\int_{B_\rho} g(P(x^*+\xi,t^*))\frac{\partial}{\partial\xi}(\xi_i h(|\xi|))\,\mathrm{d}\xi - f(P_2)\Big].$$

Now $P(x^*+\xi,t^*)$ is in $[0, P_2]$ so $g(P(x^*+\xi,t^*)) \ge 0$ and $h(\rho) \ge 0$. Therefore

$$\frac{\partial P(x^*,t^*)}{\partial t} \le P_2[2\int_0^\infty \theta(a)\frac{p(x^*,a,t^*)}{P_2}\mathrm{d}a$$
$$+\max_{P\in[0,P_2]} g(P)\sum\int_{B_\rho}\left|\frac{\partial}{\partial\xi}(\xi_i h(|\xi|))\right|\mathrm{d}\xi - f(P_2)] < 0$$

by (69), a contradiction, and the proof of (a) is complete. The proof of (b) is similar. □

Remark. Condition (68) is satisfied if $f(P_2) > 2\bar{\theta}$ and similarly for (69)–(71). Biologically it says that at close packing cell division is overcome by cell loss.

3.7 Global Existence

We now show that the solutions are global. We work in X_0^k, for some $k \ge 2$, to get sufficient regularity for Propositions 1 and 2 to hold. However, Proposition 2 only gives boundedness of P in X. To prove global existence of the mild solution of (11)–(13) in Y^k we have to show that $||v(t)||_{Y^k}$ is bounded on bounded subsets of $[0,\infty)$. This is accomplished in several steps, using Lemma 4.

Theorem 6. *Suppose that for some $k \ge 2$, $f \in C^{k+1}(\mathbb{R})$, $g \in C^{k+2}(\mathbb{R})$, $h \in C^1[0,\rho]$, $h \ge 0$, θ satisfies (H.1), $\beta_0 > 0$, and there exists $P_1 > 0$ such that $f(P_1) = \beta_0$. Let $\phi_0(a) - \beta_0 Q e^{-\beta_0 a} \in Y_0^{\alpha,k}$ where $Q = 0$ or $Q = P_1$, and $\frac{1}{2} < \alpha < 1$ and $\phi_0(a,x) \ge 0$ for all $x \in \mathbb{R}^N$ and almost all $a \ge 0$. Suppose further that either part (a) or part (b) from Proposition 2 holds.*

Then in Theorems 3–5 the solutions are global.

Proof. Take T_{max} to be the sup of the T_0 such that there is a unique mild solution $p(a,x,t)$ of (11)–(13) in Y^k, so Theorems 3 and 4 hold on $[0, T_{max})$. Thus, by Propositions 1 and 2, for $t \in [0, T_{max})$, $p(t) \ge 0$ a.e., and there exists $\bar{P}$ such that $0 \le P(t) \le \bar{P}$, so in particular, $||I(p(t))||_X \le 2\bar{\theta}||P(t)||_X$. Set $V(t) = A_k^\alpha P(t)$ and $v(t) = A_k^\alpha p(t)$. We assume $T_{max} < +\infty$.

First we will show that there exists $\bar{R}$ such that $\sup_{t\in[0,T_{max})}||V(t)||_X \leq \bar{R}$. Take $t \in [0, T_{max})$. From above $||A_k^{-\alpha}V(t)||_X \leq \bar{P}$, so from (57), using (38), (52), and (15),

$$||V(t)||_X \leq ||T_k(t)A_k^{\alpha}\Phi_0||_X + \int_0^t ||A_k^{\alpha}T_k(t-s)\{G(V(s)) - I(A_k^{-\alpha}v(s))\}||_X \mathrm{d}s$$

$$\leq ||A_k^{\alpha}\Phi_0||_X + \int_0^t C_2 e^{\beta(t-s)}(t-s)^{-\alpha}(\bar{K}_1(\bar{P}) + 2\bar{\theta}C_1)||V(s)||_X \mathrm{d}s.$$

Thus from [32] Sect. 1.2.1 there exists $\bar{K}_2$ such that

$$\sup_{t\in[0,T_{max})} ||V(t)||_X \leq \bar{K}_2||A_2^{\alpha}\Phi_0||_X = \bar{R},$$

say.

We now prove that there exists $\bar{R}_k$ such that for $t \in [0, T_{max})$

$$||v(t)||_{Y^k} \leq \bar{R}_k. \tag{74}$$

First, from (34), using (39) and (53),

$$||v(t)||_Y \leq ||S_k(t)A_k^{\alpha}\phi_0||_Y + \int_0^t ||A_k^{\alpha}S_k(t-s)G(v(s))||_Y \mathrm{d}s$$

$$\leq e^{2\bar{\theta}t}||A_k^{\alpha}\phi_0||_Y + \int_0^t C_2 e^{\bar{\beta}(t-s)}(t-s)^{-\alpha}\bar{K}_1(\bar{R})||v(s)||_Y \mathrm{d}s,$$

so that, as above, there exists $\bar{R}_0$ such that $||v(t)||_Y \leq \bar{R}_0$ for $t \in [0, T_{max})$.

Then, using (46),

$$||v(t)||_{Y^2} \leq e^{2\bar{\theta}t}||A_k^{\alpha}\phi_0||_{Y^2} + \int_0^t C_2 e^{\bar{\beta}(t-s)}(t-s)^{-\alpha}\bar{K}_3(\bar{R}_0)||v(s)||_{Y^2} \mathrm{d}s,$$

so again there exists $\bar{R}_2$ such that $||v(t)||_{Y^2} \leq \bar{R}_2$ for $t \in [0, T_{max})$. Thus (74) follows by induction, and hence by Theorem 1 the solutions are global. □

4 Local Stability for the Problem (11)–(13) in $L^1((0,\infty); L^2(\mathbb{R}^N))$

We will now consider conditions for the local stability of the nonzero stationary solution of (11)–(13). To do this we set the problem in $L^1((0,\infty); L^2(\mathbb{R}^N))$.

4.1 Existence of Solutions

For any $1 < p < \infty$ define the operator, $\tilde{A}_{L^p} : \mathscr{D}(\tilde{A}_{L^p}) \subset L^p(\mathbb{R}^N) \to L^p(\mathbb{R}^N)$ by

$$\tilde{A}_{L^p}\phi = -\delta\Delta\phi,\ \phi \in \mathscr{D}(A_{L^p}) \quad \text{with}$$
$$\mathscr{D}(\tilde{A}_{L^p}) = W^{2,p}(\mathbb{R}^N).$$

Similarly we can also define $\tilde{A}_{L^1} : \mathscr{D}(\tilde{A}_{L^1}) \subset L^1(\mathbb{R}^N) \to L^1(\mathbb{R}^N)$, for a suitable domain $\mathscr{D}(\tilde{A}_{L^1}) \supset W^{2,1}(\mathbb{R}^N)$ (see [30]). Now take $1 \le p < \infty$. For $\omega > 0$, define $A_{L^p} = \tilde{A}_{L^p} + \omega I$ with $\mathscr{D}(A_{L^p}) = \mathscr{D}(\tilde{A}_{L^p})$. It is well known (see [37]) that $-\tilde{A}_{L^p}$ generates a positive analytic semigroup $\{T_{L^p}(t)\}_{t\ge 0}$, with $\|T_{L^p}(t)\| \le 1$. Set $Y_p = L^1((0,\infty); L^p(\mathbb{R}^N))$, and let $S_{L^p}(t) : Y_p \to Y_p$ be the semigroup related to $T_{L^p}(t)$ as in (20). Thus Eqs. (38) and (39) hold also in $L^p(\mathbb{R}^N)$. Further, as in [18], if $\frac{1}{2} < \alpha < 1$,

$$\mathscr{D}(A^{\alpha}_{L^p}) \hookrightarrow W^{1,p}(\mathbb{R}^N).$$

Thus if we define $D_i : \mathscr{D}(D_i) \subset L^p(\mathbb{R}^N) \to L^p(\mathbb{R}^N)$,

$$D_i\phi = \frac{\partial\phi}{\partial x_i}, \quad \text{with } \mathscr{D}(D_i) = W^{1,p}(\mathbb{R}^N),$$

(40) and (41) also hold in $L^p(\mathbb{R})$.

Theorem 7. *Take $N \le 3$. Suppose that $f \in C^1(\mathbb{R})$, $g \in C^2(\mathbb{R})$, $h \in L^1(0,\rho)$, θ satisfies (H.1), $\beta_0 > 0$, and there exists $P_1 > 0$ such that $f(P_1) = \beta_0$. Let $\phi_0 - Q\beta_0 e^{-\beta_0 a} \in L^1((0,\infty); \mathscr{D}(A^{\alpha}_{L^2}))$, where $\frac{3}{4} < \alpha < 1$, and either $Q = 0$ or $Q = P_1$.*

Then there exists $T_0 > 0$ such that the problem (11)–(13) has a unique mild solution $p(t)$ such that $p(t) - Q\beta_0 e^{-\beta_0 a} \in C([0,T_0]; L^1((0,\infty); \mathscr{D}(A^{\alpha}_{L^2})))$. Also there is continuous dependence on the initial data in the sense of Theorem 1.

Furthermore, if in addition $\phi_0 - Q\beta_0 e^{-\beta_0 a} \in L^1((0,\infty); \mathscr{D}(A^{\alpha}_{L^1}))$, then we may choose $T_0 > 0$ such that also $p(t) - Q\beta_0 e^{-\beta_0 a} \in C([0,T_0]; L^1((0,\infty); \mathscr{D}(A^{\alpha}_{L^1})))$.

Proof. The first result follows from Theorem 1, since, as in [18] Theorem 3, $G : Y_2 \to Y_2$ and $\bar{G} : Y_2 \to Y_2$ satisfy (H.2).

For the final part note that $G, \bar{G} : Y_1 \cap Y_2 \to Y_1 \cap Y_2$ also satisfy (H.2) in the Y_1-norm. If $Q = 0$ consider the iterates,

$$v_n(t) = S_{L^p}(t)A^{\alpha}_{L^p}\phi_0 - \int_0^t A^{\alpha}_{L^p} S_{L^p}(t-s)G(v_n(s))\,ds.$$

Note that for $p = 1$ and $p = 2$ these sequences are equal, but converge in $L^1((0,\infty); L^p(\mathbb{R}^N))$ to v^1, say, and $v^2 = A^\alpha_{L^2} p$, respectively. Now, by considering the restrictions to bounded subsets of $\mathbb{R}^N$, it can be seen that $v^1 = v^2$ almost everywhere, so that $p(t) \in L^1((0,\infty); \mathscr{D}(A^\alpha_{L^1}))$ as required. Similarly for $Q = P_1$ using $\bar{G}$. □

4.2 Local Asymptotic Stability in Y_2

We now consider the local asymptotic stability of the stationary solution $q(a) = \beta_0 P_1 e^{-\beta_0 a}$. For simplicity we will consider the case where f and g are logistic. We take $f(P) = (P + \beta_0 - P_1)$ and $g(P) = P(P_2 - P)$, with $P_2 > P_1$.

If we make the substitution $p = w + \beta_0 P_1 e^{-\beta_0 a}$, $W(x,t) = \int_0^\infty w(x,a,t)\mathrm{d}a$, our equation in this case becomes

$$\frac{\partial w(x,a,t)}{\partial t} = -\frac{\partial w(x,a,t)}{\partial a} + \delta \Delta w(x,a,t) + H(w(x,a,t)), \tag{75}$$

where $H : L^1((0,\infty); D(A^\alpha_{L^2})) \to Y_2$ is defined by

$$\begin{aligned} H(\phi)(a) = -\nabla \cdot \Big((\phi(x,a) \\ + P_1 \beta_0 e^{-\beta_0 a}) \int_{B_\rho} \Phi(x+\xi)((P_2 - 2P_1 - \Phi(x+\xi))\xi h(|\xi|)\mathrm{d}\xi \Big) \\ -\phi(x,a)(\beta_0 + \Phi(x)) - P_1 {\beta_0}^2 e^{-\beta_0 a} \Phi(x), \end{aligned}$$

with $\Phi(x) = \int_0^\infty \phi(x,a)\mathrm{d}a$. So the linearized problem is

$$\frac{\partial w(x,a,t)}{\partial t} = -\frac{\partial w(x,a,t)}{\partial a} + \delta \Delta w(x,a,t) + Bw(x,a,t),$$

where the bounded linear operator $B : L^1((0,\infty); D(A^\alpha_{L^2})) \to Y_2$ is defined by

$$\begin{aligned} B\phi(a) &= -\beta_0 e^{-\beta_0 a} P_1(P_2 - 2P_1) \int_{B_\rho} \nabla\Phi(x+\xi) \cdot \xi h(|\xi|)\mathrm{d}\xi - \beta_0 e^{-\beta_0 a} P_1 \Phi - \beta_0 \phi \\ &= \beta_0 e^{-\beta_0 a} F(\Phi) - \beta_0 \phi, \text{ where,} \\ F(\Phi) &= -P_1(P_2 - 2P_1) \int_{B_\rho} \nabla\Phi(x+\xi) \cdot \xi h(|\xi|)\mathrm{d}\xi - P_1 \Phi. \end{aligned}$$

So we consider the stability of the zero solution of (75). If we write $J = H - B$, then, as in Theorem 7, H satisfies (H.2) and similarly J satisfies (H.3). So in order to apply Theorem 2 it only remains to find conditions such that there exists $\gamma > 0$ as in (27).

Let $\{R(t)\}_{t\geq 0}$ be the strongly continuous semigroup in $L^1((0,\infty); D(A^{\alpha}_{L^2}))$ associated with the solutions of the linearized problem (L), written in abstract form as

$$\begin{aligned} \frac{dw}{dt} &= -\frac{\partial w}{\partial a} - \tilde{A}_{L^2}w + Bw, \quad t > 0, \\ w(0,t) &= 2\int_0^\infty \theta(a)w(a,t)\mathrm{d}a \\ w(a,0) &= \psi_0(a), \end{aligned}$$

where $\psi_0 = \phi_0 - \beta_0 P_1 e^{-\beta_0 a}$.

From now on we shall require that θ also satisfies

(H.4) $\theta : [0, \infty) \to \mathbb{R}_+$ is Lipschitz continuous.

Write $L^1 = L^1(0,\infty)$. The following result will be used to determine conditions for the stability of (L). It exploits methods and definitions from [70].

Proposition 3. *Consider the problem (La):*

$$\frac{\mathrm{du}}{\mathrm{dt}} = -\frac{\partial u}{\partial a} - \nu u + \mu e^{-\beta_0 a}\beta_0 \int_0^\infty u(a,t)\mathrm{d}a \tag{76}$$

$$u(0,t) = 2\int_0^\infty \theta(\hat{a})u(\hat{a},t)\mathrm{d}\hat{a} \tag{77}$$

$$u(a,0) = \phi(a) \in L^1(0,\infty), \tag{78}$$

where ν and μ are real constants. This problem has a unique global solution in the sense that $u \in C([0,\infty); L^1)$ is the unique solution of the equation (Ia)

$$u(a,t) = \begin{cases} 2\int_0^\infty \theta(\hat{a})u(\hat{a},t-a)\mathrm{d}\hat{a} + \int_{t-a}^t(-\nu u(s+a-t,s) \\ \qquad +\mu e^{-\beta_0(s+a-t)}\beta_0\int_0^\infty u(\hat{a},s)\mathrm{d}\hat{a})\mathrm{d}s \quad a.a.\ 0 < a < t \\ \phi(a-t) + \int_0^t(-\nu u(s+a-t,s) \\ \qquad +\mu e^{-\beta_0(s+a-t)}\beta_0\int_0^\infty u(\hat{a},s)\mathrm{d}\hat{a})\mathrm{d}s \quad a.a.\ a > t. \end{cases}$$

Let $U(t) : L^1 \to L^1$ be the strongly continuous semigroup of solutions of (Ia), then, if ω_0 is the growth bound of $U(t)$,

$$\omega_0 \leq \max\{\beta_0 - \nu, \beta_0 - \nu + \mu\}. \tag{79}$$

Proof. Problem (La) satisfies the conditions of [70] Propositions 3.2 and 3.7. So the problem has a unique global solution for $\phi \in L^1$ and $\{U(t)\}_{t\geq 0}$ is a strongly continuous semigroup of bounded linear operators in L^1 with generator

$$\mathscr{A}\phi = -\phi' - \nu\phi + \mu e^{-\beta_0 a}\beta_0 \int_0^\infty \phi(a)\mathrm{d}a,$$

$$\mathscr{D}(\mathscr{A}) = \{\phi \in L^1 : \phi \text{ is absolutely continuous on } [0,\infty), \phi' \in L^1, \text{ and}$$
$$\phi(0) = 2\int_0^\infty \theta(\hat{a})\phi(\hat{a})\mathrm{d}\hat{a}\}.$$

The operator $\mathscr{C} : L^1 \to L^1, \mathscr{C}\phi = \mu e^{-\beta_0 a}\beta_0 \int_0^\infty \phi(a)\mathrm{d}a$ is bounded and compact so that, by [70] Proposition 4.14, $\omega_1(\mathscr{A}) = \omega_1(\mathscr{B})$, where $\mathscr{B}\phi = -\phi' - \nu\phi$, and $\omega_1(\mathscr{A})$ is the essential growth bound of $U(t)$ (see [70] for the properties of the essential growth bound). By [70] Theorem 4.6, $\omega_1(\mathscr{B}) \le -\nu$.

Let $\sigma(\mathscr{A})$ be the spectrum of $\mathscr{A}$ and let $E\sigma(\mathscr{A})$ be the essential spectrum of $\mathscr{A}$. From [70], Proposition 4.13,

$$\omega_0(\mathscr{A}) = \max\{\omega_1(\mathscr{A}), \sup_{\lambda\in\sigma(\mathscr{A})-E\sigma(\mathscr{A})} \mathrm{Re}\,\lambda\}.$$

Further, from [70] Proposition 4.11, if $\lambda \in \sigma(\mathscr{A}) - E\sigma(\mathscr{A})$ then $\lambda \in P\sigma(\mathscr{A})$, the point spectrum of $\mathscr{A}$. So we will determine the point spectrum of $\mathscr{A}$.

If k is an eigenvalue of $\mathscr{A}$, with associated eigenvector u, then

$$u'(a) = -(\nu + k)u + \mu\beta_0 e^{-\beta_0 a}\int_0^\infty u(a)\mathrm{d}a, \tag{80}$$

$$u(0) = 2\int_0^\infty \theta(a)u(a)\mathrm{d}a. \tag{81}$$

Suppose first that $\mu \neq 0$. Then, if $k = \beta_0 - \nu$, from (80),

$$u(a) = u(0)e^{-\beta_0 a} + C\mu a\beta_0 e^{-\beta_0 a},$$

where $C = \int_0^\infty u(a)\mathrm{d}a$, and (81) is only possible if $u = 0$. If, on the other hand, $k \neq \beta_0 - \nu$, then from (80)

$$u(a) = \frac{\mu\beta_0}{\nu + k - \beta_0}e^{-\beta_0 a}C + He^{-(\nu+k)a},$$

where H is a constant. Then (81) is true if and only if either $H = 0$ or $1 = 2\int_0^\infty \theta(a)e^{-(\nu+k)a}\mathrm{d}a$. If $H = 0$ then u is a nonzero solution if and only if $\mu \neq 0$ and $k = \mu + \beta_0 - \nu$. But if $1 = 2\int_0^\infty \theta(a)e^{-(\nu+k)a}da$ then $\mathrm{Im}\,k \neq 0$ and

$$1 = 2\int_0^\infty \theta(a)e^{-\nu a}e^{-(\mathrm{Re}\,k)a}\cos((\mathrm{Im}\,k)a)\mathrm{d}a < 2\int_0^\infty \theta(a)e^{-\nu a}e^{-(\mathrm{Re}\,k)a}\mathrm{d}a,$$

so Re $k \leq \beta_0 - \nu$. If on the other hand $\mu = 0$ then from (80) $u(a) = u(0)e^{-(\nu+k)a}$ and again Re $k \leq \beta_0 - \nu$. Thus (79) follows. □

Finally we have

Theorem 8. *Take $N \leq 3$. Suppose that $h \in L^2(0,\rho)$, θ satisfies (H.1) and (H.4), and $\beta_0 > 0$. Then, the stationary solution of (11)–(12) is locally asymptotically stable in Y_2 if, for all $\eta \in \mathbb{R}^N$,*

$$k(\eta) = -\delta|\eta|^2 + P_1(P_2 - 2P_1)\int_{B_\rho} \sin(\eta\cdot\xi)(\eta\cdot\xi)h(|\xi|)\mathrm{d}\xi - P_1 < 0.$$

Proof. This is equivalent to proving that the zero solution of (75) is locally asymptotically stable, so we have only to find conditions such that there exists $\gamma > 0$ such that the semigroup of solutions of (L) satisfy (27). This will be done by taking the Fourier transform of the characteristic equation of (L) and then applying Proposition 3 to the transformed equation.

First suppose that $\psi_0 \in C_b((0,\infty); L^2(\mathbb{R}^N)) \cap L^1((0,\infty); \mathscr{D}(A^\alpha_{L^2})) \cap L^1((0,\infty); \mathscr{D}(A^\alpha_{L^1}))$, with $\frac{3}{4} < \alpha < 1$. By Lemma 1, with the same argument as in Theorem 7, the problem (L) has a unique mild solution $w \in C([0,\infty); L^1((0,\infty); D(A^\alpha_{L^2}))) \cap C([0,\infty); L^1((0,\infty); D(A^\alpha_{L^1})))$. From now on take $p = 1$ or $p = 2$. Then, as in Theorem 4, $W(t) = \int_0^\infty w(a,t)\mathrm{d}a$ satisfies

$$W(t) = T_{L^p}\Psi_0 + \int_0^t T_{L^p}(t-s)\{F(W(s)) - \beta_0 W(s) + I(w(s))\}\mathrm{d}s,$$

($\Psi_0 = \int_0^\infty \psi_0 \mathrm{d}a$) and $A^\alpha_{L^p} W(t)$ and hence also $F(W(t))$ is continuous on $[0,\infty)$ and locally Hölder continuous on $(0,\infty)$.

Note that $\phi(a) = \beta_0 e^{-\beta_0 a} F(W)$ satisfies (77), so that $S_{L^p}(t)(\beta_0 e^{-\beta_0 a} F(W))$ is continuous in a. If $w(a,t)$ is the mild solution of (L) then using a similar argument to that used in Lemma 3,

$$w(a,t) = e^{-\beta_0 t} S_{L^p}(t)\psi_0 + \int_0^t e^{-\beta_0(t-s)} S_{L^p}(t-s)\beta_0 e^{-\beta_0 a} F(W(s))\mathrm{d}s,$$

and hence $w(a,t)$ is continuous in a except possibly at $a = t$.

Thus if we define $w_c(t) = w(t+c,t)$ for $c \geq -t$, then $w_c(t)$ satisfies

$$w_c(t) = e^{-\beta_0 t} T_{L^p}(t-t_c)w_c(t_c) + \int_{t_c}^t e^{-\beta_0(t-s)} T_{L^p}(t-s)\beta_0 e^{-\beta_0(s+c)} F(W(s))\mathrm{d}s,$$

where if $c \geq 0$ then $t_c = 0$ and $w_c(t_c) = \psi_0(c)$, and if $c < 0$ then $t_c = -c$ and $w_c(t_c) = 2\int_0^\infty \theta(\hat{a})w(\hat{a},-c)\mathrm{d}\hat{a}$. From [44] Theorem 4.3.3 this solution is also a classical solution, so $w_c(t) \in C([t_c,\infty); L^2(\mathbb{R}^N)) \cap C^1((t_c,\infty); L^2(\mathbb{R}^N)) \cap C([t_c,\infty); L^1(\mathbb{R}^N)) \cap C^1((t_c,\infty); L^1(\mathbb{R}^N))$ and

$$\frac{\mathrm{d}w_c(t)}{\mathrm{d}t} = \delta \Delta w_c(t) - \beta_0 w_c(t) + \beta_0 e^{\beta_0(t+c)} F(W(t)). \tag{82}$$

If we denote the Fourier transform of $\psi \in L^1(\mathbb{R}^N) \cap L^2(\mathbb{R}^N)$ (see [19], Sect. 4.3)) by

$$\hat{\psi}(\eta) = \frac{1}{(2\pi)^{\frac{N}{2}}} \int_{R^N} \exp(-i\eta \cdot x)\psi(x)\mathrm{d}x,$$

we may take the Fourier transform of (82) to get $\hat{w}_c(t) \in C([t_c,\infty)) \cap C^1((t_c,\infty))$ with

$$\frac{\mathrm{d}\hat{w}_c(t)}{\mathrm{d}t} = -\nu \hat{w}_c(t) + \beta_0 e^{\beta_0(t+c)} \mu \hat{W}(t), \tag{83}$$

where $\nu = \delta|\eta|^2 + \beta_0 > 0$ and $\mu = P_1(P_2 - 2P_1)\int_{B_p} \sin(\xi \cdot \eta)(\xi \cdot \eta)h(|\xi|)\mathrm{d}\xi - P_1$. Also, for $c > 0$, $\hat{w}_c(x,0) = \hat{\psi}_0(x,c)$, and for $c < 0$, $\hat{w}_c(x,0) = 2\int_0^\infty \theta(\hat{a})\hat{w}(\hat{a},-c)\mathrm{d}\hat{a}$. But $\hat{w}_c(t) \in C([t_c,\infty))$ and the right-hand side of (83) is in $C([t_c,\infty))$, so in fact $\hat{w}_c(t) \in C^1([t_c,\infty))$, and we may integrate (83) to get

$$\hat{w}_c(t) = \hat{w}_c(t_c) + \int_{t_c}^t -\nu \hat{w}_c(s) + \beta_0 e^{-\beta_0(s+c)} \mu \hat{W}(s)\mathrm{d}s.$$

Further $\hat{w}(a,t) = \hat{w}_{a-t}(t)$, so, putting $c = a - t$, we see that $\hat{w}(a,t)$ is the unique solution of (Ia). But $-\nu + \beta_0 < 0$ and $\mu - \nu + \beta_0 = k(\eta)$, so, using Proposition 3, if $k(\eta) < 0$ for all $\eta \in \mathbb{R}^N$, there exist $\gamma > 0$ and $M \geq 1$ such that $||\hat{w}(a,t)||_{L^2} \leq Me^{-\gamma t}||\hat{\psi}_0||_{L^2}$. Hence by Plancherel, $||w(a,t)||_{L^2} \leq Me^{-\gamma t}||\psi_0||_{L^2}$.

Finally $C_b((0,\infty); L^2(\mathbb{R}^N)) \cap L^1((0,\infty); \mathscr{D}(A^\alpha_{L^2})) \cap L^1((0,\infty); \mathscr{D}(A^\alpha_{L^1})) \supset C_c((0,\infty); C_c^\infty(\mathbb{R}^N))$ (where C_c denotes continuous functions of compact support) so it is dense in $L^1((0,\infty); L^2(\mathbb{R}^N))$, so that $||R(t)|| \leq Me^{-\gamma t}$, as required, and the result follows. □

5 Simulations

We illustrate the model (11)–(13) with simulations applicable to in vitro wound healing experiments, which are widely used in cancer research [1, 5, 6, 24–27, 45, 46, 50, 51, 72]. Our simulations are based on experiments in [67], in which MCF10A mammary epithelial cancer cells were grown in monolayer cultures with varying expression of HER2 (human epidermal growth factor receptor 2). In cultures overexpressing HER2, TGF-β (transforming growth factor beta) promotes cell motility by lamellipodia, which are cytoskeleton projections on cell surfaces. The cultures were scored in thin lines and then allowed to proliferate in replenished medium. In the cultures with uninhibited TGF-β the wounds closed completely, but in the cultures with inhibited TGF-β they did not. In earlier work [17, 18] we simulated these experiments, and here we extend this work to incorporate cell age and, thus, the cell cycle. For the numerical simulations we used a backward difference scheme for the spatial variables and a method of characteristics discretization scheme for the age variables. MATHEMATICA codes are available on request.

In [17] we examined the role of the diffusion coefficient δ and the sensing radius ρ in (11) and demonstrated that larger values of δ and smaller values of ρ lead to complete wound closure, but smaller values of δ and larger values of ρ gave rise to rippled patterns of incomplete closure. Here, as in [18], we set $g(P) = P(\lambda - P)$ and examine the role of λ, which is a non-dimensionalized cell density parameter corresponding to cell packing. As in [18], but now with cell age incorporated, we will see that there is a critical value of λ such that closure occurs below this value, but not above. We take $f(P) = (P + \beta_0 - P_1)$, $\lambda > P_1$, $h(x) = \arctan(100.0\,x)\,/(9.0\,x\,\arctan 100.0)$, and

$$\theta(a) = \begin{cases} \theta_0(a-2.0)(4.0-a)\exp(-2.0(a-2.0)) + .1, & 2.0 < a < 4.0 \\ 0.0 & a \le 2.0 \\ 0.0 & a \ge 4.0. \end{cases}$$

We choose $\delta = 1$ and $\rho = 1$. Recall that $\beta_0 > 0$ is the intrinsic growth constant, that is, the unique value β_0 such that $2\int_0^\infty \theta(a)e^{-\beta_0 a}\,da = 1$ and that if $P_1 > 0$ is such that $f(P_1) = \beta_0$, then $q(a) = \beta_0 P_1 e^{-\beta_0 a}$ is a stationary solution (independent of x and λ). We choose $\beta_0 = 0.55$, and to compare with [18] we take $P_1 = 1$, and this forces $\theta_0 = 7.2839$. Thus, the stationary solution in this case is $q(a) = 0.55\exp(-0.55a)$ (Fig. 1, left panel). By Theorem 8 the stationary solution is locally asymptotically stable in $L^1(0,\infty); \mathscr{D}(A^{\alpha}_{L^2}))$ if for all $\eta \in \mathbb{R}$

$$k(\eta) = -\delta|\eta|^2 + P_1(\lambda - 2P_1)\int_{-\rho}^{\rho} \sin(\eta\xi)\eta\xi h(|\xi|)d\xi - P_1 < 0.$$

For our parameters this condition is satisfied if $\lambda < 17.62$. We take the initial data as $\phi_0(x,a) = 0.55\exp(-0.55a) - 0.5\exp(-(0.1x)^{10})\exp(-0.6a)$ (Fig. 1, right panel), which represents the lateral scoring in the one-dimensional spatial variable x. In Figs. 2 and 3 we simulate the behavior of $P(x,t) = \int_0^\infty p_0(x,a,t)da$ (the total population density in x at time t) for values of $\lambda = 16.0, 17.0, 18.0$, and 19.0. In Fig. 2 we see that if $\lambda = 16.0$ or 17.0, then $P(x,t)$ appears to converge to the stationary solution 1.0. In Fig. 3 we see that if $\lambda = 18.0$ or $\lambda = 19.0$, then $P(x,t)$ oscillates in x about the stationary solution 1.0 with a series of peaks, symmetric to the right and left of $x = 0.0$, and rising above and falling below 1.0. The oscillation

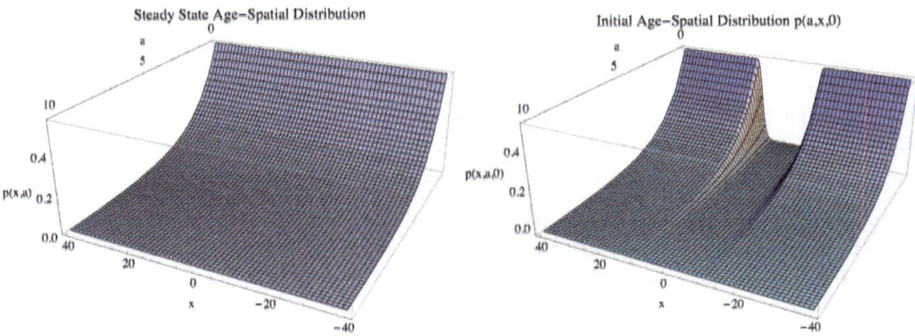

Fig. 1 *Left panel*: steady-state distribution $q(a) = 0.55\,e^{-0.55a}$ (independent of spatial position x). *Right panel*: age-space initial distribution $p(a,x,0) = \phi_0(x,a)$ at time $t = 0$. The center cavity corresponds to the wound in the spatial and age variables at time $t = 0$. Both graphs are independent of the cell density parameter λ

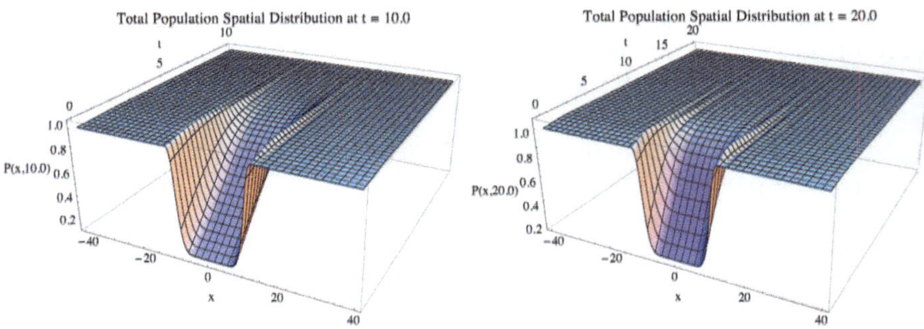

Fig. 2 Spatial distribution of the total population $P(x,t)$ for $\lambda = 16.0$, $t = 10.0$ (*left panel*) and $\lambda = 17.0$, $t = 20.0$ (*right panel*). The wound appears to close completely, but requires longer times for higher values of λ

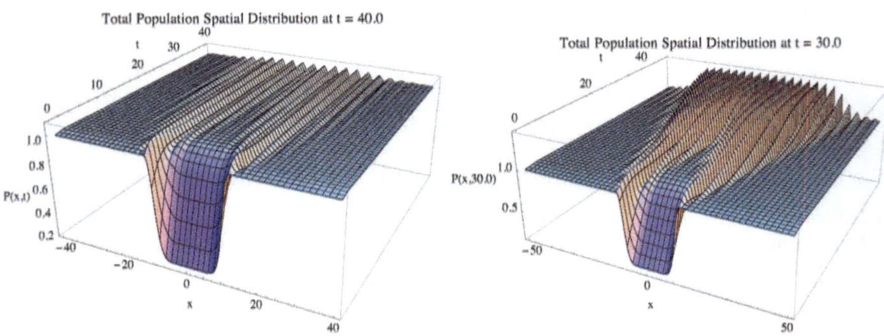

Fig. 3 Spatial distribution of the total population $P(x,t)$ for $\lambda = 18.0$, $t = 40.0$ (*left panel*) and $\lambda = 19.0$, $t = 30.0$ (*right panel*). The wound does not close, but instead develops a complex pattern of peaks and valleys which do not converge to 1.0

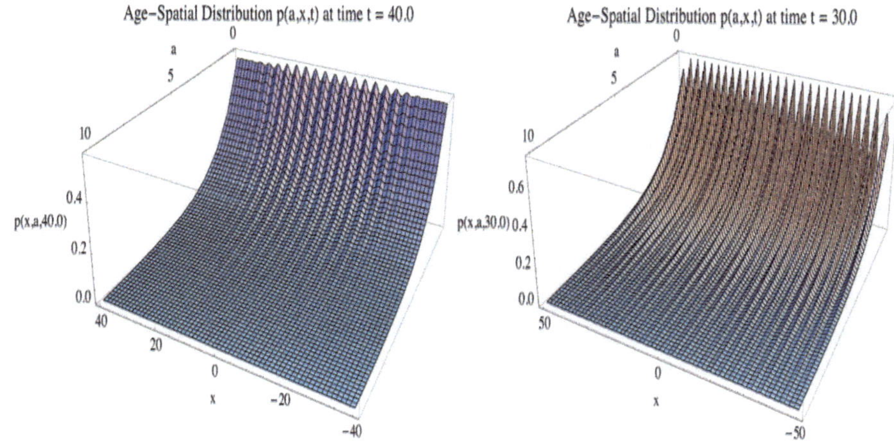

Fig. 4 Age-spatial distribution $p(x, a, t)$ for $\lambda = 18.0$, $t = 40.0$ (*left panel*) and $\lambda = 19.0$, $t = 30.0$ (*right panel*). The cell age dependence of the density $\rho(x, a, t)$ exhibits an oscillating pattern in the spatial variable x

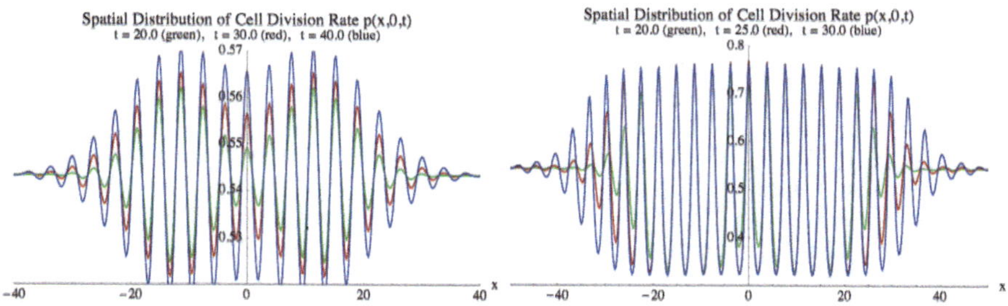

Fig. 5 Spatial distribution of the of the cell division rate $p(x, 0, t)$ at various times for $\lambda = 18.0$ (*left panel*) and $\lambda = 19.0$ (*right panel*). The pattern of oscillations widens in the x-direction and amplifies in magnitude with increasing values of λ and advancing time

pattern in the spatial variable x appears to widen as time increases and increase in amplitude as λ increases. A similar oscillation pattern in the spatial variable x is also observed in the age-space density $p(x, a, t)$ (Fig. 4) and the division rate $p(x, 0, t)$ (Fig. 5) for $\lambda = 18.0$ and 19.0. This behavior compares with Figs. 2 and 3 in [18], where the solutions should be compared with $P(x, t)$. The simulations reveal the complete or incomplete closure of the wound in terms of the parameter λ.

Summary

We have proven basic theoretical results for the model (11)–(13), which incorporates space and age into the dynamics of a proliferating cell population. This continuum model describes cell–cell adhesion processes, which are significant in cancer progression. Specifically, we proved the existence, uniqueness, positivity, and boundedness of solutions with sufficiently smooth positive initial data. Our results incorporate the cell cycle with a cell age variable reflecting heterogeneous cell division times. We examined in certain cases the asymptotic behavior of solutions and demonstrated the local stability of solutions with respect to parametric input. We illustrated these issues in simulations corresponding to wound healing experiments, in which cell–cell adhesion plays a significant role. In future work we will extend these ideas to settings in which stem cells are distinguished in cell proliferation dynamics. It is well recognized that cancer stem cells play a significant role in the pathogenesis of cancer, and continuum models of tumors incorporating spatial behavior, the cell cycle, cell differentiation, cell mutation, and stem cell lineages are important for cancer research.

References

1. J.C. Arciero, Q. Mi, M.F. Branco, D.J. Hackam, D. Swigon, Continuum model of collective cell migration in wound healing and colony expansion. Biophys. J. **100**, 535–543 (2011)
2. N.J. Armstrong, K.J. Painter, J.A. Sherratt, A continuum approach to modelling cell–cell adhesion. J. Theor. Biol. **243**, 98–113 (2006)
3. N.J. Armstrong, K.J. Painter, J.A. Sherratt, Adding adhesion to a chemical signalling model for somite formation. Bull. Math. Biol. **71**, 1–24 (2009)
4. H. Byrne, D. Draso, Individual based and continuum models of growing cell populations: a comparison. J. Math. Biol. **58**, 657–687 (2009)
5. H. Byrne, M.A.J. Chaplain, D.L. Evans, I. Hopkinson, Mathematical modelling of angiogenesis in wound healing: comparison of theory and experiment. J. Theor. Med. **2**, 175–197 (2000)
6. X. Chen, A. Friedman: A free boundary problem arising in a model of wound healing. SIAM J. Math. Anal. **32**(4), 778–800 (2000)
7. V. Christini, J. Lowengrub, *Multi-Scale Modeling of Cancer* (Cambridge University Press, Cambrigde, 2010)
8. G. Di Blasio, Mathematical analysis for an epidemic model with spatial and age structure. J. Evol. Equat. **10**(4), 929–953 (2010)
9. G. Di Blasio, A. Lorenzi, An identification problem in age-dependent population diffusion. Num. Funct. Anal. Optim. **34**(1), 36–73 (2013)
10. G.J. Doherty, H.T. McMahon, Mediation, modulation and consequences of membrane-cytoskeleton interactions. Ann. Rev. Biophys. **37**, 65–95 (2008)
11. A. Ducrot, P. Magal, S. Ruan, Travelling wave solutions in multigroup age-structured epidemic models. Arch. Ration. Mech. Anal. **195**(1), 311–331 (2010)
12. A. Ducrot, Travelling waves for a size and space structured model in population dynamics: point to sustained oscillating solution connections. J. Differ. Equat. **250**(1), 410–449 (2011)
13. R. Durrett, Cancer modeling: a personal perspective. Notices Am. Math. Soc. **60**(3), 304–309 (2013)
14. J. Dyson, E. Sánchez, R. Villella-Bressan, G.F. Webb, An age and spatially structured model of tumor invasion with haptotaxis. Discrete Contin. Dyn. Syst. -B **8**(1), 45–60 (2007)
15. J. Dyson, R. Villella-Bressan, G.F. Webb, A spatially structured model of tumor growth with cell age, cell size and mutation of cell phenotypes. Math. Model. Nat. Phenom. **2**(3), 69–100 (2007)
16. J. Dyson, R. Villella-Bressan, G.F. Webb, An age and spatially structured model of tumor invasion with haptotaxis II. Math. Pop. Stud. **15**, 73–95 (2008)

17. J. Dyson, S. Gourley, R. Villella-Bressan, G.F. Webb, Existence and asymptotic properties of solutions of a nonlocal evolution equation modelling cell–cell adhesion. SIAM J. Math. Anal. **42**(4), 1784–1804 (2010)
18. J. Dyson, S. Gourley, G.F. Webb, A nonlocal evolution equation model of cell–cell adhesion in higher dimensional space. J. Biol. Dyn. **7**(Suppl 1), 68–87 (2013). doi:10.1080/17513758.2012.755572
19. L.C. Evans, *Partial Differential Equations*. Graduate Studies in Mathematics (American Mathematical Society, Providence, 2004)
20. W.E. Fitzgibbon, M.E. Parrott, G.F. Webb, Diffusive epidemic models with spatial and age dependent heterogeneity. Discrete Contin. Dyn. Syst. **1**(1), 35–57 (1995)
21. W.E. Fitzgibbon, M.E. Parrott, G.F. Webb, A diffusive age-structured SEIRS epidemic model. Meth. Appl. Anal. **3**(3), 358–369 (1996)
22. A. Friedman, *Tutorials in Mathematical Biosciences, II: Cell Cycle, Proliferation, and Cancer*. Springer Lecture Notes in Mathematics, vol. 1872 (Springer, New York, 2005)
23. A. Friedman, Mathematical analysis and challenges arising from models of tumor growth. Math. Mod. Meth. Appl. Sci. **17**, 1751–1772 (2007)
24. A. Friedman, B. Hu, C. Xue, Analysis of a mathematical model of ischemic cutaneous wounds. SIAM J. Math. Anal. **42**, 2013–2040 (2010)
25. E.A. Gaffney, P.K. Maini, J.A. Sherratt, P.D. Dale, Wound healing in the corneal epithelium: biological mechanisms and mathematical models. J. Theor. Med. **1**, 13–23 (1997)
26. E.A. Gaffney, P.K. Maini, J.A. Sherratt, S. Tutt, The mathematical modelling of cell kinetics in corneal epithelial wound healing. J. Theor. Biol. **197**, 111–141 (1999)
27. A. Gandolfi, M. Iannelli, G. Marnoschi, An age-structured model of epidermis growth. J. Math. Biol. **62**, 15–40 (2011)
28. A. Gandolfi, M. Iannelli, G. Marinoschi, Time evolution for a model of epidermis growth. J. Evol. Equat. **13**, 509–533 (2013)
29. A. Gerisch, M.A.J. Chaplain, Mathematical modelling of cancer cell invasion of tissue: local and nonlocal models and the effect of adhesion. J. Theor. Biol. **250**, 684–704 (2008)
30. D. Guidetti, On elliptic systems in L^1. Osaka J. Math. **30**, 397–429 (1993)
31. H.J.A.M. Heijmans, *The Dynamical Behaviour of the Age-Size-Distribution of a Cell Population* (Centre for Mathematics and Computer Science, Amsterdam; Springer, Berlin, 1984)
32. D. Henry, *Geometric Theory of Semilinear Parabolic Equations*. Lecture Notes in Mathematics, vol. 840 (Springer, New York, 1981)
33. M. Kubo, M. Langlais, Periodic solutions for a population dynamics problem with age-dependence and spatial structure. J. Math. Biol. **29**(4), 363–378 (1991)
34. M. Kubo, M. Langlais, Periodic solutions for nonlinear population dynamics models with age-dependence and spatial structure. J. Differ. Equat. **109**(2), 274–294 (1994)
35. P. Laurençot, Ch. Walker, *Proteus mirabilis* swarm-colony development with drift. J. Math. Pures Appl. **92**(5), 476–498 (2009)
36. J.S. Lowengrub, H.B. Frieboes, F. Jin, Y.-L. Chuang, X. Li, P. Macklin, S.M. Wise, V. Cristin, Nonlinear models of cancer: bridging the gap between cells and tumours. Nonlinearity **23**, R1–R91 (2010)
37. L. Lorenzi, A. Lunardi, G. Metafune, D. Pallara, Analytic semigroups and reaction diffusion problems. Internet Seminar 2004–2005, http://www.math.unipr.it/~lunardi/LectureNotes.html
38. A. Lunardi, An introduction to interpolation theory. Internet Seminar January 2005, http://www.math.unipr.it/~lunardi/LectureNotes.html
39. J.A.J. Metz, O. Diekmann, *The Dynamics of Physiologically Structured Populations*. Lecture Notes in Biomathematics, vol. 68 (Springer, New York, 1986)
40. X. Mora, Semilinear parabolic problems define semiflows on C^k spaces. Trans. Am. Math. Soc. **278**, 21–55 (1983)
41. P.J. Murray, A. Walter, A.G. Fletcher, C.M. Edwards, M.J. Tindall, P.K. Maini, Comparing a discrete and continuum model of the intestinal crypt. Phys. Biol. **8**(2), 026011 (2010)

42. P.J. Murray, J-W. Kang, G.R. Mirams, S.-Y. Shin, H.M. Byrne, P.K. Maini, K.-H. Cho, Modelling spatially regulated β-catenin dynamics and invasion in intestinal crypts. Biophys. J. **99**(3), 716–725 (2010)
43. K.J. Painter, N.J. Armstrong, J.A. Sherratt, The impact of adhesion on cellular invasion processes in cancer and development. J. Theor. Biol. **264**, 1057–1067 (2010)
44. A. Pazy, *Semigroups of Linear Operators and Applications to Partial Differential Equations* (Springer, Berlin, 1983)
45. G.J. Pettet, M.A.J. Chaplain, D.S.L. McElwain, J. Norbury, A model of wound healing-angiogenesis in soft tissue. Math. Biosci. **136**, 35–63 (1996)
46. G.J. Pettet, M.A.J. Chaplain, D.S.L. McElwain, H.M. Byrne: On the role of angiogenesis in wound healing. Proc. Roy. Soc. Lond. B **263**, 1487–1493 (1996)
47. J.S. Ross, J.A. Fletcher, G.P. Linette, J. Stec, E. Clark, M. Ayers, W. Fraser Symmans, L. Pusztai, K.J. Bloom, The Her-2/neu gene and protein in breast cancer 2003: biomarker and target of therapy. Oncologist **8**(4), 307–25 (2003)
48. S. Ruan, J. Wu, Modeling spatial spread of communicable diseases involving animal hosts, in *Spatial Ecology*, S. Cantrell, C. Cosner, S. Ruan (Chapman & Hall CRC Press, Boca Raton, 2009), pp. 293–316
49. J.A. Sherratt, J.C. Dallon, Theoretical models of wound healing: past successes and future challenges. Comp. Rend. Biol. **325**(5), 557–564 (2002)
50. J.A. Sherratt, J.D. Murray, Mathematical analysis of a basic model for epidermal wound healing. Proc. Biol. Sci. **241**, 29–36 (1990)
51. J.A. Sherratt, J.D. Murray, Models of epidermal wound healing. J. Math. Biol. **31**, 703–716 (1993)
52. J.A. Sherratt, S.A. Gourley, N.J. Armstrong, K.J. Painter, Boundedness of solutions of a nonlocal reaction-diffusion model for adhesion in cell aggregation and cancer invasion. Euro. J. Appl. Math. **20**, 123–144 (2009)
53. J.W.-H. So, J. Wu, X. Zou, A reaction-diffusion model for a single species with age structure. I Travelling wavefronts on unbounded domains. Proc. Roy. Soc. Lond. Ser. A Math. Phys. Eng. Sci. **457**(2012),1841–1853 (2001)
54. S.L. Tucker, S.O. Zimmerman, A nonlinear model of population dynamics containing an arbitrary number of continuous structure variables. SIAM J. Appl. Math. **48**(3), 549–591 (1988)
55. S. Turner, J.A. Sherratt, Intercellular adhesion and cancer invasion: a discrete simulation using the extended Potts model. J. Theor. Biol. **216**, 85–100 (2002)
56. Ch. Walker, A haptotaxis model with age and spatial structure and nonlinear age-boundary conditions. PAMM **7**(1), 1040601–1040602 (2007)
57. Ch. Walker, Global well-posedness of a haptotaxis model with spatial and age structure. Differ. Integral Equat. **20**(9), 1053–1074 (2007)
58. Ch. Walker, Global existence for an age and spatially structured haptotaxis model with nonlinear age-boundary conditions. Eur. J. Appl. Math. **19**, 113–147 (2008)
59. Ch. Walker, Positive equilibrium solutions for age-and spatially-structured population models. SIAM J. Math. Anal. **41**(4), 1366–1387 (2009)
60. Ch. Walker, Global bifurcation of positive equilibria in nonlinear population models. J. Differ. Equat. **248**(7), 1756–1776 (2010)
61. Ch. Walker, Bifurcation of positive equilibria in nonlinear structured population models with varying mortality rates. Ann. Math. Pura Appl. **190**(1), 1–19 (2011)
62. Ch. Walker, On positive solutions of some systems of reaction-diffusion equations with nonlocal initial conditions. J. Reine Angew. Math. **660**, 149–179 (2011)
63. Ch. Walker, On nonlocal parabolic steady-state equations of cooperative or competing systems. Nonlinear Anal. Real World Appl. **12**(6), 3552–3571 (2011)
64. Ch. Walker, A note on a nonlocal nonlinear reaction-diffusion model. Appl. Math. Lett. **25**(11), 1772–1777 (2012)

65. Ch. Walker, Global continua of positive solutions for some quasilinear parabolic equation with a nonlocal initial condition. J. Dyn. Differ. Equat. **25**, 159–172 (2013). doi:10.1007/s10884-013-9292-7
66. Ch. Walker, Some remarks on the asymptotic behavior of the semigroup associated with age-structured diffusive populations. Monatsh. Math. **170**, 481–501 (2013). doi:10.1007/s00605-012-0428-3
67. S.E. Wang, I. Shin, F.Y. Wu, D.B. Friedman, C.L. Arteaga, HER2/Neu (ErbB2) signaling to Rac1-Pak1 is temporally and spatially modulated by transforming growth factor β. Cancer Res. **66**(19), 9591–9600 (2006)
68. G.F. Webb, An age-dependent epidemic model with spatial diffusion. Arch. Ration. Mech. Anal. **75**(1), 91–102 (1980)
69. G.F. Webb, A recovery-relapse epidemic model with spatial diffusion. J. Math. Biol. **14**(2), 77–194 (1982)
70. G.F. Webb, *Theory of Nonlinear Age-Dependent Population Dynamics*. Monographs and textbooks in Pure and Applied Mathematics (Marcel Dekker, New York, 1985)
71. G.F. Webb, Population models structured by age, size and spatial position, in *Structured Population Models in Biology and Epidemiology*, ed. by P. Magal, S. Ruan. Lecture Notes in Mathematics, vol. 1936, Mathematical Biosciences Series (Springer, Berlin, 2008), pp.1–49
72. C. Xue, A. Friedman, C.K. Sen, A mathematical model of ischemic cutaneous wounds. Proc. Natl. Acad. Sci. USA **106**, 16782–16787 (2009)

A General Framework for Multiscale Modeling of Tumor–Immune System Interactions*

Marina Dolfin, Mirosław Lachowicz, and Zuzanna Szymańska

Abstract In this paper we review methods that allow the construction of a consistent set of models that may describe the interactions between a tumor and the immune system on microscopic, mesoscopic, and macroscopic scales. The presented structures may be a basis for a description on the sub–cellular, cellular, and macroscopic levels. Important open problems are indicated.

Keywords Tumor • Immune system • Asymptotic analysis • Multiscale description • Kinetic equations • Systems of nonlinear ODEs

1 Introduction

Interactions between a tumor and the immune system are very complex and involve many factors—the players of the game. To some extent we believe that some aspects of the process may be adequately described on the macroscopic scale, but certainly there are aspects that cannot. For this reason it is important to develop a theoretical framework that allows for multiscale descriptions taking into account sub-cellular and cellular phenomena together. From a theoretical point of view this

---

*To Nicola Bellomo in occasion of his Anniversary.

M. Dolfin
Department of Civil, Computer, Construction and Environmental Engineering and of Applied Mathematics (DICIEAMA), University of Messina, Contrada Di Dio,
Vill. S. Agata, 98166 Messina, Italy
e-mail: mdolfin@unime.it

M. Lachowicz (✉)
Institute of Applied Mathematics and Mechanics, Faculty of Mathematics, Informatics and Mechanics, University of Warsaw, ul. Banacha 2, 02–097 Warsaw, Poland
e-mail: lachowic@mimuw.edu.pl

Z. Szymańska
ICM, University of Warsaw, ul. Pawińskiego 5a, 02–106 Warsaw, Poland
e-mail: mysz@icm.edu.pl

© Springer Science+Business Media New York 2014
A. d'Onofrio, A. Gandolfi (eds.), *Mathematical Oncology 2013*,
Modeling and Simulation in Science, Engineering and Technology,
DOI 10.1007/978-1-4939-0458-7_5

would involve the construction of a hierarchy of various models on the microscopic, mesoscopic, and macroscopic scales of description and a mathematical theory on possible relationships between them. Such a set of models may be related to various aspects of the competition between a tumor and the immune system as well as its various time scales.

In the present paper we discuss a theoretical framework for the construction of micro- and meso-models that may be related to macroscopic models and that are able to take into account various additional aspects of the microscopic scale.

The paper is organized as follows. In Sect. 2 we present the biological background. Section 3 reviews the general approach based on Active Particle Systems. Section 4 reviews a general macroscopic model proposed by A. d'Onofrio. Section 5 presents the general strategy that is going to be applied. The mathematical framework at the microscopic scale is given in Sect. 6. The class of models at the mesoscopic scale is contained in Sect. 7. Finally the mathematical links between various models are reviewed in Sect. 8.

2 Biological Background

Cancer as a disease exists at different biological levels. At the smallest level (or scale), cancer begins with changes in the DNA of a cell, which consequently lead to a disturbed functioning of the intracellular signaling pathways. In turn, abnormal activity of signaling pathways causes abnormal behavior at the cellular level. This means that cells

- do not respond to intercellular signals regulating proliferation and so proliferate in an uncontrolled way;
- do not differentiate normally and fully;
- do not undergo apoptosis.

Impaired processes at the cellular level lead to the formation of macroscopic abnormalities at the tissue level such as the formation of solid tumors and related pathological processes such as tumor-induced angiogenesis (new blood vessel formation). At this level we can talk about impaired functioning of whole organs, which in turn leads to the starvation and death of the organism.

Like the disease itself, the organism defense against tumor lesions includes a number of mechanisms at different levels, starting from molecular mechanisms of DNA repair, through intracellular apoptotic pathways, and finally to the activation of the body's immune system. Defense mechanisms associated with the immune system commence as soon as tumor cells appear and continue to act at a late stage in the overall process of cancer development when other host mechanisms have failed [38, 69]. However, many observations indicate a significant role for the immune system concerning the body's defense against tumors. For example, in patients who have chronic immunosuppression, one can observe a significant increase in the incidence of certain tumors in transplant recipients (about a hundredfold increase in the case of lymphomas and about a four hundredfold increase in the case of Kaposi's sarcoma [16, 38]).

The immune system protects organisms from infection with different specificity. Initially there is a physical barrier to prevent the entry of pathogens such as bacteria or viruses into the body. If a pathogen does break through these barriers, the innate immune system initially provides an immediate, but nonspecific response. If pathogens successfully avoid this nonspecific response there is a third layer of protection—the adaptive immune system. The adaptive immune response to infection relies on improved recognition of a pathogen previously eliminated and this is possible through the so-called immunological memory [55, 58]. Both types of immune response, innate and adaptive, rely on the ability of the immune system to distinguish between the host's own and foreign molecules.

Once activated, the immune system is very effective at detecting and eliminating pathogens such as bacteria or viruses. It identifies its targets by recognizing specific molecular entities—antigens—that are made by these agents. Often the immune response to a pathogen depends on previous encounters with this pathogen's antigens. The immune system has the ability to learn to recognize certain antigens and in the case of reexposure to this pathogen, the host response is faster and more efficient. This is known as the adaptive immune response. At the same time some cells of the immune system are naturally endowed with the ability to recognize certain antigens and do not require prior exposure to these antigens. In such a case we talk about the innate immune response. Having recognized the presence of foreign antigens the immune system undertakes to neutralize and destroy the infectious particle [69]. There are two basic types of immune response: the humoral immune response and the cellular immune response. In each type of immune response both of these components are present, with one of them usually being more severe.

Humoral Immunity

Responsible for the humoral immunity branch of the adaptive immune system are antibodies, which are produced by B lymphocytes. The soluble antibodies are present in the blood, lymph, and other body fluids, and they act in the intercellular spaces. They are capable to specifically recognize and bind foreign antigens, thus marking them as targets for attack by other effector mechanisms. The humoral immune response can be triggered by administration of a fluid containing molecular antigens, whereas other reactions require the administration of previously sensitized immune cells [38].

There are different ways in which the antibodies are involved in the body's immune defense. Antibodies that bind to the pathogen can prevent it from entering the target cell and neutralize it (neutralization by antibodies is also important in preventing bacterial toxins from entering cells) [42]. Virus particles, bacteria, or

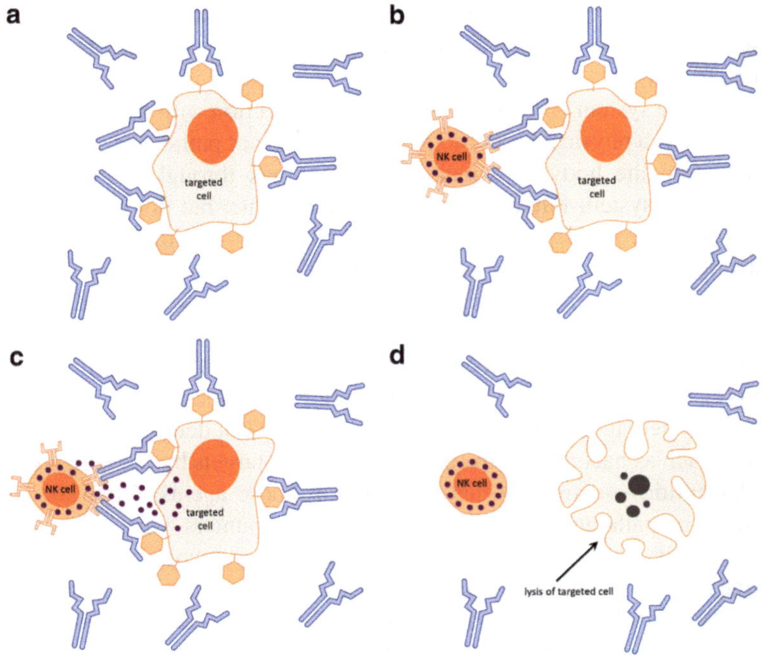

Fig. 1 Natural killer (NK) lymphocyte and its function. A target cell becomes coated by antibodies that recognize and bind antigens exposed on cell surface (plot **a**). Natural killer lymphocyte uses its cell surface receptors to bind the coating antibody molecules (plot **b**). This binding results in activation of NK cells, which proceed to destroy the target cell, using cytotoxic granules that are introduced into the targeted cell (plot **c**) and cause its death (plot **d**). Based on [69]

mammalian cell displaying antigens on its surface become covered by antibodies that are recognized by phagocytic cells such as macrophages or cytotoxic cells such as natural killer cells. In the first case the whole complex is engulfed by macrophages while in the second case the target cell is first killed by cytotoxic cells (see Fig. 1 explaining NK cells activity) [69]. Importantly cells like macrophages and natural killer do not have the ability of recognizing the specific antigens on their own, and so to recognize the antigen and for the secretion of an adequate antibody there is involved a complicated system of various cells (antigen-presenting cells, T helper cells, B lymphocytes) that are connected together in a complex network (see Fig. 2 representing B lymphocytes activation) [42, 69].

Cellular Response

In the case of cellular immune response, specialized cytotoxic cells—T lymphocytes—can on their own recognize and react directly with cells displaying certain antigens. T lymphocytes have developed their own mechanism recognizing

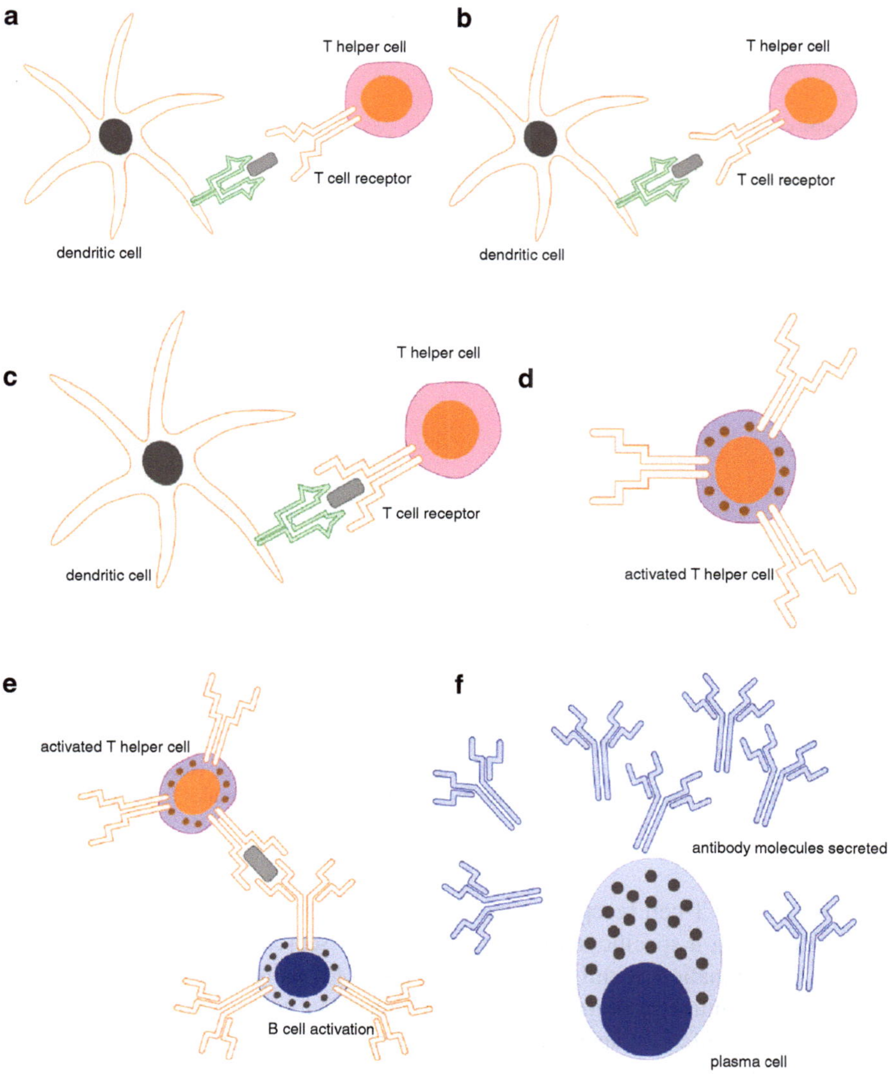

Fig. 2 Activation of B cells in the lymph nodes. A dendritic cell meets a number of T helper cells that display their own T cell receptor on their surface. In the first two encounters none of the T helper cells recognize and bind the antigen being presented (plots **a** and **b**). Finally, the dendritic cell succeeds in finding a T helper cell whose T cell receptor matches the antigen being presented by the dendritic cell (plot **c**). This results in the activation of the T helper cell to (plot d). The T helper cell leaves the dendritic cell and starts to search for B cells that also display on their surface the same antigen (plot **e**). When and if the T helper cell finds such a B cell, the B cell becomes activated, begins to proliferate and differentiates into a plasma cell that starts to release the antibody molecules that are capable of recognizing the antigen (plot **f**). Based on [69]

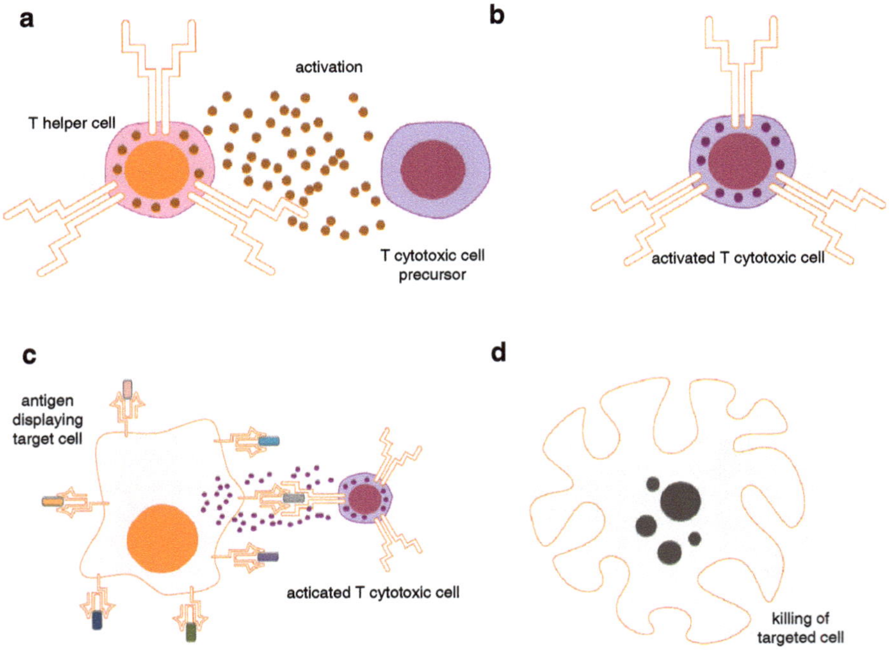

Fig. 3 Activation of cytotoxic T cells by helper T cells. In addition to inducing B cells to produce and secrete antibodies (see Fig 2), T helper cells have a second independent function, i.e., activation of precursors of cytotoxic T cells (plot **a**). Active T cytotoxic lymphocytes, with the use of their T cell receptors, are capable to recognize and bind antigens exposed on the surface of many target cells (plots **b** and **c**). As a result, T cytotoxic cells produce cytotoxic granules that cause the death of the target cell (plot **d**). Based on [69]

the antigen, the T cell receptors, which they use to identify particular antigens. In interaction with a specific antigen, cytotoxic cells secrete cytotoxic granules that cause the elimination of antigenic cells (see Fig. 3 for more details) [69].

Cellular immune response is particularly important in case of tumors and cells into which the infective agents like viruses or bacteria have entered. In such a case humoral immune response is unable to detect aberrant proteins that are embedded inside a target cell. This does not apply to T cytotoxic cells that usually can detect oligopeptides presented on the cell surface by MHC class I molecules, and can proceed to kill the cell. The whole process is mediated by T helper cells [69].

Tumor—Immune System Interaction

There is growing evidence of antitumor activity of the immune system. It is believed that the most important immunological antitumor defense mechanisms are: cytotoxic activity of NK cells, cytotoxic activity of cytotoxic T cells, activity of cytokines secreted by T lymphocytes, cytotoxicity of activated macrophages and neutrophils, activity of cytokines secreted by macrophages, and antibody-dependent cytotoxicity [38, 39, 69].

An observation that strongly supports the theory of anti-tumor activity of the immune system is that among tumor patients one may observe a weakening of the immune reactivity, especially of cellular type [38]. In particular, the risk of cancer strongly increases after solid organ transplantation. In the case of organ transplants, patients receive immunosuppressive drugs to prevent them rejecting the transplant, and the chronic immunosuppression caused by such drugs is a major risk factor of cancer. Cancer registry-based studies show that a large number of cancers (mostly those associated with oncogenic viruses) occur at increased rates relatively to the general population [67].

Moreover, new set of data, obtained from mice models and clinical epidemiology, suggests that immune system plays an important role in preventing cancers also in the case of some non-virus-induced cancers. Transplantation experiments have shown that cancer cells obtained from immunodeficient mice are often inefficient at initiating secondary tumors in syngeneic immunocompetent hosts. Vice versa, cancer cells obtained from tumors arising in immunocompetent mice are equally efficient at initiating transplanted tumors in both types of hosts [39, 66].

Known for years immune surveillance theory suggests that cells and tissues are constantly monitored by the immune system that recognizes and eliminates most of the tumor cells in the early stages of development [33]. According to this theory, clinically diagnosed solid tumors have been in some way able to avoid detection by the various arms of the immune system, or have been able to limit the scope of the immune response, thus avoiding severe damages [39]. As an extension of the immune surveillance theory recent studies indicate the existence of a so-called equilibrium phase in which tumor cells not completely eliminated by the immune system may proceed into a state where the immune system controls tumor cell growth but is not able to completely eliminate the transformed cells. In the past, the existence of an equilibrium phase was difficult to evidence because of poor experimental techniques and inability to identify and monitor early neoplastic lesions that may be subject to immunological control. As an outcome of the equilibrium phase two scenarios are possible: in the first, the immune system may eventually eliminate all tumor cells, while in the second, the constant interaction of the immune system with tumors may ultimately lead to immune selection and tumor progression [66]. Immune selection is the result of the inefficient pressure that effector immune mechanisms put on tumor cells. This results in the selective survival of tumor cells, whose sensitivity to the cytotoxic effector mechanisms is impaired, for example, due to a reduced antigenicity [38].

Counterintuitively, certain immunological processes observed in patients not only do not exert an antineoplastic activity, but may even contribute to the development of the tumor [38]. Histopathological examination of many tumors revealed their infiltration by cells of the immune system. With the development of experimental techniques it becomes apparent that any tumor lesion causes from mild to acute inflammation. For years, this phenomenon was regarded as an evidence of an attempt by the immune system to eradicate the tumor. More than ten years ago, it became apparent that the tumor associated inflammation has paradoxically enhancing effect on the tumor progression. Inflammation can contribute to tumor progression by supplying bioactive molecules such as growth, survival, and proangiogenic factors to the tumor environment [39].

2.1 *Mathematical Modeling of the Tumor-Immune System Interactions*

A number of mathematical models have recently been developed, to study the details of specific immune cell activities within a tumor, and the general implications of an immune response for tumor growth and progression.

Kuznetsov at al. [45] proposed a mathematical model of the cytotoxic T lymphocyte response to the growth of an immunogenic tumor. Owen and Sherratt [60] proposed a mathematical model for the spatial interactions of macrophages, tumor cells, and normal tissue cells, focusing on the ability of macrophages to kill mutant cells. Matzavinos at al. [57] presented a mathematical model describing the growth of a solid tumor in the presence of an immune system response, with particular attention upon the attack of tumor cells by so-called tumor-infiltrating cytotoxic lymphocytes (TICLs). Later, Matzavinos and Chaplain [56] provided a traveling wave analysis of this model. Lejeune at al. [54] studied the competition between tumor progression due to cancer cell replication and tumor regression due to immune cell cytotoxicity. Kirschner and Panetta [44], Szymańska [64], and Joshi at al. [43] developed mathematical models of immunotherapy and cancer vaccination. De Pillis at al. [61] presented a mathematical model that describes tumor–immune interactions, focusing on the role of natural killer (NK) and CD8+ T cells in tumor surveillance, with the goal of understanding the dynamics of immune-mediated tumor rejection. For an excellent review of the mathematical topics related to the modeling of the immune competition at the cellular level, see the papers of Bellomo and Preziosi [14] and Bellomo at al. [15].

Since interactions between the immune system and tumor cells are mainly based on the interaction of individual cells, in many cases the description on a microscale (or mesoscale) of interacting elements seems to be more appropriate. In fact, in tumor cells (cf. [9, 10, 13, 14, 41, 47]) competing with the immune system, the following steps of the evolution have been taken into account:

- loss of differentiation and replication,
- reproduction of cells in the form of identical descendants,
- interaction (activation or inhibition) and competition at the cellular level with immune and environmental cells, e.g., through the emission of cytokine signals.

These steps are related to cellular and subcellular interactions between tumor cells and cells and factors of immune system. The subcellular scale refers to processes that take place within the cells or at the cell membrane, e.g., synthesis of DNA and the activities of chemical signals between the cells. The cellular scale refers to various types of interactions between cells, e.g., interactions between tumor cells and immune system cells. Therefore, those scales may be connected with the microscopic level of description, i.e., the level of interacting individual elements of the system or, with some suitable reduction in the detail of the description, to the mesoscopic scale of a test-element—see Sect. 5.

3 Active Particle Systems

3.1 Modeling Complex Systems: Challenges and Perspectives

In some review papers by Bellomo and coworkers [6, 8, 11], specific features characterizing living complex systems are enlightened; some keywords may help to point out these features:

- *Strategy*: a specific feature of living entities is the ability to express strategies and structured organizations. It is worth stressing that a strategy depends strongly on the state of the surrounding environment which, in general, changes in time also due to feedback effects. A collective strategy may also be developed in particular structures.
- *Heterogeneity*: although belonging to the same structure, interacting entities can be different. In biology one could say that there are different phenotype expressions generated by the same genotype.
- *Learning*: ability to change own strategy to increment fitness, boosted by inputs coming from the environment and by mutual interactions.
- *Nonlinearity*: interactions are usually nonlinear and may involve the immediate neighbor, a geometrical, in general changing in time, distribution of a fixed number of neighbors or also distant entities. In the case of tumor–immune system interaction, a main rôle in long-distance interactions is played by the cytokines acting through chemical signaling.
- *Selection*: a Darwinian-type selection acts by means of generation of entities better fitted to the environment; its expression is related to a different time scale with respect to that of learning phenomena and may be taken into consideration or not in modeling depending on the adopted time scale. As an example, tumor cells have evolutionary abilities enhancing their survival in an environment involving hostile immune system cells. Evolutionary features related to the process of immunoevasion as a consequence of encounters among tumor cells and T lymphocytes are discussed in [1], in the context of neo-quasi-Lamarckian theories which have been the object of debate in the recent scientific literature (see [32] and references therein).

The above-listed sources of complexity force to tackle some technical aspects of modeling: a leading one is the multiscale intrinsic nature of the problem, involving microscopic, mesoscopic, and macroscopic scales, somehow (not directly and not necessarily in this order) related to sub–cellular, cellular and macroscopic phenomena. The core of the connection between modeling scales and physical phenomena is linked to the concept of size. A general starting point could be—see [8]—

> "*... focusing on biological systems, such microscopic state may refer, as a whole, to genes, cells, organs, individuals, populations. However, once specific entities have been selected, one has to consider the variables describing their possible states and their link with lower and higher scales. Only at this stage, the concept of size takes place.*"

3.2 Mesoscopic Models in the KTAP Framework

A general strategy in order to develop mathematical models for complex living systems at the mesoscopic level of description was proposed and discussed by N. Bellomo and coworkers by adopting the framework of the *kinetic theory for active particles* (KTAP for short); for a recent literature one can refer to [4, 6–8, 13] (see also references therein). A way to reduce the overall complexity of the system may be that of decomposing it into interacting subsystems, each of them having its defined mathematical rules and a lower order of complexity with respect to the whole system. This idea dates back to the *Theory of Modules* proposed by Hartwell [40], where the cells of a functional subsystem (a module in Hartwell's theory) collectively express a certain common strategy labeled by a scalar variable called *activity*. Cells may modify their activity due to interactions with other cells and this possibility may also lead to change the functional subsystem to which they belong. Moreover interactions may induce proliferative and/or destructive events which modify the overall number of involved cells. In particular, in case of proliferative events, a daughter cell may be generated presenting genetic modifications with respect to the mother cell so that the daughter cell may belong to a different functional subsystem with respect to the mother's one. In the case of tumor–immune system interactions, proliferative events are characteristic of tumor cells, but also immune system cells may proliferate triggered by the presence of tumor cells (for an example of macroscopic models of immune cell proliferation, triggered in that case by antigens, see [25, 26]). Destructive events may occur due to the ability of the immune system to suppress tumor cells. The activity, i.e., the strategy that the cells belonging to each subsystem collectively express, may be identified by the variable $u \in \mathbb{U}$, where $\mathbb{U}$ is usually an interval in $\mathbb{R}^1_+$, as it will be in the following. In general $\mathbb{U}$ may be a (Lebesgue measurable) subset of $\mathbb{R}^d$ $(d \geq 1)$ and further generalizations are straightforward. This variable takes different values from one cell to another, thus defining the microscopic state of the single cell. This idea comes from the kinetic theory of gas where every entity involved in the process (a particle), in the spatially homogeneous case, is characterized by its velocity. Similarly, other entities of the system (not necessarily cells) may also be characterized by various different characteristic parameters.

In the present paper the evolution of the system, constituted by a large number of interacting entities (factors, particles, or cells), is defined by a *probability density*

$$f = f(t) = f_j(t,u) = f(t,j,u) \; : \; \mathbb{R}^1_+ \times \mathbb{J} \times \mathbb{U} \to \mathbb{R}^1_+ \,, \tag{1}$$

where $j \in \mathbb{J}$, with $\mathbb{J} \subset \{0, 1, 2, \ldots\}$, characterizes the population to which the entity belongs (functional subsystem), $u \in \mathbb{U}$, and $\mathbb{U}$ is an (bounded or unbounded) interval in $\mathbb{R}^1_+$ as we stated before (generalizations being straightforward). The function $f = f(t)$ is a probability density in the sense that

$$\sum_{j \in \mathbb{J}_1} \int_{\mathbb{U}_1} f(t,j,u)\, du$$

is the probability of an entity at time $t \geq 0$, in a population in $\mathbb{J}_1 \subseteq \mathbb{J}$, with the activity $u \in \mathbb{U}_1$, where $\mathbb{U}_1$ is a measurable subset of $\mathbb{U}$.

As we stated before some standard approaches deal with non-normalized distribution functions that may directly describe the proliferative–destructive character of interactions. Here however we deal with the probabilistic framework referring to the moments of probability densities.

The time evolution of the probability density f is defined by a suitable kinetic equation that takes into consideration all possible interactions of the involved elements. Such a level of description is referred to one test entity of the system and therefore may be called the *mesoscopic* level of description. It requires a large (infinite) number of interacting entities. Referring to the specific features characterizing living complex systems briefly discussed at the beginning of the present section, the authors claim that, in particular, it may be worth to note the analysis of the intrinsic evolutive features of the KTAP.

One may also consider the microscopic level of description in which the physical finite number of interacting elements is taken into account. In paper [49] (see also [46] and references therein) a class of Markov jump processes was proposed and related with the *mesoscopic* and *macroscopic* levels of the description, through nonlinear equations, in the framework of the asymptotic theory. In [49] the nonlinearity of the equations at the macroscopic level was related with the interactions between multiple entities of the system at the microscopic level.

Here we review an approach that relates the nonlinear equations at the macroscopic level with the mesoscopic equations with nonlinear interaction coefficients in the spirit of papers [4, 50]. This approach goes back to the Enskog idea in kinetic theory of gas [12] where the rates of interactions between particles depend on the distribution function. We propose a class of nonlinear kinetic equations that, after averaging over the activity parameter, reduces to the class of macroscopic equations describing the competition between tumor and immune system developed in [30].

Different to [50] the present approach is not applied to the delay system at the macroscopic level (cf. [31]), although we notice that the effects of time delay may play an important rôle in the description of tumor growth.

4 Macroscopic Models of Tumor–Immune System Interaction

Our starting point here is a general class of macroscopic models introduced by A. d'Onofrio [27–31] that try to catch, in a simplified scenario, the effect of antitumoral immune surveillance taking also into account the effect of a therapeutic action represented as an external source of immune system cells. The basic idea is to deal with TCs (tumor cells) and ECs (effector cells of the immune system) as two competing populations, adopting Lotka–Volterra–type equations (the prey–predator interactions). The TCs represent the prey of ECs whose proliferation is stimulated by the presence of TCs. Moreover TCs induce a loss of ECs. This general class of models specializes to specific well-known macroscopic models in particular cases (for details see [28, 30, 31]). In the literature on the topic of antitumoral immune surveillance one can find experimental evidences of interactions among TCs and immune system cells; as an example phosphopeptides may be targets of tumor immune surveillance in humans [24]. These experimental evidences point to the importance of developing therapeutic actions called immunotherapies, although this argument presents controversial viewpoints and opinions.

One can notice that the presented macroscopic models adaptively change the interactions strength, allowing for nonlinearity features of living complex systems.

The mathematical model introduces the "*sizes*" (total number, densities, or a-dimensional quantities) of TCs and ECs, respectively, $\varrho_1(t)$ and $\varrho_2(t)$, whose dynamical evolution is represented by the following ODE system of equations:

$$\begin{aligned} \dot{\varrho}_1 &= \varrho_1\alpha_{11}(\varrho_1) - \varrho_1\beta_{11}(\varrho_1)\beta_{12}(\varrho_2) \\ \dot{\varrho}_2 &= \varrho_2\alpha_{21}(\varrho_1) - \varrho_2\beta_{21}(\varrho_1) + \gamma(\varrho_1) + S\,, \end{aligned} \tag{2}$$

where α_{jk}, β_{jk} $(j,k = 1,2)$ are given positive functions, $\gamma(\varrho_1)$ is a given function, and $S = S(t)$ is a given source term (for details see [30]).

The system of equations (2) is supplemented by the initial data

$$\varrho_1(0) = \overset{\circ}{\varrho}_1\,, \qquad \varrho_2(0) = \overset{\circ}{\varrho}_2\,, \tag{3}$$

where $\overset{\circ}{\varrho}_1 \geq 0$, $\overset{\circ}{\varrho}_2 \geq 0$ are given.

Throughout the paper we assume that, given $\overset{\circ}{\varrho}_1$ and $\overset{\circ}{\varrho}_2$, there exists a unique bounded global solution to the IVP (2), (3).

In the present paper we are going to develop a general class of models describing at the microscopic and mesoscopic levels the behavior of the entities (cells) of these two populations. Such models, by taking into consideration statistical individual properties of the entities of the system, are able to design accurate descriptions at the sub-cellular and cellular levels.

Every modeled entity is characterized by both its own population, through the parameter $j = 1, 2$, and its inner microscopic state (activity), through the microscopic parameter $u \in \mathbb{U}$.

5 General Modeling Strategy

In the present section we review the general modeling strategy in view of relating microscopic models with those at the macroscopic scale. The essential rôle is here played by the intermediate—mesoscopic—level of description (see Sect. 3).

Macroscopic models are usually written in terms of systems of ordinary (like Eq. (2)) or partial differential equations; as for instance reaction–diffusion–chemotaxis equations, see [22, 37], or systems of integro-differential equations which describe nonlocal interactions, see [23, 37, 65] and references therein. Models at the microscopic level are defined by a large number of interacting elements of the system (cells, factors etc) and are described in terms of a Markov jump process and the related linear evolution equations. The intermediate models refer to the mesoscopic level of description of test-element and are given in terms of nonlinear Boltzmann–type equations in a framework similar to *KTAP*.

The mathematical relationships between these three descriptions and explicit error estimates may be developed by the methods of asymptotic analysis and a small parameter approach.

The description of complex processes of interaction between tumor and immune system is usually carried out on a macroscopic level of interacting populations within the system, e.g., as in Eq. (2). Such an approach describes the (deterministic) evolution of densities of populations of the system rather than the individual entities. The identification of parameters is usually easier on the macroscopic scale.

If the movement in space is taken into account an important feature of the microscopic scale could be nonlocal interactions: one element may interact with another one even if the distance between them is relatively large—see [23, 37, 65].

The prototype of mathematical setting and relationships between three possible scales of description, micro, meso, and macro, can be the kinetic theory of rarefied gases, cf. [2, 19] (see also references therein). One may, however, observe an important difference: in the case of tumor—immune system interactions a basic microscopic theory, such as the classical mechanics in kinetic theory, is not available. Thus it seems reasonable to apply the following strategy. One may begin with the macroscopic models for which the identification of parameters by experiments is easier. Then one provides a theoretical framework for modeling at the microscopic scale in such a way that the corresponding models at the macro-, meso-, and microscales are asymptotically equivalent in a suitable mathematical sense. Then, if the parameters of the microscopic model are properly chosen, one may hope that it characterizes not only the macroscopic features of the system in question, but also some of its microscopic aspects. The microscopic models by their nature may be richer and can describe a wider range of phenomena.

The next sections, mostly referring to [46–49], review a general conceptual framework for the program of finding possible transitions between the different levels of description, i.e.,

- **(Mi)** interacting elements— "microscale,"
- **(Me)** statistical description of a test element—"meso-scale,"
- **(Ma)** densities of populations—"macroscale."

In mathematical terms the program may be stated in studying (asymptotic) links between the following structures:

- **(Mi)** continuous (linear) stochastic semigroups;
- **(Me)** continuous nonlinear semigroups related to the solutions of nonlinear kinetic equations;
- **(Ma)** dynamical systems related to nonlinear ODEs or PDEs.

Literature related to the rigorous (mathematical) relationships between various macroscopic, mesoscopic, and microscopic models is enormous. The interested reader is referred to the bibliography, e.g.:

- **(Mi)** and **(Me)**: [2, 19, 34, 36, 46, 48, 49, 68]
- **(Mi)** and **(Ma)**: [17, 18, 20, 21, 59]
- **(Me)** and **(Ma)**: [2, 5–7, 46, 48–51, 62, 63]

and references therein.

6 Microscopic Scale: Individually-Based Models

We follow here [49] (see also the previous approach in [46]). We consider a system composed of N interacting elements. First we assume that N is fixed and then we are going to study the limit $N \to \infty$.

This point needs an explanation. We are taking into account the complex system of all interacting elements involved in the competition between tumor and immune system. The number of elements is relatively large, and it changes in time due to proliferation and destruction processes. However we are assuming here an idealized case in which the number of elements remains fixed during the process. This is of course a mathematical simplification and it is quite unrealistic. But due to the fact that N is relatively large—it may be accepted as a model simplification. Considering the limit $N \to \infty$ (a transition from micro, to mesoscale) we understand the following sense: if N is sufficiently large, two models (micro- and meso-) should give similar results, whereas when N is not so large, they may give two different alternative ways of description of the process. We may point out that the mathematical theory that allows for changes of N is much complex (see [36]) and does not seem that may be directly applied here. This is one of the main open problems in the theory.

Each element n ($n \in \{1, 2, \ldots, N\}$) is qualified by the pair (j_n, u_n), where $j_n \in \mathbb{J}$ defines the population of the n-element and $u_n \in \mathbb{U}$, its biological state (activity). We assume that $\mathbb{J} \subseteq \{0, 1, 2, \ldots\}$ and $\mathbb{U} \subset \mathbb{R}^1_+$ is a bounded or unbounded interval.

We follow the general framework ([49], cf. [46]) that relates Markov jump processes, and more precisely the corresponding modified Liouville equations describing the evolution of probability densities, with microscopic representation of the system of N interacting elements. The modified Liouville equation defines the microscopic models that may correspond to the large class of macroscopic models related to the competition tumor–immune system. The linear generator, defining the modified Liouville equation, completely describes the time evolution of the probability density at the microscale. The corresponding stochastic semigroup may approximate the dynamical system related to the macroscopic model.

In what follows, a number N of elements of various populations is considered. The n_1-element changes its population and/or its activity at random times. It is possible [49] that

- a change occurs without any interaction;
- a change is due to the interaction with the $n_2, n_3, \ldots, n_M$ elements

$$n_2, \ldots, n_M \in \{1, \ldots, N\}, \qquad n_i \neq n_j, \quad \text{for all} \quad i \neq j \quad i, j = 1, \ldots, M,$$

  where M can be any integer satisfying $1 \leq M \leq N$.

We consider the interactions of a given element with $m - 1$ elements, where $m = 1, \ldots, M$. The rate of interaction between the element of the j_{n_1}th population with activity u_{n_1} and the elements of $j_{n_2}, \ldots, j_{n_m}$ populations with activities $u_{n_2}, \ldots, u_{n_m}$, is given by a measurable function $a^{[m]} = a^{[m]}(j_1, u_1, \ldots, j_m, u_m)$ such that

$$0 \leq a^{[m]}\Big((j_{n_1}, u_{n_1}), (j_{n_2}, u_{n_2}), \ldots, (j_{n_m}, u_{n_m})\Big) \leq a^{[m]}_+, \tag{4}$$

for all $j_{n_1}, \ldots, j_{n_m} \in \mathbb{J}$ and all $u_{n_1}, \ldots, u_{n_m}$ in $\mathbb{U}$,
where $0 < a^{[m]}_+ < \infty$ are constants, $m = 1, 2, \ldots, M$.

The transition into the kth population with the activity v of an element of the j_{n_1}-th population with activity u_{n_1}, due to the interaction with elements of $j_{n_2}, \ldots, j_{n_m}$ populations with activities $u_{n_2}, \ldots, u_{n_m}$, respectively, is defined by the measurable function $A^{[m]}$ such that

$$A^{[m]} = A^{[m]}\Big((k, v); (j_{n_1}, u_{n_1}), \ldots, (j_{n_m}, u_{n_m})\Big) \geq 0, \tag{5}$$

for all $k, j_{n_1}, \ldots, j_{n_m} \in \mathbb{J}$ and all $v, u_{n_1}, \ldots, u_{n_m}$ in $\mathbb{U}$,

$$\sum_{k\in\mathbb{J}} \int_{\mathbb{U}} A^{[m]}\Big((k,v)\,;\,(j_{n_1},u_{n_1}),(j_{n_2},u_{n_2}),\,\ldots\,,(j_{n_m},u_{n_m})\Big)\,\mathrm{d}v = 1\,, \tag{6}$$

for all $j_{n_1}\,,\,\ldots\,,j_{n_m} \in \mathbb{J}$ and all $u_{n_1}\,,\,\ldots\,,u_{n_m} \in \mathbb{U}$ such that

$$a^{[m]}\Big(\,(j_{n_1}\,,\,u_{n_1})\,,\,\ldots\,,\,(j_{n_m}\,,\,u_{n_m})\,\Big) > 0\,.$$

The (microscopic) stochastic model is completely determined by the functions $a^{[m]}$ and $A^{[m]}$, $m = 1, \ldots, M$. Different choices of $a^{[m]}$ and $A^{[m]}$ give rise to different microscopic models (Markov jump processes).

Given N, M, $a^{[m]}$ and $A^{[m]}$, $m = 1, \ldots, M$, we assume that the stochastic system is defined by the Markov jump process of N elements through the following generator Λ_N acting on a suitable test function ϕ (a real-valued, Borel measurable, bounded function)

$$\begin{aligned}
&\Lambda_N\phi\Big((j_1,u_1),\ldots,(j_N,u_N)\Big)\\
&= \sum_{m=1}^{M} \frac{1}{(m-1)!\binom{N}{m-1}} \sum_{\substack{1\le n_1,\ldots,n_m\le N\\ n_i\neq n_j\ \forall i\neq j}} a^{[m]}\Big((j_{n_1},u_{n_1}),(j_{n_2},u_{n_2}),\ldots,(j_{n_m},u_{n_m})\Big)\\
&\times\Bigg(\sum_{k\in\mathbb{J}}\int_{\mathbb{U}} A^{[m]}\Big((k,v);(j_{n_1},u_{n_1}),(j_{n_2},u_{n_2}),\ldots,(j_{n_m},u_{n_m})\Big)\\
&\times\phi\Big((j_1,u_1),\ldots,(j_{n_1-1},u_{n_1-1}),(k,v),(j_{n_1+1},u_{n_1+1}),\ldots,(j_N,u_N)\Big)\mathrm{d}v\\
&-\phi\Big((j_1,u_1),\ldots,(u_N,u_N)\Big)\Bigg)\,.
\end{aligned}$$

Λ_N is the generator for a Markov jump process in $\big(\mathbb{J}\times\mathbb{U}\big)^N$ that can be constructed as in Ref. [35], Sect. 4.2 — cf. [49].

However, in the present paper, as in [46–49], we refer only to the evolution of probability densities.

Let the system be initially distributed according to the probability density $\overset{\circ}{f}{}^N \in L_1^{(N)}$. $L_1^{(N)}$ is the space equipped with the standard norm

$$\|f\|_{L_1^{(N)}} = \sum_{j_1\in\mathbb{J}}\int_{\mathbb{U}} \cdots \sum_{j_N\in\mathbb{J}}\int_{\mathbb{U}} \Big|f\Big((j_1,u_1),\ldots,(j_N,u_N)\Big)\Big|\,\mathrm{d}u_1\ldots\mathrm{d}u_N\,.$$

Time evolution is described by the following linear equation, the modified Liouville equation,

$$\frac{\partial}{\partial t} f^N = \Lambda_N^* f^N\,; \qquad f^N\Big|_{t=0} = \overset{\circ}{f}{}^N\,, \tag{7}$$

where Λ_N^* is the generator

$$
\begin{aligned}
&\Lambda_N^* f^N\Big(t, (j_1, u_1), (j_2, u_2), \ldots, (j_N, u_N)\Big) \\
&= \sum_{m=1}^{M} \frac{1}{(m-1)!\binom{N}{m-1}} \sum_{\substack{1 \le n_1, \ldots, n_m \le N \\ n_i \ne n_j \ \forall i \ne j}} \\
&\Bigg(\sum_{k \in \mathbb{J}} \int_{\mathbb{U}} A^{[m]}\Big((j_{n_1}, u_{n_1}); (k, v), (j_{n_2}, u_{n_2}), \ldots, (j_{n_m}, u_{n_m})\Big) \\
&\times a^{[m]}\Big((k, v), (j_{n_2}, u_{n_2}), \ldots, (j_{n_m}, u_{n_m})\Big) \\
&\times f^N\Big(t, (j_1, u_1), \ldots, (j_{n_1-1}, u_{n_1-1}), (k, v), (j_{n_1+1}, u_{n_1+1}), \ldots, (j_N, u_N)\Big)\, dv \\
&- a^{[m]}\Big((j_{n_1}, u_{n_1}), (j_{n_2}, u_{n_2}), \ldots, (j_{n_m}, u_{n_m})\Big) f^N\Big(t, (j_1, u_1), \ldots, (j_N, u_N)\Big)\Bigg).
\end{aligned}
$$

The generator is the difference between two terms:

- the *gain term* that is the sum of terms describing changes from state (k, v) of the n_1-element into (j_{n_1}, u_{n_1}) due to the interaction with the $n_2, \ldots, n_m$ elements with states $(j_{n_2}, u_{n_2}), \ldots, (j_{n_m}, u_{n_m})$, respectively, for $2 \le m \le M$, and the term $(m = 1)$ describing the direct changes of state (k, v) of the n_1-element into (j_{n_1}, u_{n_1}) without interactions;
- the *loss term* that is the sum of terms describing changes from state (j_{n_1}, u_{n_1}) of the n_1-element into another state due to the interaction with the $n_2, \ldots, n_m$ elements with states $(j_2, u_2), \ldots, (j_m, u_m)$, respectively, for $2 \le m \le M$, or without interactions for $m = 1$.

The operator Λ_N^*, under Assumptions (4), (5), and (6), is a bounded linear operator in the space $L_1^{(N)}$. For this reason the Cauchy Problem (7) has the unique solution given by a continuous semigroup according to the formula

$$f^N(t) = e^{t\Lambda_N^*} \mathring{f}^N$$

in $L_1^{(N)}$ for all $t \ge 0$. Furthermore, standard arguments show that the solution is nonnegative for nonnegative initial data and the $L_1^{(N)}$–norm is preserved

$$\| f^N(t) \|_{L_1^{(N)}} = \| \mathring{f}^N \|_{L_1^{(N)}} = 1\,, \qquad \text{for } t > 0\,. \tag{8}$$

From this, $\left(e^{t\Lambda_N^*}\right)_{t \ge 0}$ is a continuous semigroup of Markov operators, that is, a continuous stochastic semigroup.

We consider symmetric functions, that is,

$$f^N\Big((j_1,u_1),\ldots,(j_N,u_N)\Big) = f^N\Big((j_{r_1},u_{r_1}),\ldots,(j_{r_N},u_{r_N})\Big), \tag{9}$$

for all $j_1,\ldots,j_N$ in $\mathbb{J}$, all $u_1,\ldots,u_N$ in $\mathbb{U}$, and for any permutation $\{r_1,\ldots,r_N\}$ of the set $\{1,\ldots,N\}$.

We introduce the s-individual marginal density $(1 \leq s < N)$

$$\begin{aligned} & f^{N,s}\Big((j_1,u_1),\ldots,(j_s,u_s)\Big) \\ &= \sum_{j_{s+1}\in\mathbb{J}}\int_{\mathbb{U}}\cdots\sum_{j_N\in\mathbb{J}}\int_{\mathbb{U}} f^N\Big((j_1,u_1),\ldots,(j_N,u_N)\Big)\,\mathrm{d}u_{s+1}\ldots\mathrm{d}u_N\,, \end{aligned} \tag{10}$$

$f^{N,N} = f^N$, and $f^{N,s'} \equiv 0$ if $s' > N$.

The function f^N satisfies Eq. (7) if and only if $f^{N,s}$ satisfy the following finite hierarchy of equations:

$$\frac{\partial}{\partial t} f^{N,s} = \mathfrak{A}^s_{N,M}\Big(f^{N,s}, f^{N,s+1},\ldots,f^{N,s+M-1}\Big), \tag{11}$$

for $s = 1,2,\ldots,N$, where the linear operator $\mathfrak{A}^s_{N,M}$ is given by

$$\begin{aligned} & \mathfrak{A}^s_{N,M}\Big(f^{N,s},\ldots,f^{N,s+M-1}\Big) \\ &= \sum_{m=1}^{M}\frac{(N-m+1)!}{N!}\,\mathfrak{A}^{s,m}_{N,M}\Big(f^{N,s},\ldots,f^{N,s+m-1}\Big), \end{aligned}$$

$$\begin{aligned} & \mathfrak{A}^{s,m}_{N,M}\Big(f^{N,s},\ldots,f^{N,s+m-1}\Big)\Big((j_1,u_1),\ldots,(j_s,u_s)\Big) \\ &= \sum_{k=1}^{m}\frac{(N-s)!}{(N-s-m+k)!}\sum_{\substack{1\leq n_1,\ldots,n_k\leq s\\ n_i\neq n_j\ \forall i\neq j}}\ \sum_{\{\}}\underbrace{\sum_{j_{s+1}\in\mathbb{J}}\int_{\mathbb{U}}\cdots\sum_{j_{s+m-k}\in\mathbb{J}}\int_{\mathbb{U}}}_{(m-k)\text{ times}} \\ & \Bigg(\sum_{k\in\mathbb{J}}\int_{\mathbb{U}} A^{[m]}\Big((j_{n_1},u_{n_1});(k,v),\{\mathbf{u}_{n_2},\ldots,\mathbf{u}_{n_k},\mathbf{u}_{s+1},\ldots\mathbf{u}_{s+m-k}\}\Big) \\ & \times a^{[m]}\Big((k,v),\{\mathbf{u}_{n_2},\ldots,\mathbf{u}_{n_k},\mathbf{u}_{s+1},\ldots\mathbf{u}_{s+m-k}\}\Big) \\ & \times f^{N,s+m-k}\Big((j_1,u_1),\ldots,(j_{n_1-1},u_{n_1-1}), \\ & (k,v),(j_{n_1+1},u_{n_1+1}),\ldots,(j_s,u_s),\ldots,(j_{s+m-k},u_{s+m-k})\Big)\,\mathrm{d}v \\ & -a^{[m]}\Big((j_{n_1},u_{n_1}),\{\mathbf{u}_{n_2},\ldots,\mathbf{u}_{n_k},\mathbf{u}_{s+1},\ldots\mathbf{u}_{s+m-k}\}\Big) \\ & \times f^{N,s+m-k}\Big((j_1,u_1),\ldots,(j_{s+m-k},u_{s+m-k})\Big)\Bigg)\,\mathrm{d}u_{s+1}\ldots\mathrm{d}u_{s+m-k}\,, \end{aligned}$$

where $\mathbf{u_n} = (j_n, u_n)$, and $\sum_{\{\}}$ means the sum over all permutation of variables within $\{\}$. We adhere to the convention that if $m - k < 1$, then

$$\sum_{\{\}} \underbrace{\sum_{j_{s+1}\in\mathbb{J}} \int_{\mathbb{U}} \cdots \sum_{j_{s+m-k}\in\mathbb{J}} \int_{\mathbb{U}}}_{(m-k)\text{ times}},$$

$\{\mathbf{u}_{s+1}, \ldots \mathbf{u}_{s+m-k}\}$ and $\mathrm{d}\mathbf{u}_{s+1} \ldots \mathrm{d}\mathbf{u}_{s+m-k}$ do not appear in the corresponding term.

Taking sufficiently large N, we may expect that the solution of the finite hierarchy (11) approximates the solution of the following infinite hierarchy of equations

$$\frac{\partial}{\partial t} f^s = \mathfrak{B}^s_M\Big(f^s, f^{s+1}, \ldots, f^{s+M-1}\Big), \qquad s = 1, 2, \ldots, \tag{12}$$

where the linear operator $\mathfrak{B}^s_M$ is given by

$$\mathfrak{B}^s_M\Big(f^s, f^{s+1}, \ldots, f^{s+M-1}\Big) = \sum_{m=1}^{M} \mathfrak{B}^{s,m}_M\, f^{s+m-1},$$

and

$$\mathfrak{B}^{s,m}_M\, f^{s+m-1}\Big((j_1, u_1), \ldots, (j_s, u_s)\Big) = \sum_{n=1}^{s} \sum_{\{\}} \underbrace{\sum_{j_{s+1}\in\mathbb{J}} \int_{\mathbb{U}} \cdots \sum_{j_{s+m-1}\in\mathbb{J}} \int_{\mathbb{U}}}_{(m-1)\text{ times}}$$

$$\Bigg(\sum_{k\in\mathbb{J}} \int_{\mathbb{U}} A^{[m]}\Big((j_n, u_n); (k, v), \{\mathbf{u}_{s+1}, \ldots \mathbf{u}_{s+m-1}\}\Big)$$
$$\times a^{[m]}\Big((k, v), \{\mathbf{u}_{s+1}, \ldots \mathbf{u}_{s+m-1}\}\Big)$$
$$\times f^{s+m-1}\Big((j_1, u_1), \ldots, (j_{n-1}, u_{n-1}), (k, v),$$
$$(j_{n+1}, u_{n+1}), \ldots, (j_s, u_s), \ldots, (j_{s+m-1}, u_{s+m-1})\Big)\, \mathrm{d}v$$
$$-a^{[m]}\Big((j_n, f u_n), \{\mathbf{u}_{s+1}, \ldots \mathbf{u}_{s+m-1}\}\Big)$$
$$\times f^{s+m-1}\Big((j_1, u_1), \ldots, (j_{s+m-1}, u_{s+m-1})\Big)\Bigg)\mathrm{d}u_{s+1} \ldots \mathrm{d}u_{s+m-1}.$$

The integral versions of hierarchies (11) and (12) read

$$f^{N,s}(t) = \overset{\circ}{f}{}^{N,s} + \int_0^t \mathfrak{A}^s_{N,M}\Big(f^{N,s}, f^{N,s+1}, \ldots, f^{N,s+M-1}\Big)(t_1)\, \mathrm{d}t_1, \tag{13}$$

where $s = 1, \dots, N$, and

$$f^s(t) = \overset{\circ}{f}{}^s + \int_0^t \mathfrak{B}^s_M\Big(f^s, f^{s+1}, \dots, f^{s+M-1}\Big)(t_1)\, \mathrm{d}t_1 \,, \tag{14}$$

with $s = 1, 2, \dots$, respectively.

A hierarchy $\{f^s\}_{s=1,2,3,\dots}$ is called the *admissible hierarchy*, if it is a sequence of functions f^s satisfying, for $s = 1, 2, \dots$:

(i) f^s is a probability density on $(\mathbb{J} \times \mathbb{U})^s$;
(ii) $f^s\big((j_1, u_1), \dots, (j_s, u_s)\big) = f^s\big((j_{r_1}, u_{r_1}), \dots, (j_{r_s}, u_{r_s})\big)$ for all $j_1, \dots, j_s$ in $\mathbb{J}$ and a.a. $u_1, \dots, u_s$ in $\mathbb{U}$ and for any permutation $\{r_1, \dots, r_s\}$ of the set $\{1, \dots, s\}$;
(iii) $f^s\big((j_1, u_1), \dots, (j_s, u_s)\big) = \sum_{j_{s+1} \in \mathbb{J}} \int_{\mathbb{U}} f^{s+1}\big((j_1, u_1), \dots, (j_{s+1}, u_{s+1})\big)\, \mathrm{d}u_{s+1}$ for all $j_1, \dots, j_s$ in $\mathbb{J}$ and a.a. $u_1, \dots, u_s$ in $\mathbb{U}$.

We have [49]the following:

Theorem 6.1. *Let Assumptions* (4)*,* (5)*,* (6) *be satisfied and* $\{\overset{\circ}{f}{}^s\}_{s=1,2,\dots}$ *be an admissible hierarchy. Then, for all* $t > 0$*, there exists a unique hierarchy* $\{f^s(t)\}_{s=1,2,\dots}$*,* $f^s(t) \in L_1^{(s)}$ $(s = 1, 2, \dots)$ *that is a solution of Eq.* (14) *with initial data* $f^s(0) = \overset{\circ}{f}{}^s$ $(s = 1, 2, \dots)$*. Moreover* $\{f^s(t)\}_{s=1,2,\dots}$*, for all* $t > 0$*, is an admissible hierarchy.*

Proof. See [49].

7 Mesoscopic model

In this section we are going to consider a general model (cf. [50]) containing the following terms:

- the term $\mathcal{A}$ describing stochastic change of the parameter u due to the interactions between two elements—the Enskog-type kinetic operator;
- the term $\mathcal{B}$ describing stochastic change of the parameter u without taking into consideration the interactions between elements.

The general class of bilinear systems of Boltzmann-like integro-differential equations describes the dynamics of elements undergoing kinetic binary interactions—see [3,46]. This type of equations can model interactions between pairs of elements of various populations at the mesoscopic scale. In the Enskog-type description [12] we consider the binary interactions between the elements of the system but the rates of interaction (contrary to the Boltzmann-type description) are modified by "*pair correlation functions*".

The general mesoscopic model reads

$$\partial_t f_j(t,u) = \mathcal{A}_j[f](t,u) + \mathcal{B}_j[f](t,u), \quad j \in \mathbb{J}. \tag{15}$$

We consider the Banach space X equipped with the norm

$$\|f\| = \sum_{j \in \mathbb{J}} \int_{\mathbb{U}} (1+u)|f_j(u)|\,du$$

that is, we consider f such that f_j are integrable on $\mathbb{U} \subseteq \mathbb{R}^1_+$ and their first moments (with respect to u variable) are bounded, $j \in \mathbb{J}$. If $\mathbb{U}$ is a bounded subset of $\mathbb{R}^1_+$, then $(1+u)$ may be substituted by 1.

Moreover we consider the positive cone X_+

$$X_+ = \left\{ f \in X \; : \; f_j \geq 0, \qquad j \in \mathbb{J} \right\}.$$

First we define the Enskog-type terms $\mathcal{A}_j$.

The rate of interaction between an element in jth population with the activity u and the element in kth population with the activity v, $j,k \in \mathbb{J}$, $u,v \in \mathbb{U}$ is given by the function

$$a_{jk} = a_{jk}\left(\bar{f}_j, u, \bar{f}_k, v\right) \tag{16}$$

that depends on $\bar{f}_j$, $\bar{f}_k$, where

$$\bar{f}_j(t) = \int_{\mathbb{U}} u f_j(t,u)\,du, \qquad j \in \mathbb{J}, \tag{17}$$

is the first moment of f_j.

In (16) here we assume the dependence on the first moment only, but the generalization into more complex dependence on various higher moments is straightforward.

The transition into the jth population with activity u due to the interaction of element in kth population and with activity v and the element in lth population with activity w, $j,k,l \in \mathbb{J}$, $u,v,w \in \mathbb{U}$ is given by the function

$$A^j_{kl} = A^j_{kl}\left(\bar{f}_j, u; \bar{f}_k, v, \bar{f}_l, w\right)$$

that depends on $\bar{f}_j$, $\bar{f}_k$, and $\bar{f}_l$.

Now we consider the following term of the Enskog type:

$$\mathcal{A}_j[f](t,u) = \mathcal{A}^+_j[f](t,u) - \mathcal{A}^-_j[f](t,u), \qquad j \in \mathbb{J}, \tag{18}$$

where

$$\mathcal{A}_j^+[f](t,u) = \sum_{k,l=1}^{2} \int_{\mathbb{U}^2} A_{kl}^j(\bar{f}_j, u; \bar{f}_k, v, \bar{f}_l, w) \times$$
$$a_{kl}(\bar{f}_k, v, \bar{f}_l, w) f_k(t,v) f_l(t,w) \mathrm{d}v\mathrm{d}w,$$

$$\mathcal{A}_j^-[f](t,u) = f_j(t,u) \sum_{k\in\mathbb{J}} \int_{\mathbb{U}} a_{jk}(\bar{f}_j, u, \bar{f}_k, v) f_k(t,v)\, \mathrm{d}v\,.$$

We assume that A_{kl}^j is a transition probability:

$$A_{kl}^j \geq 0\,, \qquad \int_{\mathbb{U}} A_{kl}^j(\bar{f}_j, u; \bar{f}_k, v, \bar{f}_l, w)\, \mathrm{d}u = 1\,, \tag{19}$$

for all $j, k, l \in \mathbb{J}$ and each $f_j = f_j(t) \in X_+, t > 0$.

Moreover, let

$$0 \leq a_{kl}(\bar{f}_k, v, \bar{f}_l, w) \leq \text{const}\,, \tag{20}$$

for all $k, l \in \mathbb{J}$ and each $f_j = f_j(t) \in X_+, t > 0$.

Next in a similar way we define the term $\mathcal{B}$

$$\mathcal{B}_j[f](t,u) = \mathcal{B}_j^+[f](t,u) - \mathcal{B}_j^-[f](t,u)\,, \qquad j \in \mathbb{J}\,, \tag{21}$$

where

$$\mathcal{B}_j^+[f](t,u) = \sum_{k\in\mathbb{J}} \int_{\mathbb{U}} B_k^j(\bar{f}_j, u, \bar{f}_k(t), v) b_k(\bar{f}_k, v) f_k(t,v)\, \mathrm{d}v\,,$$

$$\mathcal{B}_j^-[f](t,u) = b_j(\bar{f}_j, u) f_j(t,u)\,.$$

We assume that B_k^j is a transition probability, so that

$$B_k^j \geq 0\,, \qquad \int_{\mathbb{U}} B_k^j(\bar{f}_j, u; \bar{f}_k, v)\, \mathrm{d}u = 1\,, \tag{22}$$

for all $j, k \in \mathbb{J}$ and each $f_j = f_j(t) \in X_+, t > 0$.

Moreover, let

$$0 \leq b_k(\bar{f}_k, v) \leq \text{const}\,, \tag{23}$$

for all $k \in \mathbb{J}$ and each $f_j = f_j(t) \in X_+, t > 0$.

The time evolution of the probability density $f = f_j(t, u)$ is obtained by solving the IVP for Eq. (15) with the initial data

$$f_j\Big|_{t=0} = \overset{\circ}{f}_j\,, \qquad j \in \mathbb{J}\,. \tag{24}$$

Under suitable assumption similar to those in [4] the operator appearing at the right-hand side of (15) is Lipschitz continuous in X and thus it gives a local existence result.

Theorem 7.1. *Consider the IVP* (15), (24) *for* $\overset{\circ}{f} = (\overset{\circ}{f}_1, \overset{\circ}{f}_2) \in X_+$. *Assume that conditions* (19), (20), (22), *and* (23) *are satisfied and that*

- A^j_{kl} *and* B^j_k *satisfy the following conditions:*

$$\int_{\mathbb{U}} (1+u) A^j_{kl}(\bar{f}_j, u; \bar{f}_k, v, \bar{f}_l, w)\, du \leq \text{const} \tag{25}$$

$$\int_{\mathbb{U}} (1+u) B^j_k(\bar{f}_j, u; \bar{f}_k, v)\, du \leq \text{const} \tag{26}$$

  for all $j,k,l \in \mathbb{J}$ *and each* $f_j = f_j(t) \in X_+$, $t > 0$, *with positive constants (independent of all variables) denoted by "const";*
- A^j_{kl} *and* B^j_k *are Lipschitz continuous in* X *with respect to* f_j, f_k, f_l *and* f_j, f_k, *respectively, i.e., if* $\|f\|, \|g\| \leq M$, *then*

$$\begin{aligned} &\int_{\mathbb{U}} (1+u) \Big| A^j_{kl}(\bar{f}_j, u; \bar{f}_k, v, \bar{f}_l, w) \\ &- A^j_{kl}(\bar{g}_j, u; \bar{g}_k, v, \bar{g}_l, w)\Big|\, du \leq L\, \|f - g\|\,, \end{aligned} \tag{27}$$

  and

$$\begin{aligned} &\int_{\mathbb{U}} (1+u) \Big| B^j_k(\bar{f}_j, u; \bar{f}_k, v) \\ &- B^j_k(\bar{g}_j, u; \bar{g}_k, v)\Big|\, du \leq L\, \|f - g\|\,, \end{aligned} \tag{28}$$

  with L*—a positive constant depending on* M.

Then, there exists $T > 0$ *and a unique strong solution* $f = f(t)$ *in* X *of the IVP* (15), (24) *on the time interval* $[0, T]$. *Moreover* $f(t) \in X_+$ *for any* $t \in [0, T]$.

The conditions assure that the operator appearing at the right-hand side of (15) is Lipschitz continuous in X which proves the existence of a unique solution local in time. Positivity follows in a standard way.

This result together with the conservative properties leads to the following global existence result (cf. [4]):

Theorem 7.2. *Consider the IVP* (15), (24) *for initial data* $\overset{\circ}{f} \in X_+$ *under the assumptions of Theorem* (7.1). *Then there exists a unique strong solution* $f = f(t)$ *in* X *of the IVP* (15), (24)*, for any* $t > 0$. *Moreover,* $f(t) \in X_+$ *for any* $t > 0$.

The proof is obtained by observing that the probability density properties (19) and (23) assure that

$$\sum_{j \in \mathbb{J}} \int_{\mathbb{U}} f_j(t,u)\, \mathrm{d}u = \sum_{j \in \mathbb{J}} \int_{\mathbb{U}} \overset{\circ}{f}_j(u)\, \mathrm{d}u \qquad \forall\, t \in [0,T]. \tag{29}$$

Using the conservative property (29) together with (25) and (26) one can see that the X-norm is bounded on every compact time set. Consequently, the solution can be extended for arbitrary time interval.

Finally, let us mention that it is important looking for equilibrium solutions of the equations. This means finding f such that

$$\mathcal{A}_j[f] + \mathcal{B}_j[f] = 0, \quad \forall\, j \in \mathbb{J}. \tag{30}$$

The conditions guaranteeing the existence of solutions of such problem may be stated similarly as in [4]. This result does not provide any information on the number of possible solutions of Eq. (30). It is an important open question to identify its solution. We may point out that for the periodic structures such an identification was given in Ref. [53] (see also [52]).

8 Micro–Macro Links

In order to derive the nonlinear equations resulting in the limit $N \to \infty$ (i.e., at the mesoscopic level) from Eq. (7) the approach of [49] (c.f. [46]) may be used.

We assume that the process starts with a factorized probability density

$$\overset{\circ}{f}{}^{N} = \overset{\circ}{f}{}^{N\otimes} := \underbrace{\overset{\circ}{f} \otimes \ldots \otimes \overset{\circ}{f}}_{N\times}, \tag{31}$$

where

$$\underbrace{\overset{\circ}{f} \otimes \ldots \otimes \overset{\circ}{f}}_{N\times}\Big((j_1,u_1),\ldots,(j_N,u_N)\Big) = \prod_{n=1}^{N} \overset{\circ}{f}(j_n,u_n),$$

i.e., N-fold outer product of the probability density $\overset{\circ}{f}$.

In the limit $N \to \infty$, the (linear) modified Liouville equation (7) yields, [49], a nonlinear Boltzmann-like integro-differential equation that can be related to the mesoscopic description. In fact we may see that the propagation of chaos is held and the solution $f^s(t)$ to Eq. (14) is the s-product of solution $f(t)$ of the following nonlinear kinetic equation; see [49]:

$$\frac{\partial}{\partial t} f(t,j,u) = \mathfrak{G}[f](t,j,u) - f(t,j,u)\mathfrak{L}[f](t,j,u)\,, \qquad j \in \mathbb{J}\,, \quad u \in \mathbb{U}\,, \tag{32}$$

where $\mathfrak{G}[f]$ is the *gain term*, given by

$$\mathfrak{G}[f](t,j,u) = \sum_{m=1}^{M} \sum_{\{\}} \underbrace{\sum_{j_2 \in \mathbb{J}} \int_{\mathbb{U}} \cdots \sum_{j_m \in \mathbb{J}} \int_{\mathbb{U}}}_{(m-1)\times} \sum_{k \in \mathbb{J}} \int_{\mathbb{U}} A^{[m]}\Big((j,u);(k,v),\{\mathbf{u}_2,\dots \mathbf{u}_m\}\Big)$$
$$\times a^{[m]}\Big((k,v),\{\mathbf{u}_2,\dots \mathbf{u}_m\}\Big) f(t,k,v) f(t,j_2,u_2)\dots f(t,j_m,u_m)\, \mathrm{d}v\, \mathrm{d}u_2 \dots \mathrm{d}u_m\,,$$

and $f\,\mathfrak{L}[f]$ is the *loss term*, defined as

$$\mathfrak{L}[f](t,j,u) = \sum_{m=1}^{M} \sum_{\{\}} \underbrace{\sum_{j_2 \in \mathbb{J}} \int_{\mathbb{U}} \cdots \sum_{j_m \in \mathbb{J}} \int_{\mathbb{U}}}_{(m-1)\times} a^{[m]}\big(j,u,\{\mathbf{u}_2,\dots \mathbf{u}_m\}\big)$$
$$\times f(t,\mathbf{u}_2)\dots f(t,\mathbf{u}_m)\, \mathrm{d}u_2 \dots \mathrm{d}u_m\,,$$

and as before, $\mathbf{u}_n = (j_n,u_n)$, we adhere to the convention that if $m = 1$ then $\sum_{\{\}} \underbrace{\sum_{j_2 \in \mathbb{J}} \int_{\mathbb{U}} \cdots \sum_{j_m \in \mathbb{J}} \int_{\mathbb{U}}}_{(m-1)\times}$, as well as $\{\mathbf{u}_2,\dots \mathbf{u}_m\}$ and $f(t,\mathbf{u}_2)\dots f(t,\mathbf{u}_m)$ do not appear in the corresponding term. By Theorem 6.1 we have

Corollary 8.1. *Let Assumptions* (4), (5), *and* (6) *be satisfied and* $\overset{\circ}{f}$ *be a probability density on* $\mathbb{J}\times\mathbb{U}$. *Then, for each* $T > 0$, *there exists an admissible hierarchy* $\{f^s\}_{s=1,2,\dots}$ *such that*

(i) *it is a unique solution of Eq.* (14) *with factorized initial data* (31),
(ii) $f^s(t)$ *is factorized,*

$$f^s(t) = \Big(f(t)\Big)^{s\otimes}\,, \tag{33}$$

for all $0 < t \le T$ *and* $s = 1,2,\dots$, *where* $f(t)$ *is the unique solution in* $L_1^{(1)}$ *of Eq.* (32) *with the initial datum* $\overset{\circ}{f}$.

As a by-product, we obtain the existence (and uniqueness) of solutions to Eq. (32).

We may now state the theorem (cf. [49]) that defines the links between the solutions to Eq. (7) and to Eq. (32) or, in other words, that defines the transition from the microscopic level to the mesoscopic level.

Theorem 8.2. *Let Assumptions* (4), (5), *and* (6) *be satisfied and* $\overset{\circ}{f}$ *be a probability density on* $\mathbb{J} \times \mathbb{U}$. *Then, for each* $T > 0$, *there exists* N_0 *such that for* $N \geq N_0$

$$\sup_{t \in [0,T]} \|f^{N,1} - f\|_{L_1^{(1)}} \leq \frac{c}{N^{\zeta}}, \tag{34}$$

where $f^N \in L_1^{(N)}$ *is the unique nonnegative solution of Eq.* (7) *corresponding to the initial datum* (31); $f \in L_1^{(1)}$ *is the unique nonnegative solution of Eq.* (32) *corresponding to the initial datum* $\overset{\circ}{f}$; *and* ζ, c *are positive constants that depend on* T.

Proof. See [49].

Now, under the assumptions that the terms A_{kl}^j and B_k^j analytically depend on the moments one may find a relationship between the linear microscopic equation (7) and the nonlinear kinetic equations (15) in an analogous way to [49].

Then we obtain

Theorem 8.3. *Let Assumptions* (4), (5), *and* (6) *be satisfied and* $\overset{\circ}{f}$ *be a probability density on* $\mathbb{J} \times \mathbb{U}$ *that satisfies the condition of Theorem* (7.1). *Then the terms* $A^{[m]}$ *and* $a^{[m]}$ *of the Eq.* (7) *may be chosen in such a way that, for each* $T > 0$, *there exists* N_0 *such that for* $N \geq N_0$

$$\sup_{t \in [0,T]} \|f^{N,1} - f\|_{L_1^{(1)}} \leq \frac{c}{N^{\zeta}} + \xi_M, \tag{35}$$

where $f^N \in L_1^{(N)}$ *is the unique nonnegative solution of Eq.* (7) *corresponding to the initial datum* (31); $f \in L_1^{(1)}$ *is the unique nonnegative solution of Eq.* (15) *corresponding to the initial datum* $\overset{\circ}{f}$; *and* ζ, c *are positive constants that depend on* T *and* ξ_M *is such that* $\lim_{M \to \infty} \xi_M \to 0$.

Theorem (8.3) states that the solution to the (nonlinear) mesoscopic equation (15) may be approximated by the solutions of the (linear) microscopic equation (7), with properly chosen $A^{[m]}$ and $a^{[m]}$, if both N and M are sufficiently large. Approximation may be realized by many possible microscopic equations (various $A^{[m]}$ and $a^{[m]}$ may be taken into account).

If, on the other hand, N is not large, then the linear equation (7) related to the microscopic description, and the nonlinear equation (15) at the mesoscopic scale, may independently play important roles in the mathematical description of the complex processes, presumably giving different results. In such a case one may expect that the microscopic model gives results that are closer to reality.

The last step is to state the relationship between the mesoscopic (Eq. (15)) and macroscopic (Eq. (2)) descriptions. There are various possible ways to relate the mesoscopic equation (15) with the macroscopic equation (2). We have

Theorem 8.4. *Let* $\mathbb{J} = \{1, 2\}$. *There exist* $\mathbb{U} \subset \mathbb{R}_+$, A^j_{kl}, *and* B^j_k *such that given be a probability density* $\overset{\circ}{f}$ *on* $\mathbb{J} \times \mathbb{U}$ *that satisfies the condition of Theorem* (7.1), *there exists, for any* $t \geq 0$, *the unique solution* $f = f(t)$ *to Eq.* (15) *(with initial data* $\overset{\circ}{f}$*) such that* $\bar{f}_j = \bar{f}_j(t)$, $j = 1, 2$, *satisfies Eq.* (2).

There exists a large class of mesoscopic equations that corresponds to the given macroscopic system. The choice of the proper mesoscopic (and then microscopic) model should base on the experimental data.

Finally we may mention that the problem of stability of some mesoscopic equations corresponding to the macroscopic equation, whose solutions are stable, was studied in the papers [51] and [62].

9 Concluding Remarks

In this chapter we have presented a general framework that may unify the various phenomena of the competition between tumor and immune system, starting from the subcellular and cellular scales to the macroscopic scale, in one general model that may result at the macroscopic scale in given models for which the identification of parameters is realizable. We have in fact not one model but a class of models on the microscopic scale, and when the experimental data provide information on the behavior of microscopic elements of the system, we may choose the adequate model. Therefore the presented approach defines new perspectives of modeling for such a complex system. It leads to many open problems on the theoretical and practical level. In this chapter we attempted to indicate some of them, and we stressed possible directions of further research.

Acknowledgment M.D. acknowledges a support from the National Group for Mathematical Physics through the grant Visiting Professors 2013. M.L. acknowledges a support from the National Science Centre of Poland through grant N N201 605640. Z.S. acknowledges a support from the National Science Centre of Poland through grant DEC-2011/01/D/ST1/04133.

References

1. M. Al-Tameemi, M. Chaplain, A. d'Onofrio, Evasion of tumours from the control of the immune system: consequences of brief encounters. Biol Direct. **7**(31), 1–22 (2012)
2. L. Arlotti, N. Bellomo, E. De Angelis, M. Lachowicz, *Generalized Kinetic Models in Applied Sciences* (World Scienctific, New Jersay (2003)
3. L. Arlotti, N. Bellomo, M. Lachowicz, Kinetic equations modelling population dynamics. Transport Theory Statist. Phys. **29**, 125–139 (2000)

4. L. Arlotti, E. De Angelis, L. Fermo, M. Lachowicz, N. Bellomo, On a class of integro–differential equations modeling complex systems with nonlinear interactions. Appl. Math. Lett. **25**, 490–495 (2012)
5. N. Bellomo, A. Bellouquid, J. Nieto, J. Soler, Multiscale biological tissue models and flux–limited chemotaxis for multicellular growing systems. Math. Models Methods Appl. Sci. **20**, 1179–1207 (2010)
6. N. Bellomo, A. Bellouquid, J. Nieto, J. Soler, On the asymptotic theory from microscopic to macroscopic growing tissue models: An overview with perspectives. Math. Models Methods Appl. Sci. **22**, 1130001 (2012)
7. N. Bellomo, A. Bellouquid, J. Nieto, J. Soler, Modelling chemotaxis from L^2–closure moments in kinetic theory of active particles. Discrete Contin. Dyn. Systems B **18**, 847–863 (2013)
8. Bellomo, N., Carbonaro, B.: Toward a mathematical theory of living system focusing on developmental biology and evolution: A review and prospectives. Physics of Life, Reviews **8**, 1–18 (2011)
9. N. Bellomo, G. Forni, Dynamics of tumor interaction with the host immune system. Math. Comput. Modelling **20**, 107–122 (1994)
10. N. Bellomo, G. Forni, Looking for new paradigms towards a biological-mathematical theory of complex multicellular systems. Math. Models Methods Appl. Sci. **16**, 1001–1029 (2006)
11. N. Bellomo, D. Knopoff, J. Soler, On the difficult interplay between life, "complexity" and mathematical sciences. Math. Models Methods Appl. Sci. **23**, 1861–1913 (2013)
12. N. Bellomo, M. Lachowicz, J. Polewczak, G. Toscani, *Mathematical Topics in Nonlinear Kinetic Theory II: The Enskog Equation* (World Scientific, Singapore 1991)
13. N. Bellomo, N.K. Li, P.K. Maini, On the foundations of cancer modelling: Selected topics, speculations and perspectives. Math. Models Methods Appl. Sci. **18**, 593–646 (2008)
14. N. Bellomo, L. Preziosi, Modelling and mathematical problems related to tumour evolution and its interaction with immune system. Math. Comput. Model. **32**, 413–452 (2000)
15. N., Bellomo, A. Bellouquid, E. de Angelis, The modelling of immune competition by generalised kinetic (Boltzmann) models: review and research perspectives. Math. Comput. Modelling **37**, 65–86 (2003)
16. M., Bodnar, U., Foryś, Z. Szymańska, Model of AIDS–related tumour with time delay, Appl. Math. (Warsaw) **36**, 263–278 (2009)
17. V. Capasso, D. Morale, in *Rescaling Stochastic Processes: Asymptotics*. eds. by J. Banasiak, V. Capasso, M.A.J. Chaplain, M. Lachowicz, J. Miękisz, Multiscale Problems in the Life Sciences. From Microscopic to Macroscopic, Lecture Notes Math. vol. 1940, (Springer, Berlin 2008) pp. 91–146.
18. Capasso, V., Morale, D.: Asymptotic behavior of a system of stochastic particles subject to nonlocal interactions. Stoch. Anal. Appl. **27**, 574–603 (2009)
19. C. Cercignani, R. Illner, *Pulvirenti, M. The Mathematical Theory of Dilute Gases* (Springer, New York 1994)
20. N. Champagnat, R. Ferrière, S. Méléard, Unifying evolutionary dynamics: from individual stochastic processes to macroscopic models. Th. Popul. Biol. **69**, 297–321 (2006)
21. N. Champagnat, R. Ferrière, S. Méléard, From individual stochastic processes to macroscopic models in adaptive evolution. Stochastic Models **24** 2–44 (2008)
22. M.A.J. Chaplain, in *Modelling aspects of cancer growth: insight from mathematical and numerical analysis and computational simulation*, eds. by J. Banasiak, V. Capasso, M.A.J. Chaplain, M. Lachowicz, J. Miękisz, Multiscale Problems in the Life Sciences. From Microscopic to Macroscopic, Lecture Notes Math. vol. 1940, (Springer, Berlin 2008) pp. 147–200.
23. M.A.J. Chaplain, M. Lachowicz, Z. Szymańska, D. Wrzosek, Mathematical modelling of cancer invasion: The importance of cell–cell adhesion and cell-matrix adhesion. Math. Models Methods Appl. Sci. **24**, 719–743 (2011)
24. M. Cobbold, H. De La PeÃśa, A. Norris, J.M. Polefrone, J. Qian, A.M. English, K.L. Cummings, S. Penny, J.E. Turner, J. Cottine, J.G. Abelin, S.A. Malaker, A.L. Zarling, H.W. Huang, O. Goodyear, S.D. Freeman, J. Shabanowitz, G. Pratt, C. Craddock, M.E Williams, D.F. Hunt, V.H. Engelhard, MHC Class I-Associated Phosphopeptides Are the Targets of Memory-like Immunity in Leukemia. Sci. Transl. Med. **203**, 203–125 (2013)

25. Criaco, D., Dolfin, M.,. Restuccia, L: Approximate smooth solution pf a mathematical model for the activation and clonal expansion of T cells. Math. Biosci. Engineer. **10**, 59–73 (2013)
26. D. Criaco, M. Dolfin, A phenomenological approach to the dynamics of activation and clonal expansion of T cells. Math. Computer Model. **53**, 314–329 (2011)
27. A. d'Onofrio, A general framework for modeling tumour–immune system competition and immunotherapy: mathematical analysis and biomedical inferences. Physica D **208**, 220–235 (2005)
28. d'Onofrio, A.: Tumor–immune system interaction: modeling the tumor–stimulated proliferation of effectors and immunotherapy. Math. Models Methods Appl. Sci. **16**, 1375–1401 (2006)
29. A. d'Onofrio, Tumor evasion from immune control: Strategies of a MISS to become a MASS. Chaos, Solitons and Fractals **31**, 261–268 (2007)
30. A. d'Onofrio, Metamodeling tumor–immune system interaction, tumor evasion and immunotherapy. Math. Comput. Model. **47**,614–637 (2008)
31. A. d'Onofrio, F. Gatti, P. Cerrai, L. Freschi, Delay–induced oscillatory dynamics of tumour–immune system interaction. Math. Comput. Model. **51**, 572–591 (2010)
32. A. d'Onofrio, A. Ciancio, Simple biophysical model of tumor evasion from immune system control. Phys Rev E Stat Nonlin Soft Matter Phys. **84** (2011)
33. P. Ehrlich, Über den jetzigen Stand der Karzinomforschung. Ned. Tijdschr. Geneeskd. **5**, 273–290 (1909)
34. A. Eibeck, W. Wagner, Stochastic interacting particle systems and nonlinear kinetic equations. Annals Appl. Prob. **13**, 845–889 (2003)
35. S.N., Ethier, T.G. Kurtz, Markov Processes, Characterization and Convergence. Wiley, New York (1986)
36. D., Finkelshtein, Y., Kondratiev, O. Kutoviy, Semigroup approach to birth–and–death stochastic dynamics in continuum. J. Funct. Anal. **262**, 1274–1308 (2012)
37. A. Gerisch, M.A.J. Chaplain, Mathematical modelling of cancer cell invasion of tissue: local and non-local models and the effect of adhesion. J. Theoret. Biol. **250**, 684–704 (2008)
38. J. Gołab, M. Jakóbisiak, W. Lasek (eds.), Immunologia, Wydawnictwo Naukowe PWN, Warszawa (2002), in Polish
39. D., Hanahan, R.A. Weinberg, Hallmarks of cancer: the next generation. Cell **144**, 646–674 (2011)
40. H.L. Hartwell, J.J. Hopfield, S. Leibner, A.W. Murray, From molecular to modular cell biology. Nature **402**, c47–c52 (1999)
41. R. B. Herberman (ed.), *NK Cells and Other Natural Effector Cells* (Academic Press, New York 1982)
42. C.A. Janeway, P. Travers, M. Walport, *Immunobiology: The Immune System in Health and Disease* (Garland Science, New York 2001)
43. B., Joshi, X., Wang, S., Banerjee, H., Tian, A., Matzavinos, M.A.J. Chaplain, On immunotherapies and cancer vaccination protocols: A mathematical modelling approach. J. Theor. Biol. **259**, 820–827 (2009)
44. D., Kirschner, J.C. Panetta, Modeling immunotherapy of the tumor - immune interaction. J. Math. Biol. **37**, 235–252 (1998)
45. V.A., Kuznetsov, I.A. Makalkin, M.A. Taylor, A.S. Perelson, Nonlinear dynamics of immunogenic tumours: Parameter estimation and global bifurcation analysis. Bull. Math. Biol. **56**, 295–321 (1994)
46. M. Lachowicz, in *Links between microscopic and macroscopic descriptions* eds. by J. Banasiak, V. Capasso, M.A.J. Chaplain, M. Lachowicz, J. Miękisz Multiscale Problems in the Life Sciences. From Microscopic to Macroscopic, Lecture Notes Math vol. 1940, (Springer, Berlin 2008) pp. 201–268
47. Lachowicz, M.: in *Towards microscopic and nonlocal models of tumour invasion of tissue* eds. by In: N. Bellomo, M. Chaplain, E. De Angelis, (Birkhauser, Boston 2008) pp. 49–63.
48. M. Lachowicz, Microscopic, mesoscopic and macroscopic descriptions of complex systems. Prob. Engin. Mech. **26**, 54–60 (2011)

49. Lachowicz, M.: Individually–based Markov processes modeling nonlinear systems in mathematical biology. Nonlinear Analysis Real World Appl. **12**, 2396–2407 (2011)
50. M. Lachowicz, A. Quartarone, A general framework for modeling tumor–immune system competition at the mesoscopic level. Appl. Math. Letters **25**, 2118–2122 (2012)
51. Lachowicz, M., Quartarone, A., Ryabukha, T.V.: Stability of solutions of kinetic equations corresponding to the replicator dynamics, Kinetic Relat. Models (to appear)
52. M. Lachowicz, T.V. Ryabukha, Equilibrium solutions for microscopic stochastic systems in population dynamics. Math. Biosci. Engin. **10**, 777–786 (2013)
53. M.Lachowicz, D. Wrzosek, Nonlocal bilinear equations. Equilibrium solutions and diffusive limit. Math. Models Methods Appl. Sci. **11**, 1375–1390 (2001)
54. O. Lejeune, M.A.J. Chaplain, I. El Akili, Oscillations and bistability in the dynamics of cytotoxic reactions mediated by the response of immune cells to solid tumours. Math. Comput. Modelling **47** 649–662 (2008)
55. G.W. Litman, J.P., Cannon, L.J. Dishaw, Reconstructing immune phylogeny: new perspectives. Nat. Rev. Immunol. **5**, 866–879 (2005)
56. A. Matzavinos, M.A.J. Chaplain, Travelling wave analysis of a model of the immune response to cancer. Biologies **327**, 995–1008 (2004)
57. A. Matzavinos, M.A.J. Chaplain, V. Kuznetsov, Mathematical modelling of the spatio–temporal response of cytotoxic T–lymphocytes to a solid tumour. Math. Med. Biol. **21**, 1–34 (2004)
58. G. Mayer, *Microbiology and Immunology On–Line Textbook* (USC School of Medicine, 2006)
59. D. Morale, V. Capasso, K. Oelschläger, An interacting particle system modelling aggregation behaviour: from individuals to populations. J. Math. Biol. **50**, 49–66 (2005)
60. M.R. Owen, J.A. Sherratt, Pattern formation and spatiotemporal irregularity in a model for macrophage–tumour interaction. J. Theor. Biol. **189**, 63–80 (1997)
61. L.G. de Pillis, A.E. Radunskaya, C.L. Wiseman, A validated mathematical model of cell–mediated immune response to tumor growth. Cancer Res. **65**, 7950–7958 (2005)
62. A. Quartarone, Stability of solutions of some mesoscopic equations in Biomathematics. Ph.D. Thesis. University of Messina. Department of Mathematics and Informatics. Messina (2013)
63. L. Saint–Raymond, *Hydrodynamic Limits of the Boltzmann Equation, 1 Lecture Notes Mathematics*, vol. 1971, (Springer, Berlin 2009)
64. Z. Szymańska, Analysis of immunotherapy models in the context of cancer dynamics. Int. J. Appl. Math. Comput. Sci. **13**, 407–418 (2003)
65. Z. Szymańska, C. Morales Rodrigo, M., Lachowicz, M.A.J. Chaplain, Mathematical modelling of cancer invasion of tissue: The role and effect of nonlocal interactions. Math. Models Methods Appl. Sci. **19**, 257–281 (2009)
66. M.W. Teng, J.B. Swann, C.M. Koebel, R.D. Schreiber, M.J. Smyth, Immune–mediated dormancy: an equilibrium with cancer. J. Leukoc. Biol. **84**, 988–983 (2008)
67. C.M. Vajdic, van Leeuwen, M.T.: Cancer incidence and risk factors after solid organ transplantation. Int. J. Cancer. **125**, 1747–1754 (2009)
68. W. Wagner, A functional law of large numbers for Boltzmann type stochastic particle systems. Stochastic Anal. Appl. **14**, 591–636 (1996)
69. R.A. Weinberg, *The Biology of Cancer. Garland Sciences* (Taylor & Francis Group, New York 2007)

The Power of the Tumor Microenvironment: A Systemic Approach for a Systemic Disease

Irina Kareva, Kathleen P. Wilkie, and Philip Hahnfeldt

Abstract Cancer is increasingly recognized as not solely a disease of the genes and chromosomes but as a systemic disease that affects numerous components of the host including blood vessel formation, immune cell function, and nutrient recycling. This review summarizes a variety of time-dependent mathematical models that focus on the consequences of tumor growth within an evolving microenvironment, represented by a dynamic carrying capacity. Transcending the specifics of each model, their overview reveals that the key to tumor control really lies in controlling the support furnished the tumor by its microenvironment.

Keywords Tumor microenvironment • Mathematical modeling • Cancer systems biology • Tumor-promoting inflammation • Metronomic chemotherapy

1 Introduction

Despite the increase in targeted efforts that followed the National Cancer Act of 1971, where President Nixon declared a "war on cancer," the cure for cancer remains elusive. In the following 40 years, significant advances have been made for certain genetically simple cancers, such as acute lymphoblastic leukemia (ALL), a rare case where the cancerous clone can be therapeutically eradicated [1]. More genetically complex cancers, such as those of the breast, prostate, pancreas, and lung, however, are still not responsive to the majority of available treatments [1]. The limited therapeutic success of these cancers may be due to the fact that genetically complex cancers continually engage their microenvironment, creating a niche within the host, where abnormal cells have a competitive advantage over normal cells of the surrounding tissue [2].

Tumor microenvironments are dynamic heterogeneous systems influenced by many factors, including space and nutrient availability. Additionally, continuous

I. Kareva • K.P. Wilkie • P. Hahnfeldt (✉)
Center of Cancer Systems Biology, GeneSys Research Institute, Tufts University School of Medicine, Boston, MA 02135, USA
e-mail: Philip.Hahnfeldt@tufts.edu

© Springer Science+Business Media New York 2014
A. d'Onofrio, A. Gandolfi (eds.), *Mathematical Oncology 2013*,
Modeling and Simulation in Science, Engineering and Technology,
DOI 10.1007/978-1-4939-0458-7_6

interactions between somatic cells and immune cells influence the environment through both pro- and antitumor actions, depending on their context [3]. In this review we focus on three main biological processes that affect tumor growth: angiogenesis, tumor-promoting inflammation, and nutrient availability. Tumor growth is always accompanied by new blood vessel growth, or angiogenesis [4]. Since this discovery, angiogenesis has become a target for anticancer therapies [5, 6]. Inflammatory responses also accompany tumor growth and can promote tumor development from initiation to progression and metastasis [3, 7]. In light of these developments, anti-inflammatory drugs have been proposed as anticancer agents [8]. And finally, competition for nutrients such as phosphorus—used in ribosome synthesis—can modulate tumor growth and development [9, 10], and thus dialysis has been proposed after cytotoxic therapy to aid tumor control [11].

In order to successfully treat a cancerous lesion, the tumor's influence over its microenvironment needs to be understood, requiring thorough investigation of the constituent parts, both individually and as a whole. Mathematical modeling provides a useful conceptual tool to evaluate the relative importance of various components of the microenvironment in tumor growth dynamics, which can then be translated into more targeted and fiscally responsible experimental, and eventually clinical, investigations.

A frequent criticism of the mathematical modeling toolkit is that it results in little more than an intellectual exercise, lacking practical use since it cannot compare to the weight and importance of wet-lab experiments. One must remember, however, that the purpose of mathematical modeling is not to replace such experiments, but rather to extend them. A carefully constructed conceptual model can aid in evaluating the relative importance of major players within the experimentally observed system. Not only can modeling provide a more financially responsible framework to validate theories and design experiments, but it can also provide valuable negative results. Since the basic assumptions underlying a model are predetermined, if predicted results qualitatively contradict experimental observations, then the assumptions must be reevaluated. Such is usually easier to address in silico rather than in vivo or even in vitro, which makes mathematical modeling an important and powerful complement to experimental work, and it should be treated as such.

Mathematical models of tumor growth and cancer-immune interactions provide a framework within which the complex biological system can be analyzed. Several approaches have been applied to quantify this system, with perhaps the most common approach being ordinary differential equations [12–20]. Such time-dependent models provide valuable insight into the complex dynamics of the system and can be modified to consider stochastic effects [21, 22] and the evolutionary nature of the system [23, 24]. Other approaches allow incorporation of spatial effects with either partial differential equations [25, 26] or agent-based simulations [27, 28].

Here, we review several mathematical models of tumor growth, selected because of their ability to describe the continuous evolution of the tumor microenvironment. In these models, the microenvironment is represented as a dynamic carrying capacity that supports, and ultimately limits, tumor growth. Noticeably, rather than discuss specific details of each model, we present the general functional forms originally proposed by the corresponding authors and instead focus our discussions on the major features and conclusions derived from each framework. We then use these frameworks to explore the significance of various components of the tumor microenvironment, in particular, of angiogenic factors, tumor-promoting inflammation, and nutrient availability. To conclude, we discuss how these results could be used to influence and augment current approaches to cancer therapy.

2 A Model of Angiogenesis-Dependent Tumor Growth

We start by discussing a model of tumor growth by Hahnfeldt et al. [6], where the effective vascular support of the tumor microenvironment, or carrying capacity, grows in a time-dependent manner with the tumor mass. The tumor-associated vasculature is controlled by the production of angiogenic stimulators and inhibitors, in a similar fashion to healing wounds or organogenesis. That is, the growing mass of cancer and stromal cells produce both pro- and antiangiogenic factors that initially stimulate angiogenesis and ultimately limit the angiogenic capacity of the tumor environment.

To study this process quantitatively, they proposed a mathematical framework composed of two compartments: the tumor volume $V(t)$ and the carrying capacity of the tumor environment $K(t)$. The two compartments are coupled in that the tumor cannot exceed the size allowed by the capacity and the capacity is controlled by the tumor volume.

Mathematically, the tumor volume is assumed to grow according to a generalized logistic law. That is,

$$\frac{\mathrm{d}V}{\mathrm{d}t} = P(t)V(t) \qquad \text{where} \qquad P(t) = \frac{\lambda}{\alpha}\left[1 - \left(\frac{V(t)}{K(t)}\right)^{\alpha}\right].$$

This general law captures both of the most commonly assumed tumor growth models: logistic growth with $P(t) = \lambda\left(1 - \frac{V(t)}{K(t)}\right)$ if $\alpha = 1$ and Gompertzian growth with $P(t) = -\lambda_1 \ln\left(\frac{V(t)}{K(t)}\right)$ in the limit as $\alpha \to 0$.

The rate of change of the variable carrying capacity, $\frac{\mathrm{d}K}{\mathrm{d}t}$, should depend on the current level of environmental support, $K(t)$, the current tumor volume, $V(t)$, and time, t. Four proposed factors that may influence the capacity are an intrinsic loss rate, stimulatory and inhibitory angiogenic signals produced by cancer-stroma interactions, and inhibition due to antiangiogenic treatments. Mathematically, these can be interpreted as the four terms written below on the right-hand side:

$$\frac{\mathrm{d}K}{\mathrm{d}t} = f(K, V, t) = -\lambda_2 K + bS(V, K) - dI(V, K) - eKg(t).$$

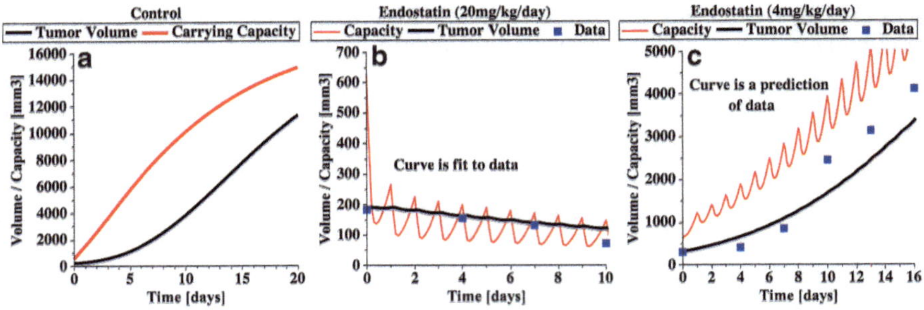

Fig. 1 A demonstration of the Hahnfeldt et al. [6] model for tumor growth in a dynamically growing carrying capacity. In (**a**), the carrying capacity grows with the tumor mass in the control (untreated) condition. In (**b**), parameters for the antiangiogenic drug endostatin were estimated from the experimental data at the dosage of 20 mg/kg/day. In (**c**) these same parameter values were used to predict the outcome of an additional experiment where the dosage was changed to 4 mg/kg/day. Excellent agreement of the model prediction to the experimental data is observed

Under the assumption that angiogenic stimulators have fast clearance rates while angiogenic inhibitors have slow clearance rates, the final functional form of

$$\frac{\mathrm{d}K}{\mathrm{d}t} = -\lambda_2 K + bV - dKV^{\frac{2}{3}} - eKg(t) \tag{1}$$

was proposed. While the exact form of the stimulator (bV) and inhibitor ($dKV^{\frac{2}{3}}$) terms can vary, the main conclusion of the work was that the ratio of the two terms should be proportional to the volume raised to the power $\frac{2}{3}$. Note that both tumor volume and the carrying capacity have units of volume. Details of the derivation can be found in [6].

The effects of antiangiogenic drugs on tumor growth are well characterized by this model because it specifically describes the angiogenic support of the tumor environment. As such, model validation by a series of experiments involving Lewis lung carcinoma cells in C57BL/6 mice with antiangiogenic treatments of TNP-470, angiostatin, or endostatin was performed. Assuming a Gompertzian growth law, tumor growth parameters (λ_1, λ_2, b, d, and $K(0)$) as well as treatment-specific parameters (e and $g(t)$) for the three drugs were estimated from experimental data.

The predictive power of this model was demonstrated by the excellent agreement between simulated growth dynamics and experimental data for drug dosages and combination therapies not used in the parameter estimation process (see Fig. 1). This quantitative theory may be useful for clinical determination of optimal drug and dosing protocols. Additionally, it has led to many theoretical investigations including generalizations of the theory for antiangiogenic therapy [29, 30], prediction of optimal antiangiogenic [31, 32] and combination of antiangiogenic-cytotoxic [33, 34] treatment protocols, spatial analyses for tumors with varying chemotherapeutic sensitivities [35], predictions of metastatic spread including angiogenesis [36], and investigations of the environmental regulatory effects on tumor growth and cancer-immune dynamics [37]. Overall, the concept of a dynamic tumor microenvironment, as captured by this model, has led to advances in our understanding of tumor growth dynamics and the consequences of cancer treatment.

3 Immune Predation in the Dynamic Tumor Microenvironment

The microenvironment of a tumor provides sustainability signals, such as nutrient and space availability, to the growing neoplasm. Cancer cells, however, can have varying sensitivities to these signals [38], especially across cancer type. Recent work by Wilkie and Hahnfeldt [37] demonstrates how varying sensitivities to these environmental regulatory signals can affect tumor dynamics and cancer-immune interactions, as well as result in significant variation in therapeutic outcome.

To do so, generalized logisitic growth is used in the mathematical formulation instead of Gompertzian or logistic growth. This introduces another parameter that represents the strength of the connection between the growing tumor and the carrying capacity. Additionally, they consider immune predation of cancer cells, resulting in the three-compartment model described below:

$$\frac{\mathrm{d}C}{\mathrm{d}t} = \frac{\mu}{\alpha}(1 + \Psi(I,C))C\left(1 - \left(\frac{C}{K_C}\right)^{\alpha}\right)$$

$$\frac{\mathrm{d}I}{\mathrm{d}t} = \lambda(1 + rC)\left(1 - \frac{I}{K_I}\right)$$

$$\frac{\mathrm{d}K_C}{\mathrm{d}t} = pC - qK_C C^{\frac{2}{3}}$$

Here, $C(t)$ represents the cancer population, $I(t)$ the immune population, K_I the constant carrying capacity of the immune population, and $K_C(t)$ the carrying capacity of the cancer population, which is considered to be either constant or dynamic (according to the Hahnfeldt et al. [6] model) in their analyses [37]. Immune predation of cancer cells occurs through the growth-modulating function $\Psi(I,C) < 0$.

With a constant cancer carrying capacity, parameter fitting to experimental data determines a fixed parameter set specifying the level of sensitivity to the regulatory signals provided by the environment. With a dynamic cancer carrying capacity, however, which has the ability to stimulate support at various rates, several sets of parameters were found specifying varying degrees of sensitivity, yet all fit the growth data equally well.

Environmental growth regulatory signals typically grow and evolve with, and in response to, the growing neoplasm. This process requires a conversion of the microenvironment from a normal state to a tumor-supporting niche. To expedite this conversion process, Hu et al. [39] co-injected cancer cells with either normal fibroblasts, tumor-associated fibroblasts, or pro-inflammatory arthritis-associated fibroblasts. They found that tumor weight may be enhanced by the tumor- or arthritis-associated fibroblasts compared to the normal fibroblasts or control, suggesting that inflammatory pro-tumor stromal cells may accelerate the conversion process within the microenvironment. In other words, the addition of pro-tumor inflammatory fibroblasts altered the growth regulatory signals produced by the environment and sensed by the cancer cells, resulting in accelerated growth.

Another example of dysregulation in environmental regulatory signaling can occur after chemotherapy or radiation therapy. Following such treatments, cancer regrowth can occur at rates up to 15–20 times faster than the rate observed pre-treatment [40, 41]. This accelerated regrowth may be partially due to tumor mass de-bulking without simultaneously targeting the environment, essentially leaving the pro-tumor behaviors of the stroma unaltered. Hence, residual cancer cells do not have to face the challenge and initial resistance associated with converting the environment to a tumor-supporting niche and thus can flourish in the already existing tumor-supporting environment.

With the above mentioned mathematical model, the phenomenon of accelerated regrowth was demonstrated by disrupting the environmental regulatory signals from those that grow with the cancer (a dynamic carrying capacity) to those that are already highly tumor supporting (a constant capacity equal to the maximum value for each parameter set). Interestingly, the sensitivity of each cancer (or parameter set) determined how the tumor would regrow following this disruption. Some sets predicted accelerated regrowth while others closely matched the original growth rate.

In terms of cancer-immune interactions, the immune-induced dormant state was shown to be essentially eliminated for cancers (or parameter sets) associated with high sensitivity to environmental regulation, and simulated immunotherapy treatments were shown to result in different outcomes depending on this sensitivity. The same therapy was predicted to result in the elimination for low-sensitivity cancers but growth and escape for high-sensitivity cancers. Disruption of the regulatory signals further altered the predicted outcomes of simulated immunotherapy, causing all cancers to grow large but at rates dependent on their level of sensitivity.

Important implications of this work are the potential explanation for why the same treatment may work for some patients but not others (cancer cell sensitivity to environmental regulation), and that primary versus follow-up treatments should not be expected to achieve the same outcomes (due to disruption of environmental signals). This work further supports the idea that to control a cancer, both the tumor and the surrounding environment should be targeted, simultaneously, to remove both the cancerous cells and the tumor-supporting environment.

4 Immune-Mediated Tumor Stimulation via the Dynamic Tumor Microenvironment

It is now well accepted that various immune cell types can stimulate cancer at every step from initiation to progression to metastasis [7, 42–45]. Immune cells are recruited to the tumor site by cytokines and danger signals that initiate an inflammatory response and thus can promote angiogenesis and tumor growth [3, 7, 46]. These dichotomous behaviors that immune cells exhibit confound the already complex system of cellular interplay that evolves in the tumor microenvironment.

A first attempt to mathematically investigate the dichotomous roles of immune cells within the tumor environment was recently undertaken by Wilkie and Hahnfeldt [47]. Based on the model for dynamic carrying capacity described above, the new framework incorporates both the cytotoxic effects and the proangiogenic effects of immune cells in a tumor microenvironment. The polarization of the grouped action for all immune cell types within the microenvironment is classified as either pro-tumor or antitumor, referring to the relative production of factors controlling angiogenesis, immunosuppression, and cancer cell predation.

The model modifies the Wilkie and Hahnfeldt framework described above to allow both immune and cancer population growth with dynamic carrying capacities [37]. The model equations below describe the dynamics of the four compartments: cancer cells $C(t)$, immune cells $I(t)$, the cancer carrying capacity $K_C(t)$, and the immune carrying capacity $K_I(t)$:

$$\frac{\mathrm{d}C}{\mathrm{d}t} = \frac{\mu}{\alpha}(1+\Psi(I,C))C\left(1-\left(\frac{C}{K_C}\right)^{\alpha}\right)$$

$$\frac{\mathrm{d}I}{\mathrm{d}t} = \frac{\lambda}{\gamma}(1+rC)\left(1-\left(\frac{I}{K_I}\right)^{\gamma}\right)$$

$$\frac{\mathrm{d}K_C}{\mathrm{d}t} = p(1+I)^a C^{1-a} - qK_C(1+I)^b C^{\frac{2}{3}-b}$$

$$\frac{\mathrm{d}K_I}{\mathrm{d}t} = xI^{\frac{1}{2}}C^{\frac{1}{2}} - yK_I I^{\frac{1}{3}}C^{\frac{1}{3}} - z(K_I - I_e).$$

Notice that the cancer carrying capacity now considers the angiogenic actions of immune cells, that is, $K_C(t) = f(K_C, C, I, t)$, and the cytotoxic actions are incorporated into the growth-modulating function $\Psi(I,C)$.

Parameters a and b control the immune polarization as discussed above. Pro-tumor immunity polarization is described by $a > b$ so that more weight is placed on the pro-angiogenic actions of immune cells than the antiangiogenic actions. Similarly, antitumor immunity polarization is described by $a < b$. In the immune carrying capacity, equal weight is placed on the immune and cancer cell actions to control the environmental recruitment signals for the immune response.

The antitumor and pro-tumor immunity polarizations predict different outcomes based solely on the level of angiogenesis-promoting inflammation. This is easily

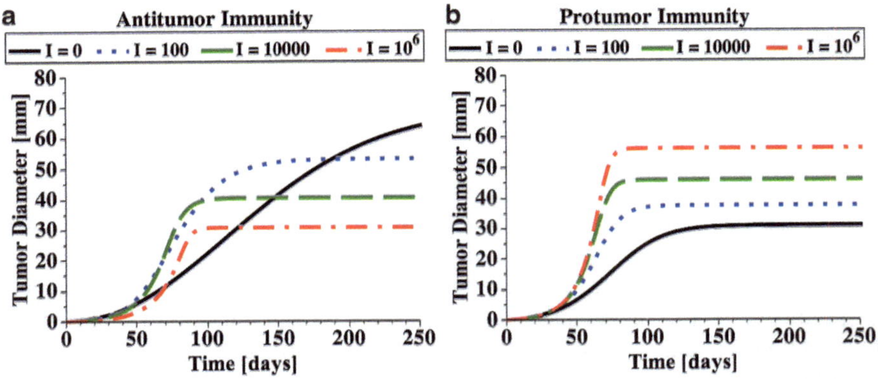

Fig. 2 A demonstration of the effect that antitumor (**a**) and pro-tumor (**b**) immunity can have on tumor growth. When present at constant levels, antitumor immunity causes a decrease in final tumor burden whereas pro-tumor immunity causes an increase in final tumor burden. Before these final states are achieved, however, both polarizations cause a period where immune presence increases tumor burden compared to the control (zero-immune) case. These results highlight the importance of time dynamics in the evolution of cancer-immune interactions as antitumor immunity may appear to enhance tumor growth in the short term while actually suppressing it in the long term

seen in the simulations without immune predation ($\Psi = 0$) in Fig. 2. With more weight placed on antiangiogenic functions in antitumor immunity, the final tumor burden is reduced compared to the control case where no immune cells are present. Conversely, with more weight placed on proangiogenic functions in pro-tumor immunity, the final tumor burden is enhanced compared to the control case. In both situations, however, there is an early period of growth where the presence of inflammation stimulates the tumor to grow faster than the control, regardless of whether the final tumor burden is reduced or enhanced.

This work has great clinical importance as it demonstrates how stimulating the immune response can result in an early period of enhanced tumor growth even if a later-stage reduction in growth is obtained. Such contradictory results have been observed in clinical trials for immunotherapies [48] where responses were classified into four distinct patterns: shrinkage of the tumor, stable disease, response after initial increase in burden, and response in the presence of new lesions. With chemotherapeutic treatments, shrinkage of the tumor is expected, differing substantially from the response patterns described above for immunotherapies. As such, different evaluation criteria are required for immunotherapies versus chemotherapies, which have only recently been established. The predictions of this model suggest that a response after initial increase should not be a surprising consequence of immunotherapy, but rather an expected outcome of stimulating an antitumor immune response. Furthermore, the results suggest that targeting the polarization of the microenvironment to transform a pro-tumor environment into an antitumor environment will lead to improved tumor suppression. Insights into cancer-immune dynamics gained through mathematical modeling of both tumor-promoting and tumor-inhibiting immune effects will lead to a greater understanding of the disease, of treating the disease, and of how therapeutic success should be evaluated.

5 Nutrient Availability in the Tumor Microenvironment

The dynamic carrying capacity can also be analyzed from the point of view of nutrient availability. Within such a paradigm, tumors are similar to ecological systems where heterogeneous populations of cells, such as can be found in most solid tumors, compete with somatic cells and with each other not only for space but also for nutrients. Moreover, in the context of limited nutrient availability, it is the appropriate allocation of the limited nutrients that may prove to be a crucial factor in whether the tumor will progress to malignancy or not.

Based on experimental observations, Elser [9, 10] has proposed what has become known as the growth rate hypothesis (GRH), which suggests that variations not in absolute amounts of carbon, phosphorus, and nitrogen in the cell's, or organism's microenvironment, but their ratios may affect the growth rate of the organism. Specifically, the GRH predicts that increased phosphorus availability can select in favor of more rapidly growing phenotypes, since additional nutrients can be available for phosphorus-rich ribosomal RNA, which is a requirement for rapid growth. This prediction was verified experimentally [49], as mice that were fed a phosphorus-rich diet had more advanced lung tumor progression and growth compared to mice kept on a phosphorus-poor diet.

The GRH also predicts that highly proliferative cells, such as actively growing tumor cells, should be characterized by a low intracellular carbon:phosphorus ratio, since most of the phosphorus would be allocated in favor of replicative machinery and specifically ribosomes. Interestingly, these predictions were confirmed experimentally [50] for cancers of the colon and the lung but not in the kidney or liver. These results raised the possibility that variations in microenvironmental conditions can provide a selective force towards either highly replicative cell clones in some cases, or slower-growing but perhaps more apoptosis-resistant cell clones in other cases.

In this case, phosphorus becomes a dynamic resource that may determine whether or not a more proliferative cancer clone comes to dominate the population. Phosphorus, much like many other nutrients that are consumed by growing cells, can be recycled through cell death, suggesting that tumors with high cell mortality rates may create conditions that naturally favor high proliferation. Moreover, it is also possible that changing phosphorus availability in the tumor microenvironment may be an additional factor that can drive cells out of the dormant state. Therefore, the implications of this hypothesis need to be thoroughly investigated, as they can have pivotal implications for determining the aggressiveness of a growing tumor.

Several models have been introduced to look at the effects of phosphorus as a dynamic resource on tumor dynamics. Kuang and colleagues [51] introduced a model where they study the dynamic interactions of a population of healthy cells $x(t)$ within the organ, tumor cells $y_i(t)$, where i represents the parenchyma cell type, tumor microvessels $z(t)$, particularly mature vascular endothelial cells, extracellular phosphorus $P_e(t) = g(P(t), x(t), y_i(t), z(t))$ and total phosphorus $P(t)$. The details of the derivation and full explanation and analysis of this particular complex and thorough model can be found in [51], however, the functional forms can be summarized as follows:

$$\frac{dx}{dt} = f_1(x(t), P_e(t)) - \text{death terms}$$

$$\frac{dy_i}{dt} = f_2(y_i(t), P_e(t), z(t)) - \text{death terms}$$

$$\frac{dz}{dt} = f_3(y_i(t), P_e(t)) - \text{death terms}$$

$$\frac{dP}{dt} = \text{inflow} - f_4(x(t), y_i(t), z(t)).$$

Similar to the previously discussed models, the model proposed by Kuang et al. considers a situation where the support of the tumor is dynamic. Unlike other models, however, the dynamic carrying capacity depends not only on the dynamics of the tumor cells, the stroma, and the angiogenic support but also on fluctuating nutrient availability, specifically phosphorus.

Through a number of simulations, the authors observed that, interestingly, the ultimate size of the tumor is insensitive to its predetermined carrying capacity but instead depends primarily on phosphorus supply. The authors emphasize the importance of stoichiometric constrains imposed by limiting nutrients by demonstrating a common feature of a variety of model variations that they explored: phosphorus supply plays a key role in affecting tumor growth dynamics and final size, even compared to cell birth or death rates, which, incidentally, are the primary focus of most therapeutic intervention strategies. Moreover, the authors demonstrate that over time, it is the slower-growing tumors with lower phosphorus demands that come to dominate rather than faster-growing tumors, lending further support to the idea of focusing less on cytotoxic treatments and more on targeting the microenvironment in which the tumor cells exist. The authors also raised the possibility that excessive phosphorous that may accumulate in the organ after any cytotoxic treatment could explain a phenomenon known as "tumor lysis syndrome," where the concentration of plasma phosphorus concentration can become toxic to the patient, in addition to providing remaining tumor cells with ribosome-building materials.

6 Nutrient Availability and Tumor Heterogeneity

The effects of phosphorus availability on the growth dynamics of heterogeneous tumors were further explored by Kareva [52]. More specifically, the analysis focused on the effects of changes in nutrient availability in the microenvironment and initial composition of the tumor on overall tumor dynamics. In this particular model, the author looked at the dynamics of tumor cells $x_\alpha(t)$, where each cell is characterized by a value of α from some initial distribution and where α represents a choice of strategy in terms of resource allocation. Other variables included were intracellular phosphorus $P_{in}(t)$ and extracellular phosphorus $P_{ex}(t)$, both of which directly influence cell growth and death rates. The functional form of the full system is given by

$$\begin{aligned}\frac{\mathrm{d}x_\alpha}{\mathrm{d}t} &= f_5(x_\alpha(t), P_{in}(t))\\ \frac{\mathrm{d}P_{ex}}{\mathrm{d}t} &= P_{\text{inflow}} - P_{\text{outflow}} - f_6(x_\alpha(t), P_{in}(t))\\ \frac{\mathrm{d}P_{in}}{\mathrm{d}t} &= f_6(x_\alpha(t), P_{in}(t)).\end{aligned}$$

This modeling framework focused on distinguishing the microenvironmental conditions that can lead to selection of corresponding alternative resource allocation strategies, expanding on the hypothesis proposed by Elser et al. [50]. Specifically, the analysis focuses on whether significant changes in phosphorus availability can influence selection towards or away from a more proliferative cell phenotype. Numerical simulations predict that tumor composition can evolve to be different depending on the initial state of the microenvironment and which might not be reflected in final tumor size. This suggests that tumor size is not necessarily a good predictor of final tumor composition and hence potential aggressiveness. It was also shown that modification of parameters pertaining to nutrient uptake rates did not affect tumor composition as much as expected, suggesting that blocking nutrient transporters might not be an effective therapeutic intervention strategy. Finally, sensitivity analysis revealed two important conclusions: phosphorus availability is a major factor in determining tumor growth dynamics, and different parameters gain relative importance at different stages of tumor growth. Specifically, while growth and nutrient consumption rates were of highest importance in the initial stages of tumor growth, cell death rates gain the highest importance at later stages of growth, potentially due to increased cell turnover and increased phosphorous availability through nutrient recycling.

While approaching the question of phosphorus availability in the tumor microenvironment from different viewpoints, together, these two models highlight the importance of nutrient availability as a major factor influencing tumor growth. It is particularly important that this type of dynamic resource is contained within the cells themselves and gets replenished through cell death, making tumors a self-sustainable ecosystem. To address this issue, the administration of dialysis after cytotoxic therapy was proposed [11] as a means to reduce an otherwise self-renewing carrying capacity.

All of these models, regardless of the manner in which they describe the dynamic tumor microenvironment (whether it be a space-, nutrient-, and/or angiogenesis-dependent viewpoint), suggest that extensive cell death may not be the optimal way to achieve long-term remission. Rather, they suggest that such strategies may select for more dangerous or aggressive cancers. In light of this collection of work, a rational conclusion is that the best chance of cancer elimination is through treatments simultaneously targeting both the cancer cells and the surrounding microenvironment.

7 Therapeutic Implications

Despite the large amounts of financial and intellectual resources that have been employed in the area of cancer research, and despite the large number of therapeutic agents that are currently available on the market as a result of these efforts, with a few notable exceptions, cancer mortality rates have not significantly decreased over the past 40 years [53]. One possible explanation for these results is a lack of recognition of the fact that cancer is a systemic disease, which cannot be successfully managed without considering its nature not solely as a disease of the genes but also of the microenvironment and host. Such an approach has been confirmed theoretically with several mathematical models, some of which were discussed above. Interestingly, these conclusions are supported by a variety of mathematical models, both descriptive and conceptual, regardless of the level of their complexity. These models suggest that improved clinical outcome requires particular attention be paid to not only the cancer cells but to the different components of the tumor microenvironment.

One approach, initially proposed by Folkman [4], is to target the tumor microenvironment via antiangiogenic therapy, which attempts to block the formation of new blood vessels in the hopes of starving the tumor. Unfortunately, patient response to such treatments has been more modest than anticipated [54]. Importantly, while antiangiogenic therapy has proven to be a relatively safe treatment, it has not demonstrated an increase in patient survival [54] despite promising results in preclinical testing. One possible explanation for these unexpected results is that deactivating the angiogenic factors by blocking the molecular receptors does not remove the ultimate source of the factors, namely, the tumor cells and associated stroma. Indeed, a combination of antiangiogenic therapy with chemotherapy has been shown to improve treatment response [55], possibly due to the simultaneous targeting of both the angiogenic signals and the source of the signals (the cancer cells themselves).

The most promising therapeutic approach proposed to date, however, is to alter the dosage and timing of standard chemotherapeutics to target the tumor-associated endothelial cells, which effectively determine the tumor carrying capacity. Such an approach has been termed "metronomic chemotherapy" and is characterized by more frequent administration of lower doses of cytotoxic agents. Hahnfeldt et al. [56] evaluated the predicted effectiveness of metronomic chemotherapy compared to the standard maximum tolerated dose protocol and demonstrated that, indeed, frequent administration of lower doses of cytotoxic drugs is the most efficient approach to achieve a smaller tumor burden. Specifically, metronomic chemotherapy reduces the emergence of therapeutic resistance since all the sub-populations within the tumor, whether initially sensitive to chemotherapy or not, are equally affected by the diminishing carrying capacity. Furthermore, they suggest that continuous administration of chemotherapy is superior to other forms of dose administration, an idea that has recently been challenged.

According to Doloff and Waxman [57], continuous administration of chemotherapy, while effective against the tumor, may damage the host's natural cytotoxic immune response and thus potentially diminish overall treatment effectiveness. Their experiments demonstrate that a 6-day cycle may yield the highest overall therapeutic effectiveness, first by achieving a slower but longer-lasting reduction in tumor volume due to the frequent administration of cytotoxic agents and second by preserving antitumor immunity due to an appropriate period of recovery between doses.

Finally, the work presented by Elser and colleagues [11, 50, 51] focuses on nutrient availability in the tumor microenvironment, specifically the amount of phosphorus required by growing cells to construct molecular machinery such as ribosomes. They suggest that cytotoxic therapies should be accompanied by treatments such as dialysis. This approach may serve the dual purpose of avoiding tumor lysis syndrome, where concentrations of liberated intracellular phosphorus may reach toxic levels, and removing recycled nutrients that may otherwise become available to remaining cancer cells.

One commonality amongst the studies discussed here is that cancer should be thought of as a systemic disease that becomes increasingly difficult to manage due to its ability to engage both the tumor microenvironment, via endothelial and other stromal cells, and the host (via the immune response). The microenvironment, which ultimately transforms to support tumor growth, can be polarized into a pro-tumor niche in a variety of ways, including pro-tumor inflammation, nutrient recycling, and recruitment of angiogenic stimulators. Therefore, the successful management of cancer, a complex and systemic disease, requires a systemic multifaceted treatment approach. It may be that only by addressing the changes that occur in the microenvironment and host as a result of cancer presence can one hope for improved tumor suppression and positive clinical outcomes.

Grant Support

This work was financially supported by the National Cancer Institute under Award Number U54CA149233 (L. Hlatky) and by the Office of Science (BER), US Department of Energy, under Award Number DE-SC0001434 (P. Hahnfeldt).

References

1. V.T. DeVita Jr., T.S. Lawrence, S.A. Rosenberg (eds.), *DeVita, Hellman, and Rosenberg's Cancer: Principles and Practice of Oncology*, 9th edn. (Lippincott, Williams & Wilkins, Philadelphia, 2011)
2. I. Kareva, What can ecology teach us about cancer? Transl. Oncol. **4**(5), 266–270 (2011)
3. K.E. deVisser, A. Eichten, L.M. Coussens, Paradoxical roles of the immune system during cancer development. Nat. Rev. Cancer **6**(1), 24–37 (2006)
4. J. Folkman, Tumor angiogenesis: therapeutic implications. New Engl. J. Med. **285**(21), 1182–1186 (1971)
5. M.S. O'Reilly, T. Boehm, Y. Shing, N. Fukai, G. Vasios, W.S. Lane et al., Endostatin: an endogenous inhibitor of angiogenesis and tumor growth. Cell **88**, 277–285 (1997)
6. P. Hahnfeldt, D. Panigraphy, J. Folkman, L. Hlatky, Tumor development under angiogenic signaling: a dynamical theory of tumor growth, treatment response, and postvascular dormancy. Cancer Res. **59**(19), 4770–4775 (1999)
7. S.I. Grivennikov, F.R. Greten, M. Karin, Immunity, inflammation, and cancer. Cell **140**(6), 883–899 (2010)
8. M.J. Thun, S.J. Henley, C. Patrono, Nonsteroidal anti-inflammatory drugs as anticancer agents: mechanistic, pharmacologic, and clinical issues. J. Natl. Cancer Inst. **94**(4), 252–266 (2002)
9. M. Boersma, J.J. Elser, Too much of a good thing: on stoichiometrically balanced diets and maximal growth. Ecology **87**(5), 1325–1330 (2006)
10. J. Elser, Biological stoichiometry: a chemical bridge between ecosystem ecology and evolutionary biology. Am. Nat. **168**(Suppl 6), S25–S35 (2006)
11. J.J. Elser, J.D. Nagy, Y. Kuang, Biological stoichiometry: an ecological perspective on tumor dynamics. Bioscience **53**(11), 1112–1120 (2003)
12. A.K. Laird, Dynamics of tumor growth. Br. J. Cancer **18**(3), 490–502 (1964)
13. C. DeLisi, A. Rescigno, Immune surveillance and neoplasia–1 a minimal mathematical model. Bull. Math. Biol. **39**(2), 201–221 (1977)
14. V.A. Kuznetsov, I.A. Makalkin, M.A. Taylor, A.S. Perelson, Nonlinear dynamics of immunogenic tumors: parameter estimation and global bifurcation analysis. Bull. Math. Biol. **56**(2), 295–321 (1994)
15. D. Kirschner, J.C. Panetta, Modeling immunotherapy of the tumor-immune interaction. J. Math. Biol. **37**(3), 235–252 (1998)
16. L.G. de Pillis, A.E. Radunskaya, C.L. Wiseman, A validated mathematical model of cell-mediated immune response to tumor growth. Cancer Res. **65**(17), 7950–7958 (2005)
17. A. d'Onofrio, A general framework for modeling tumor-immune system competition and immunotherapy: mathematical analysis and biomedical inferences. Physica D **208**(3–4), 220–235 (2005)
18. A. Cappuccio, M. Elishmereni, Z. Agur, Cancer immunotherapy by interleukin-21: potential treatment strategies evaluated in a mathematical model. Cancer Res. **66**(14), 7293–7300 (2006)
19. R. Eftimie, J.L. Bramson, D.J. Earn, Interactions between the immune system and cancer: a brief review of non-spatial mathematical models. Bull. Math. Biol. **73**(1), 2–32 (2011)

20. K.P. Wilkie, A review of mathematical models of cancer-immune interactions in the context of tumor dormancy. Adv. Exp. Med. Biol. **734**, 201–234 (2013). doi:10.1007/978-1-4614-1445-2_10
21. R. Lefever, W. Horsthemke, Bistability in fluctuating environments. implications in tumor immunology. Bull. Math. Biol. **41**, 469–490 (1979)
22. A. d'Onofrio, Bounded-noise-induced transitions in a tumor-immune system interplay. Phys. Rev. E Stat. Nonlin. Soft Matter Phys. **81**(2), 021923, 1–7 (2010)
23. A. d'Onofrio, A. Ciancio, Simple biophysical model of tumor evasion from immune system control. Phys. Rev. E Stat. Nonlinear Soft Matter Phys. **84**(3), 031910 (2011). doi:10.1103/PhysRevE.84.031910
24. K.P. Wilkie, P. Hahnfeldt, Mathematical models of immune-induced cancer dormancy and the emergence of immune evasion. Interface Focus **3**, 20130010 (2013)
25. A. Matzavinos, M.A.J. Chaplain, V.A. Kuznetsov, Mathematical modelling of the spatio-temporal response of cytotoxic T-lymphocytes to a solid tumour. Math. Med. Biol. **21**(1), 1–34 (2004)
26. T. Roose, S.J. Chapman, R.K. Maini Mathematical models of avascular tumor growth. SIAM Rev. **49**(2), 179–208 (2007)
27. H. Enderling, L. Hlatky, P. Hahnfeldt, Immunoediting: evidence of the multifaceted role of the immune system in self-metastatic tumor growth. Theor. Biol. Med. Model. **9**, 31 (2012). doi:10.1186/1742-4682-9-31
28. T. Takayanagi, H. Kawamura, A. Ohuchi, Cellular automaton model of a tumor tissue consisting of tumor cells, cytotoxic T lymphocytes (CTLs), and cytokine produced by CTLs. IPSJ Trans. Math. Model. Appl. **47**(1),61–67 (2006). doi:10.2197/ipsjdc.2.138
29. A. d'Onofrio, A. Gandolfi, Tumor eradication by antiangiogenic therapy: analysis and extensions of the model by Hahnfeldt et al. (1999). Bath Biosci. **191**(2), 159–184 (2004)
30. A. d'Onofrio, A. Gandolfi, A family of models of angiogenesis and antiangiogenesis anti-cancer therapy. Math. Med. Biol. **26**(1), 63–95 (2009)
31. U. Ledzewicz, H. Schättler, Antiangiogenic therapy in cancer treatment as an optimal control problem. SIAM J. Control Optim. **46**(3), 1052–1079 (2007)
32. U. Ledzewicz, H. Schättler, Optimal and suboptimal protocols for a class of mathematical models of tumor anti-angiogenesis. J. Theor. Biol. **252**(2), 295–312 (2008)
33. U. Ledzewicz, A. d'Onofrio, H. Schättler, Tumor development under combination treatments with anti-angiogenic therapies, in *Mathematical Methods and Models in Biomedicine*, ed. by U. Ledzewicz, H. Schättler, A. Friedman, E. Kashdan. Lecture Notes on Mathematical Modeling in the Life Sciences (Springer, Heidelberg, 2012), pp. 301–327
34. S. Benzekry, G. Chapuisat, J. Ciccolini, A. Erlinger, F. Hubert, A new mathematical model for optimizing the combination between antiangiogenic and cytotoxic drugs in oncology. C. R. Math. Acad. Sci. Paris **350**(1–2), 23–28 (2012)
35. T.L. Jackson, Vascular tumor growth and treatment: consequences of polyclonality, competition and dynamic vascular support. J. Math. Biol. **44**(3), 201–226 (2002)
36. S. Benzekr, Mathematical analysis of a two-dimensional population model of metastatic growth including angiogenesis. J. Evol. Equat. **11**(1), 187–213 (2010)
37. K.P. Wilkie, P. Hahfeldt, Tumor-immune dynamics regulated in the microenvironment inform the transient nature of immune-induced tumor dormancy. Cancer Res. **73**(12), 3534–3544 (2013)
38. G.N. Naumov, E. Bender, D. Zurakowski, S.Y. Kang, D. Sampson, E. Flynn et al., A model of human tumor dormancy: an angiogenic switch from the nonangiogenic phenotype. J. Natl. Cancer Inst. **98**(5), 316–325 (2006)
39. M. Hu, J. Yao, D.K. Carroll, S. Weremowicz, H. Chen, D. Carrasco et al., Regulation of in situ to invasive breast carcinoma transition. Cancer Cell **13**(5), 394–406 (2008)
40. H. Withers, Treatment-induced accelerated human tumor growth. Semin. Radiat. Oncol. **3**(2), 135–143 (1993)
41. S.Y. El Sharouni, H.B. Kal, J.J. Battermann, Accelerated regrowth of non-small-cell lung tumours after induction chemotherapy. Br. J. Cancer **89**(12), 2184–2189 (2003)

42. S. Kraus, N. Arber, Inflammation and colorectal cancer. Curr. Opin. Pharmacol. **9**(4), 405–410 (2009)
43. A. Mantovani, P. Romero, A.K. Palucka, F.M. Marincola, Tumour immunity: effector response to tumour and role of the microenvironment. Lancet **371**(9614), 771–783 (2008)
44. F.R. Balkwill, L.M. Coussens, Cancer: an inflammatory link. Nature **431**(7007), 405–406 (2004)
45. J. Condeelis, J.W. Pollard, Macrophages: obligate partners for tumor cell migration, invasion, and metastasis. Cell **124**(2), 263–266 (2006)
46. D. Nelson, R. Ganss, Tumor growth or regression: powered by inflammation. J. Leukocyte Biol. **80**(4), 685–690 (2006)
47. K.P. Wilkie, P. Hahnfeldt, Modeling the dichotomy of the immune response to cancer: cytotoxic effects and tumor-promoting inflammation (2013). ArXiv:1305.3634
48. J.D. Wolchok, A. Hoos, S. O'Day, J.S. Weber, O. Hamid, C. Lebbé et al., Guidelines for the evaluation of immune therapy activity in solid tumors: immune-related response criteria. Clin. Cancer Res. **15**(23), 7412–7420 (2009)
49. H. Jin, C. Xu, H. Lim, S. Park, J. Shin, Y. Chung et al., High dietary inorganic phosphate increases lung tumorigenesis and alters AKT signaling. Am. J. Respir. Crit. Care Med. **179**(1), 59–68 (2009)
50. J.J. Elser, M.M. Kyle, M.S. Smith, J.D. Nagy, Biological stoichiometry in human cancer. PLoS ONE **2**(10), e1028 (2007). doi:10.1371/journal.pone.0001028
51. Y. Kuang, J.D. Nagy, J.J. Elser, Biological stoichiometry of tumor dynamics: mathematical models and analysis. Discrete Contin. Dyn. B **4**(1), 221–240 (2004)
52. I. Kareva, Biological stoichiometry in tumor micro-environments. PLoS ONE **8**(1), e51844 (2013)
53. S.M. Gapstur, M.J. Thun, Progress in the war on cancer. J. Am. Med. Assoc. **303**(11), 1084–1085 (2010)
54. J.M.L. Ebos, R.S. Kerbel, Antiangiogenic therapy: impact on invasion, disease progression, and metastasis. Nat. Rev. Clin. Oncol. **8**(4), 210–221 (2011)
55. L. Bello, G. Carrabba, C. Giussani, V. Lucini, F. Ceruutti, F. Scaglione et al., Low-dose chemotherapy combined with an antiangiogenic drug reduces human glioma growth in vivo. Cancer Res. **61**(20), 7501–7506 (2001)
56. P. Hahnfeldt, J. Folkman, L.R. Hlatky, Minimizing long-term tumor burden: the logic for metronomic chemotherapeutic dosing and its antiangiogenic basis. J. Theor. Biol. **220**, 545–554 (2003)
57. J.C. Doloff, D.J. Waxman, VEGF receptor inhibitors block the ability of metronomically dosed cyclophosphamide to activate innate immunity-induced tumor regression. Cancer Res. **72**(5), 1103–1115 (2012)

Part III
Anti-Tumor Therapies

Modeling Immune-Mediated Tumor Growth and Treatment

Lisette de Pillis and Ami Radunskaya

Abstract The immune response is an important factor in the progression of cancer, and this response has been harnessed in a variety of treatments for a range of cancers. In this chapter we develop mathematical models that describe the immune response to the presence of a tumor. We then use these models to explore a variety of immunotherapy treatments, both alone and in combination with other therapies.

Keywords Tumor-immune interactions • Effector cell kill rate • Therapy optimization • Agent-based models • Immune response kinetics

1 Introduction

The simplest model of tumor growth assumes that cells undergo mitosis at a constant rate, resulting in a tumor population that grows exponentially. However, it is quickly apparent that this model is not consistent with clinical observation. As a thought experiment, consider a breast cancer cell, which is approximately 20 microns in diameter. If we assume a doubling time of two days, then after 26 doublings or 52 days, this single cell will have produced a mass of approximately 67 million cells, with a diameter of 8 mm—in other words a detectable tumor mass. After another 18 doublings or 88 days after the single cancer cell started dividing, the mass would be the size of a beach ball (of radius 25 cm).

Clinicians knew from experience that, in general, this was not a correct description of tumor growth, even in the absence of treatment. Tumor growth could be limited by many factors, an obvious one being a limited supply of nutrients, and several early mathematical models were proposed that account for the slowing

L.G. de Pillis
Harvey Mudd College, Claremont, CA 91711, USA
e-mail: depillis@hmc.edu

A.E. Radunskaya (✉)
Pomona College, Claremont, CA 91711, USA
e-mail: aradunskaya@pomona.edu

© Springer Science+Business Media New York 2014

A. d'Onofrio, A. Gandolfi (eds.), *Mathematical Oncology 2013*,
Modeling and Simulation in Science, Engineering and Technology,
DOI 10.1007/978-1-4939-0458-7_7

of tumor growth as a result of limits on the ability of the vasculature to deliver nutrients [10, 42, 55]. There was also ample evidence that the immune system plays a significant role in the containment of tumors.

The exact role of the immune system in fighting cancer is not known, although as early as 1908, scientists proposed that the immune system could prevent the progression of many cancers. In particular, in that year, Nobel laureate Paul Ehrlich deduced that, without the immune system's intervention, there would be many more cases of cancer than we observe [25]. Throughout the past century, the "immunosurveillance" hypothesis was tested and retested, with experimental results sometimes supporting the hypothesis, sometimes rejecting it [24]. In the past two decades, the overwhelming majority of evidence is in favor of Ehrlich's hypothesis, and researchers are now seeking ways to enhance the ability of the immune system to stop the progression of the disease [26].

One of the earliest attempts to harness the immune system's response was made by an oncologist, William B. Coley, in the late 1800s, who noticed that some of his patients with what he thought was incurable cancer would improve when they simultaneously had an infection. He manufactured a mixture of dead bacteria, and experimentally administered the brew, known as "Coley's toxins," to patients with inoperable tumors. This treatment was successful enough to encourage other doctors to follow suit throughout the following decades [56]. Other immunotherapy treatments for cancer include stem cell transplants, introduced in the 1950's, and the administration of immune-stimulating cytokines. A stem cell transplant involves harvesting immune cells from the bone marrow of healthy individuals and transferring them to patients with leukemia. The administration of immune-stimulating cytokines is a technique that was pioneered and developed by Dr. Steven Rosenberg to treat patients with melanoma [50]. For an excellent review of cancer immunotherapies, see [3].

The role of the immune response in the control of cancerous cells also caught the interest of the mathematical community. Over the past twenty years, physicists and applied mathematicians have developed mathematical models that describe the interactions between tumor cells and immune cells in an attempt to understand the mechanisms behind observed behavior and to help clinicians design effective treatments. The earlier models consider tumor cells and immune cells at the population level [33, 36, 44, 52]. For an excellent survey of these early models see the book [1]. Later models include spatial effects [4, 5, 41] or focus on optimization of specific immunotherapy treatments [6, 35]. General frameworks have been developed from a systems perspective that are applicable to a variety of specific situations [19, 20]. This chapter is not intended to be an overview of this impressive body of work; the interested reader is referred to the texts cited here and the references therein. Rather, we follow our own trajectory of investigation and discovery, presenting several models of tumor–immune interactions that illustrate a variety of approaches to understanding the progression of the disease and to harnessing the immune response in the context of treatment.

This chapter is organized as follows. In Sect. 2 we develop the simplest model of the immune response, which uses two ordinary differential equations to describe two competing populations: the immune cells and the tumor cells. We add chemotherapy to this simple model to illustrate the complexity in the resulting dynamics and to demonstrate *in silico* the importance of including the immune response in the design of treatment strategies. We add more realism to the model in Sect. 3 by distinguishing between the innate and the adaptive immune responses and describe several types of modeling techniques that can be used to explore this distinction. In the final section of the chapter, Sect. 4, we discuss immunotherapies, and give several examples of mathematical models that can be used to investigate immunotherapeutic protocols.

2 The Immune Response as One Population of Effector Cells

The immune system is a complex network of interacting cells, proteins and chemicals. This network consists of excitatory and inhibitory connections, positive and negative feedback loops, and delays. In the simplest mathematical model of tumor-immune interactions, we only consider those immune cells that have the ability to destroy antigen, or foreign cells. These include natural killer (NK) cells, cytotoxic T-cells (CTL) such as $CD8^{+}$ cells, macrophages, and other scavenger cells. As a first model, we lump all of these killer cells into one population called **effector cells**. We imagine that we are considering a small volume of tissue containing a tumor, we consider the tumor to be one homogeneous population of cells, and we assume that the interaction between tumor and effector cells can be described as an *average* affect. If the number of cells in each of the populations is large, we can describe the population as a continuum, and we can describe the evolution of the average using differential equations. We also include a population of normal host cells in this model, as a proxy for overall "well-being." Since a tumor cannot grow without bound, we assume that, in the absence of an immune response, the tumor will grow to some maximum size determined by the available nutrients. We assume the same for the normal cells. Several functional forms are used to model self-limiting growth in the literature, for example, logistic, Gompertz, or von Bertalanffy. In this formulation we use a logistic growth law for both normal and tumor cells. We note, however, that other growth laws produce qualitatively similar results. Further details of this model and an analysis of its long-term behavior can be found in [14] and [15].

We let $I(t)$ denote the number of effector immune cells at time t, $T(t)$ the number of tumor cells at time t, and $N(t)$ the number of normal, or host, cells at time t. A graphical representation of the model interactions is shown in Fig. 1.

The source of the immune cells is considered to be outside of the system, and we let s denote the constant influx of innate effector cells that would be present in the absence of a tumor. Furthermore, in the absence of any tumor, the cells will die off at a per capita rate d_1, resulting in a long-term population size of s/d_1 cells. Thus, immune cell proliferation will never suffer from crowding.

The presence of tumor cells stimulates the immune response, represented by the dashed arrow in the diagram. For biological realism, we assume here a saturation

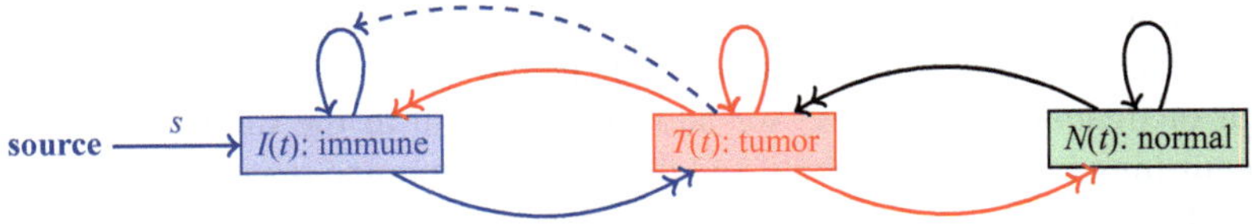

Fig. 1 A graphical representation of a population model of tumor-immune interactions, with three populations: immune effector cells (I), tumor cells (T), and normal cells (N). *Solid lines* indicate direct interactions, and *dashed lines* indicate indirect interactions. *Single arrow heads* denote a positive interaction, while double arrow heads denote an inhibitory interaction

limited effect. Furthermore, the reaction of immune cells and tumor cells can result in either the death of tumor cells or the inactivation of the immune cells, represented by two double-headed arrows.

The closed loop arrows on the tumor and normal cell population nodes represent normal growth and decay, which follows a logistic law. In addition there are two terms representing the competition between tumor and host cells, shown also as double-headed arrows in the diagram. Putting all the terms together gives the following system of ordinary differential equations:

$$\begin{aligned} \dot{N} &= r_2 N(1 - b_2 N) - c_4 T N, \\ \dot{T} &= r_1 T(1 - b_1 T) - c_2 I T - c_3 T N, \\ \dot{I} &= s + \frac{\rho I T}{\alpha + T} - c_1 I T - d_1 I. \end{aligned} \tag{1}$$

As shown in [15], this system has one "tumor-free" equilibrium at $(1/b_2, 0, s/d_1)$ and two "dead" equilibria, where the normal cell population is zero. Furthermore, the system can have one, two, or three "co-existing" equilibria, where all of the cell populations are nonzero, depending on the values of the parameters. Thus, in some parameter regimes, the system is *multistable*, where several stable equilibria exist at the same time, so that the long-term behavior of the system is determined by the initial conditions. We note that the concept of multistability is one of the few new ideas that biomathematics was able to offer the biomedical research community.

If the tumor-free equilibrium is stable, then small tumors will be eradicated by the immune system. A linearized stability analysis shows that this occurs when the *resistance coefficient* is larger than the intrinsic growth rate of the tumor, i.e., when

$$\frac{c_2 s}{d_1} + c_3 > r_1.$$

If a patient has a detectable tumor that is progressing, then we can assume that the tumor-free equilibrium is unstable. Two bifurcation diagrams are shown in Fig. 2.

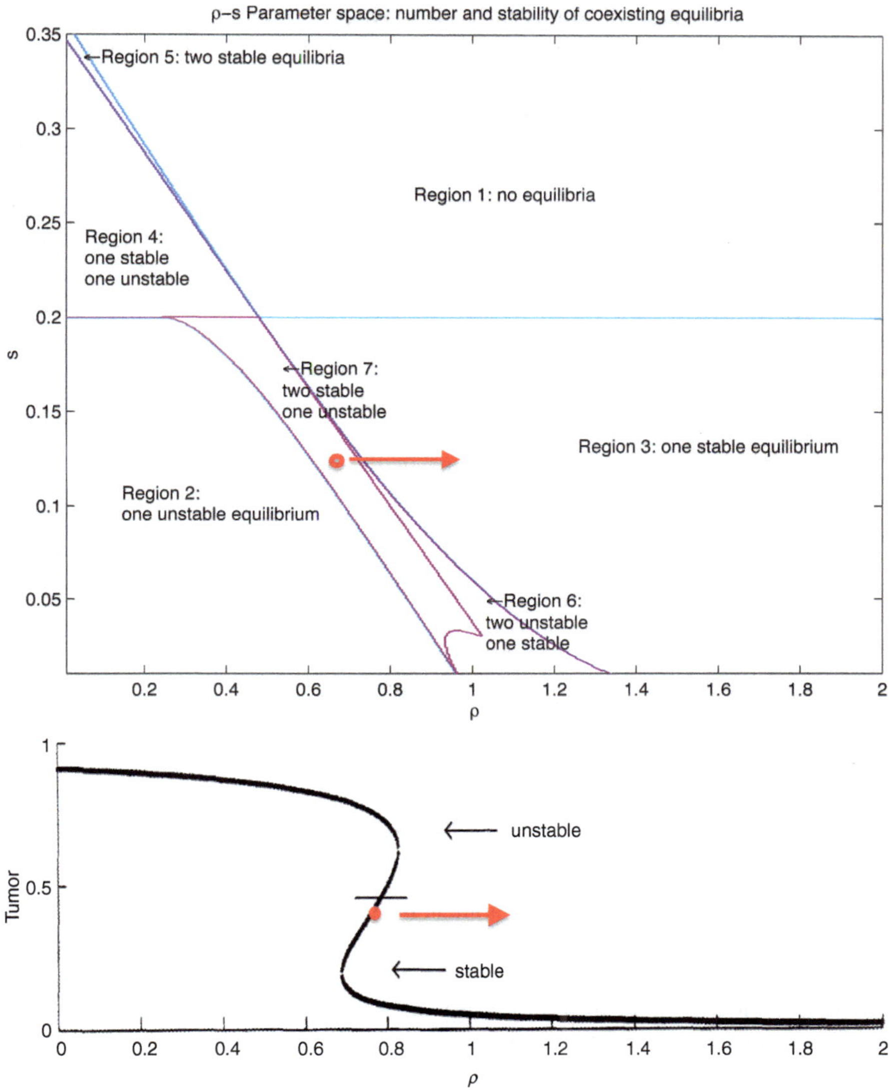

Fig. 2 Bifurcation diagram showing how changes in the immune source (s) and recruitment (ρ) parameters affect the number and stability of equilibria with nonzero tumor values. Note that the tumor-free equilibrium is not shown here: it is assumed to be unstable if there is a tumor. The red arrow indicates movement through the diagram as a result of a hypothetical treatment that enhances the immune response, such is the administration of interleukin 2 (IL2). Upper graph: Number and type of co-existing equilibria as a function of source rate, s, and immune response, ρ. Lower graph: Tumor cell populations at the equilibria as a function of the immune response rate, ρ. Stability of equilibria is indicated. Movement is from Region 2 through Regions 7 and 6 and finally into Region 3: as ρ increases from 0.1 to 2.0. Source rate $s = .12$, tumor populations as fraction of carrying capacity. See [15] for details

We can interpret these diagrams in the context of immunotherapy treatments as follows: treatment should move the system into a regime where it is attracted to a small, presumably harmless, tumor. If a patient has a detectable tumor that is progressing, we can assume that it is *not* in the basin of attraction of a stable, small-tumor equilibrium. Suppose the system is in Region 7, where it will be attracted to a relatively large tumor equilibrium (the dot in both graphs in Fig. 2). By administering cytokines that increase the immune response or by giving a vaccine that increased the immunogenicity of the tumor, the parameter ρ could be increased, moving the system to the right in the bifurcation diagrams (denoted by the arrows in both graphs). The system would then be in the basin of attraction of a relatively small tumor equilibrium, and the tumor would regress without further treatment.

We can also learn something about the effects of uncertainty in the environment by looking at the bifurcation diagram. For example, suppose the system is near the right boundary of Region 7 in Fig. 2, for example, near the point $s = .17$, $\rho = .6$. In this case, small fluctuations in the parameter s, the influx rate of effector cells in the absence of a tumor, could cause the system to move into Region 3. A reverse saddle-node bifurcation occurs where one stable equilibrium and one unstable equilibrium disappear, and the system would move towards the one remaining stable equilibrium. In this case, this would be beneficial, since the remaining equilibrium is at a point in state space with a small tumor population. The effect of stochastic fluctuations in the parameters has been discussed in the context of tumor–immune models in, for example, [7], and the effects of random fluctuations on resistance to chemotherapy is treated nicely in [21].

2.1 The Immune Response and Chemotherapy

In a scenario known as "Jeff's phenomenon," it has been clinically observed that tumors treated with cytotoxic chemotherapy can respond in a non-intuitive way. For some patients, after one treatment the tumor will shrink, and after another it might continue to grow, resulting in a temporal oscillation that is asynchronous with the chemotherapy. This phenomenon is reported in [59], where it is argued that this asynchronicity cannot be explained solely by acquired drug resistance. We hypothesize that it is the interaction of the chemotherapy with the immune response that could explain Jeff's phenomenon and test the hypothesis by adding a chemotherapeutic term to the model.

We assume that the rate of change of the concentration of drug at the tumor site, $u(t)$, can be described by a time-varying input function, $v(t)$, representing the administration of the drug, and by an elimination rate, d_1. We also assume that the drug kills all three types of cells in the model at a saturating rate, but that it acts preferentially on the more quickly dividing tumor cells and immune cells than on the normal host cells. A graphical representation of the system is shown in Fig. 3, with edges terminating in open circles denoting an inhibitory (killing) effect.

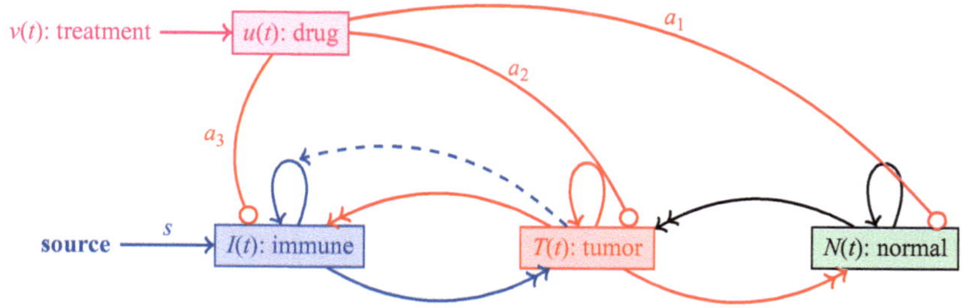

Fig. 3 A graphical representation of the model with chemotherapy. *Solid lines* represent direct interactions, with single arrowheads representing cooperative interactions, double arrowheads representing competitive interactions, and open circles representing a killing effect. *Dashed lines* represent interactions that affect the rate of another interaction

These assumptions result in the following system of equations (see [15] for parameter values and more details):

$$
\begin{aligned}
\dot{N} &= r_2 N(1 - b_2 N) - c_4 T N - a_1(1 - e^{-u})N,\\
\dot{T} &= r_1 T(1 - b_1 T) - c_2 I T - c_3 T N - a_2(1 - e^{-u})T,\\
\dot{I} &= s + \frac{\rho I T}{\alpha + T} - c_1 I T - d_1 I - a_3(1 - e^{-u})I,\\
\dot{u} &= v(t) - d_1 u.
\end{aligned}
\tag{2}
$$

where $a_1 < a_3 < a_2$. A simulation of this model demonstrating Jeff's phenomenon is shown in Fig. 4. Thus, the immune response could play a role in the delayed response of some patients to cytotoxic chemotherapy. This simple model also suggests that a close monitoring of the state of the cellular immune response could help in designing more effective treatment protocols.

With the relatively simple model given by Equation 2, we can attempt to answer the question: what is the best treatment regimen for a patient with a specific parameter set? As a first step, we must define what we mean by "best." One criterion might be "the one that minimizes tumor size at the end of treatment" and another might be "the one that is least toxic." Once the criteria are settled, optimization techniques can be applied to the system to propose effective treatment protocols. For example, suppose we wish to minimize the tumor burden after 45 days of treatment, while keeping the tumor population as low as possible and keeping the

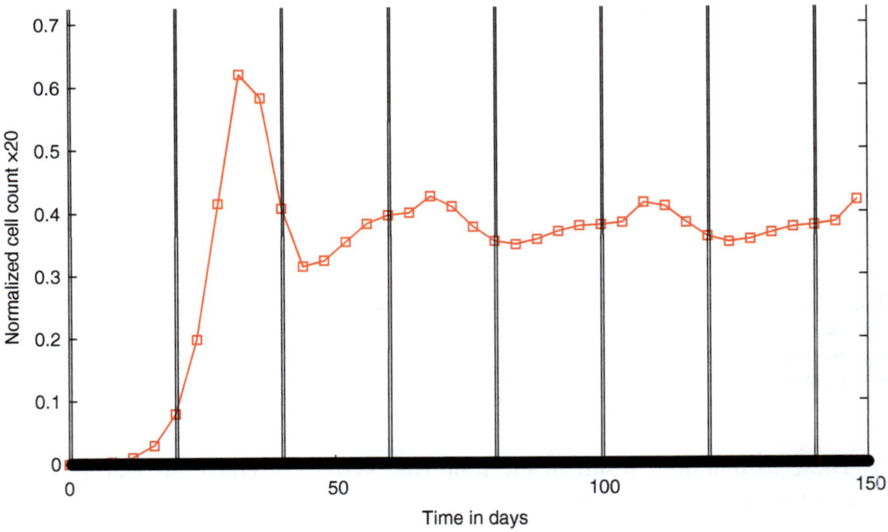

Fig. 4 Simulation of the model in system 2 demonstrates a possible role of the immune response in an asynchronous response to chemotherapy. Vertical lines show the simulated bolus injections of the drug, administered every 21 days

level of normal cells above 75 % of their normal value (a measure of toxicity). In terms of an optimization problem, we want to find the function $v(t)$, representing the administration of the drug, that minimizes the following *cost* function, where t_f is the total time of treatment.

$$J(v(t)) = w_1 T(t_f) + w_2 \int_0^{t_f} T(t)dt + w_3 \max_{t \in (t_0, t_f)} T(t), \tag{3}$$

where w_i are weighting constants. Note that three terms were required in the cost function: the first reflects our desire to minimize the tumor at the end of treatment, t_f. The second reflects our desire to minimize the total tumor present over the course of treatment, and the third term puts a penalty on any treatment that results in a large tumor at any point. The omission of either of the final two terms yields solutions with tumor populations that grow very large for a short period of time. The weighting of the three terms also yields qualitatively different results. For the set of experiments we present here, we set $w_1 = 1500, w_2 = 150, w_3 = 1000$, but other weightings might be preferred, depending on the type of tumor.

To reflect our desire to avoid excessive toxicity, we introduce a *constraint* function:

$$N(t) \geq 0.75, \quad 0 \leq t \leq t_f, \tag{4}$$

where the host cell population, $N(t)$, is scaled to a fraction of its normal value. Additional constraints are that all state variables must satisfy Equation 2 and that both the rate of drug input and the total amount of drug administered are bounded: $0 \leq v(t) \leq \max_v$, $0 \leq \int_0^{t_f} v(t)\,dt \leq v_{TOT}$.

This optimization problem can be solved using a variety of available techniques. In Fig. 5 we compare simulations using a traditional, "pulsed" protocol, where the drug is administered over short (12 h) periods, repeated every 2 days for 40 treatments (so the last treatment ends midday on Day 80). In this experiment, we simulate a patient with a relatively weak initial immune system ($I(0) = .1$) and the simulation shows that the traditional pulsed treatment is ineffective in the long term: once treatment stops, the tumor continues to grow, and the disease progresses. In the right panel of Fig. 5 we show a solution obtained using a direct collocation method, DIRCOL [57]. We required that the total amount of drug administered be no more than the total in the traditional case, so $v_{TOT} = \sum_{n=1}^{40} \int_0^{.5} 1\,dt = 20$. The optimized protocol suggests that the drug be administered over longer periods of time, on the order of days, with irregularly spaced treatments. In fact, it suggests one very short pulse of chemo at Day 125. With this treatment the tumor burden is driven to near zero by Day 70, and it remains there for the duration of the simulation. It is worth emphasizing that the only difference between the two treatments is the *timing* of the doses: the total amount of drug, and the maximum drug given are the same.

There are many possible optimization questions that could be asked in this setting. For example, it is possible that, by adding the total amount of drug used to the cost functional, one could find treatment protocols that are equally effective but that use less drug. Or it might be desirable to introduce a penalty term that curbs the destruction of the immune population. In Sect. 4 we will explore other optimization techniques and results in the context of designing cancer vaccines.

3 The Innate and Adaptive Immune Response

The human body has a huge army of defender cells, generally known as *white blood cells* (WBC) or *leukocytes*. It creates approximately a billion of these cells each day. A subset of these leukocytes are *lymphocytes*, which comprise 20–30 % of the WBC. In this section we will focus on two types of lymphocytes: the natural killer (NK) lymphocytes and the cytotoxic T lymphocytes (CTLs). Both of these cells are *cytotoxic*, meaning that they kill antigen or "nonself" cells. However, they belong to two different arms of the immune response: NK cells belong to the *innate* immune system. They form part of the immune system's regular patrol, and they are activated to *lyse*, or kill, cells that they encounter when that cell does not have a high expression level of certain molecules known as MHC I (major histocompatibility complex class I). The CTLs are part of the *adaptive* immune system. These cells originate from stem cells that then migrate to the thymus (hence the "T").

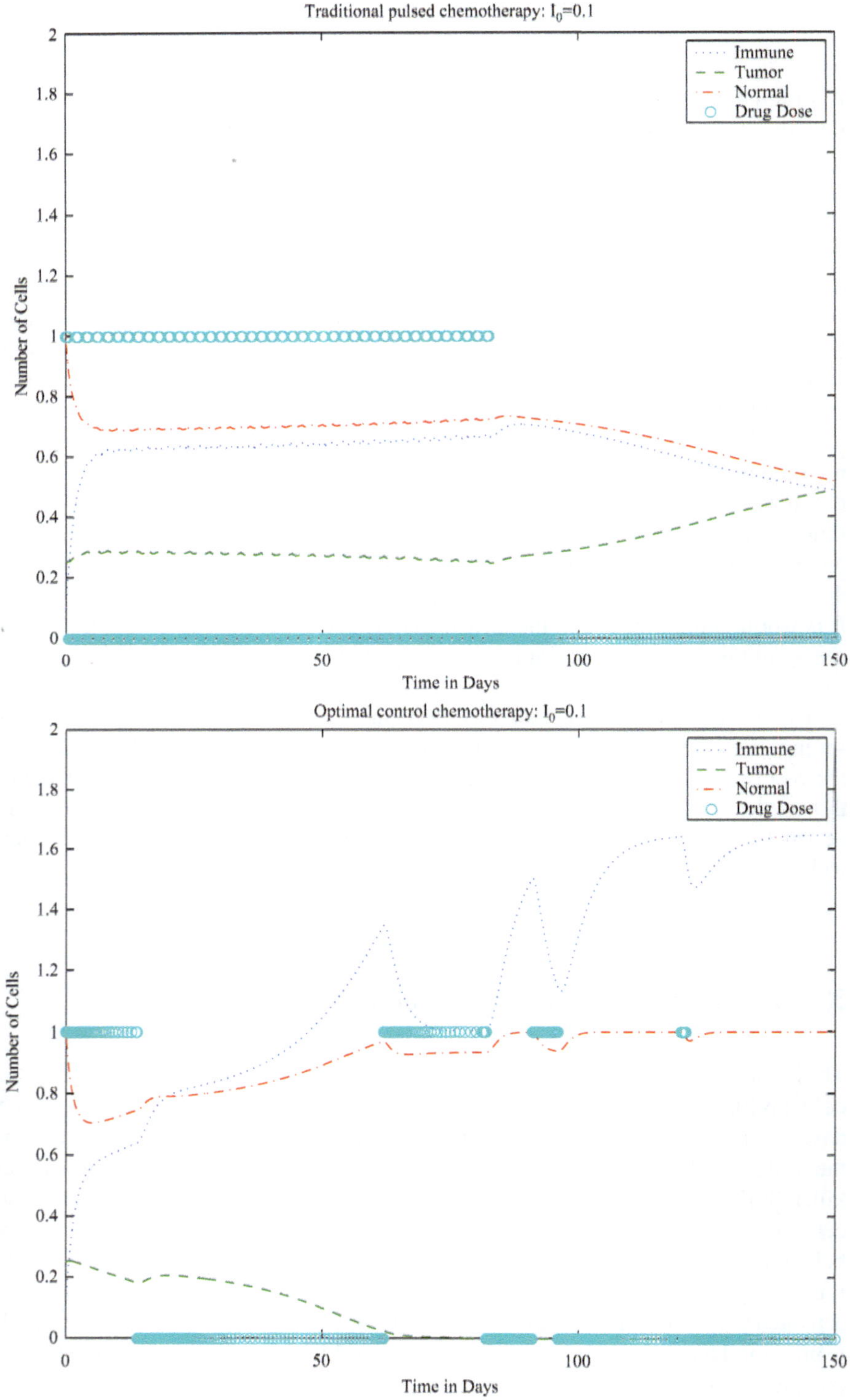

Fig. 5 (continued)

From there they are recruited to various lymph organs. When a nonself cell, or antigen, is encountered by a certain type of roving immune cells called *antigen-presenting* cells (APCs), they are engulfed, and pieces of the foreign cell are "presented" to the T-cells, activating them and causing them to proliferate into one population that can recognize and kill that particular type of foreign cell. Figure 6 gives a sketch of this process, where the antigen-presenting cell is a *macrophage*. Once the CTL is activated, it will seek out the specific antigen for which it is trained. Some activated CTLs will become memory cells, providing immunity for a second attack by the same foreign cells.

Note that other T-cells known as "helper T-cells" are also activated in this process, and these T-cells participate in the activation of the CTLs or "killer T-cells." Helper T-cells will appear in our models later, in Sect. 4.1. Helper T-cells also activate B-cells, which are key players in the *humoral response*, that part of the immune response that is mediated by antibodies. Another important class of APCs is the *dendritic cells* (DCs), which are now being used in the development of cancer vaccines. These will be discussed in Sect. 4.2. In terms of the body's fight against cancer, both NK cells and CTLs act like predators, but their methods of recognizing—and killing—their prey are different. As part of the innate immune system, natural killer cells are cytotoxic cells that are highly effective in lysing multiple (but specific) tumor cell lines [43]. Unlike cells of the specific immune system, which are drawn to a location due to the presence of antigen, the natural killer cells are constantly present guarding the body from infection and disease.

On the other hand, cytotoxic T lymphocytes are able either to lyse or to induce apoptosis in cells presenting *specific* antigens, such as tumor cells [43]. Unlike NK cells, CTLs are only able to recognize a specific antigen or tumor cell line. It is known, however, that these cells are able to destroy more than one tumor cell during their life cycle while a single natural killer cell generally kills very few [36]. After destroying the target cell, the CTLs move on in search of other antigen-presenting cells.

3.1 The dePillis–Radunskaya Law

In the fight against cancer, both the innate and the adaptive arms of the immune response are important. In fact, laboratory experiments show that without both NK cells and CTLs, tumors injected into mice will escape the immune surveillance

Fig. 5 *Left*: A patient with a relatively weak initial immune population ($I(0) = .1$ in normalized units) shows progressive disease after a series of pulsed chemotherapy treatments. The bolus treatments are simulated as injections at the maximum rate (normalized to $\max_v = 1$) for 12 h, repeated every 2 days for the course of the treatment. *Right*: A solution to the optimization problem given in Equation 3. The total amount of drug is the same in both the left (traditional treatment) and right (optimized treatment) simulations. The optimized treatment protocol is successful in eliminating the tumor

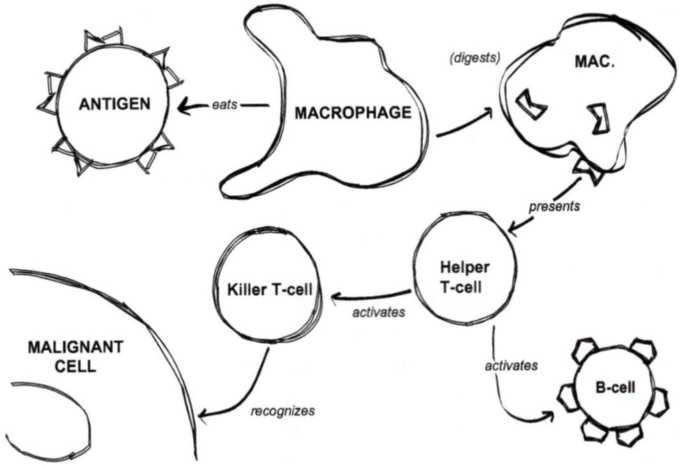

Fig. 6 A graphical representation of the activation of the adaptive immune response. A macrophage, a type of antigen-presenting cell (APC), recognizes a particular cell as "nonself" or antigen and engulfs it. The APC then presents bits of the engulfed, or "phagocytosed" cell to immature T-cells, which then begin to proliferate, activating other immune cells and, ultimately, the killer T-cells, or CTLs, and recognize and kill malignant cells of the same type as the initial antigen

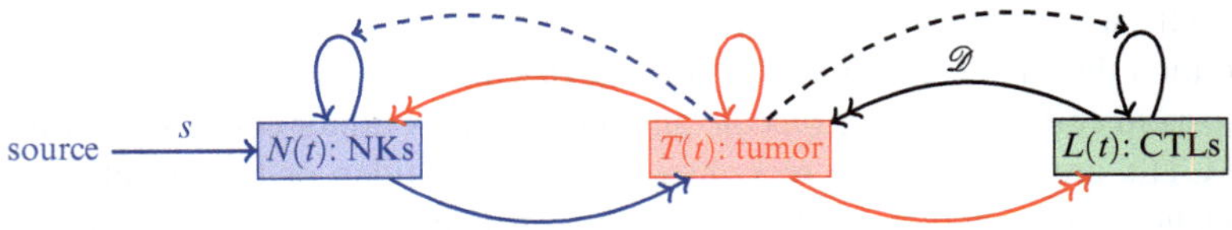

Fig. 7 Schematic of the model with two types of effector cells: natural killer cells (N), representing the innate immune response, and cytotoxic T lymphocytes (L), representing the adaptive immune response. As before, the solid lines represent direct interactions, with a single arrow-head denoting a cooperative interaction, and a double arrowhead denoting a competitive interaction. The dashed lines represent indirect interactions, where one population affects the *rate* of another interaction

(e.g., [18]). We therefore separate the effector cell population from the previous model into two subpopulations: the NK cells and the CTLs. Without the host cells, the model diagram becomes that shown in Fig. 7.

In developing the model, we assume again that the tumor grows logistically, that the NK cells, as part of the innate immunity, have a constant source, that immune cell proliferation is enhanced by the presence of the tumor, and that immune cells and tumor cells interact competitively. Furthermore, we know that the destruction of tumor cells by NK cells results in an increased uptake of antigen by antigen-presenting cells and, hence, an increase in the number of tumor-specific CTLs that are produced.

In the previous model given by System 1, competition between effector immune cells and tumor cells was represented by a mass action term of the form $-cIT$.

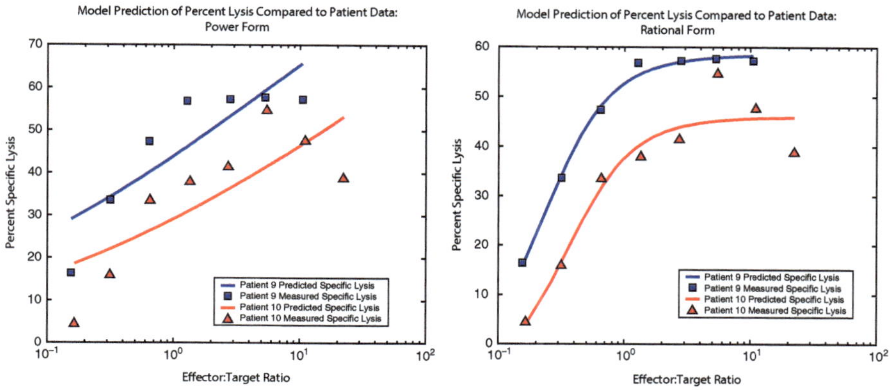

Fig. 8 *Left*: best fit kill rates using a power law for the per-cell kill rate by CTLs. *Right*: best fit kill rates using a ratio-dependent per-cell kill rate by CTLs. *Solid lines*: model simulations, ▲ and ■: data for two patients from [23], Figure from [16].

This term reflects an assumption that the number of encounters between the two cell populations is proportional to the product of the two populations, i.e., that all immune–tumor cell pairs are equally likely to occur, and that each of these encounters has an equal chance of resulting in the death of the tumor cell. A more general assumption might be that it takes n immune cells to kill one tumor cell, in which case the competition term would take the form $-cI^nT$. However, in trying to fit experimental data to a power law for the kill rate, it became apparent that the kill rate by CTLs does not follow this function form. Figure 8 shows the best fit curves to data from [23] using both a power law and a saturating *ratio-dependent* kill rate of the form: $\mathscr{D} = -d\dfrac{(L/T)^{\lambda}}{k_{1/2} + (L/T)^{\lambda}}T$. The kill rate for NKs does, on the other hand, fit a power law quite well. This gives a mathematical distinction between the functional forms for the kill rates of the two types of effector cells and has become known as the *dePillis–Radunskaya Law*. See [16] for details.

In this model we continue to denote the tumor cell population by T. We now let N denote NK cells (rather than normal cells) and L denote the number of cytotoxic T lymphocytes. We include in our equations one more distinction between the innate and the adaptive response. Since the NK cells recognize antigen directly, the response term in the NK equation (Equation 6) depends explicitly on the tumor population: $\dfrac{gT^2}{h + T^2}N$. However, as depicted in the graphic in Fig. 6, CTLs are activated by a cascade of immune events, including antigen presentation by APCs and activation of helper T-cells. Since tumor cells must be lysed before the APCs can initiate the activation cascade, the response term in the CTL equation (Equation 7) depends on the kill rate of tumor cells which in turn is given by the sum of the ratio-dependent term from the dePillis–Radunskaya Law, $\mathscr{D}$, and the power law kill rate by NK cells, denoted by rNT.

To complete the model, we include a saturating response term. The model system is

$$\frac{dT}{dt} = aT(1-bT) - cNT - \mathscr{D} \tag{5}$$

$$\frac{dN}{dt} = s - fN + \frac{gT^2}{h+T^2}N - pNT \tag{6}$$

$$\frac{dL}{dt} = -mL + \frac{j\mathscr{D}^2}{k+\mathscr{D}^2}L - qLT + rNT \tag{7}$$

where

$$\mathscr{D} = d\frac{\left(\frac{L}{T}\right)^{\lambda}}{k_{1/2} + \left(\frac{L}{T}\right)^{\lambda}}T, \tag{8}$$

With this model, we can now study the response of the system to variations in the two types of immune response. A sensitivity analysis shows that the final size of the tumor is most sensitive to the model parameter, d, which represents the maximum kill rate by CTLs. On the other hand, the size of the tumor after 40 days is affected by small changes in the parameter c, representing the strength of the kill rate by NK cells. This suggests that treatments should focus on increasing the number of CTLs, and enhancing their effectiveness and that the innate immune system cannot, by itself, control tumor growth.

The mechanisms that lead to the different kill rate laws are still unknown. It is likely that spatial effects are important, since CTLs are trained to "seek and destroy" specific antigens, while NK cells move randomly through the body. In the next section we introduce a spatial component into the model.

3.2 Adding a Spatial Component: Agent Based Models

One way to address questions about the effects of the spatial distribution of tumor and immune cells is to employ a hybrid cellular automata–partial differential equation modeling approach. The model we describe in this section is detailed in [39], and accounts for the spatiotemporal and stochastic interactions between individual tumor cells and populations of CD8+T (also known as CTLs) and NK cells while the tumor is in the pre-vascular stage.

We initialize the model with a cluster of tumor cells in a two-dimensional space of extra-cellular matrix (ECM). Nutrient diffuses from nearby blood vessels through the space to the tumor cells. In subsequent hybrid CA models [12, 22], we incorporated nutrient delivery through blood vessel sources interspersed throughout the computational domain. In the model presented here, we focus on the early pre-vascular stage, representing a small tumor burden, such as a postoperative remnant or a small satellite colony originating from a resected tumor.

As a result of the external nutrient supply, avascular tumors often develop into compact, nearly spherical structures. In these cases, the growing tumor generally develops three distinct layers—the proliferative rim, which is an outer shell of dividing cells that have direct access to nutrients that have diffused through the tissue; the quiescent layer, which is an inner layer of cells that have insufficient nutrient to allow them to divide, but enough to keep the cells alive; and a central core of necrotic cells that have died because nutrient concentrations are too low to maintain cell life. Our model simulations are able to produce a variety of tumor growth outcomes, including spherical and papillary (branchy) growth, stable and unstable oscillatory growth, proliferating and quiescent layers with a necrotic core, and lymphocyte-infiltrated growth. Lymphocyte infiltration is of particular interest given the experimental research that suggests improved survival rates for patients with intratumoral immune cells [60]. Infiltration of T-cells into the tumor mass can also lead to fibrosis and necrosis and subsequently reductions in tumor size [51, 54]. Numerical simulations of this model are in qualitative agreement with the experimental results demonstrated by, for example, Zhang et al. [60], Schmollinger et al. [51], and Soiffer et al. [54].

3.2.1 Hybrid PDE–CA Model Overview

Our model tumor grows on a two-dimensional square domain representing a patch of tissue that is supplied with nutrients by blood vessels that occupy the top and bottom boundaries, as shown in Fig. 9. The remainder of the space is partitioned into a regular grid in which the various cell types reside, and through which the nutrients diffuse. The grid is partitioned in such a way that each cellular automata grid

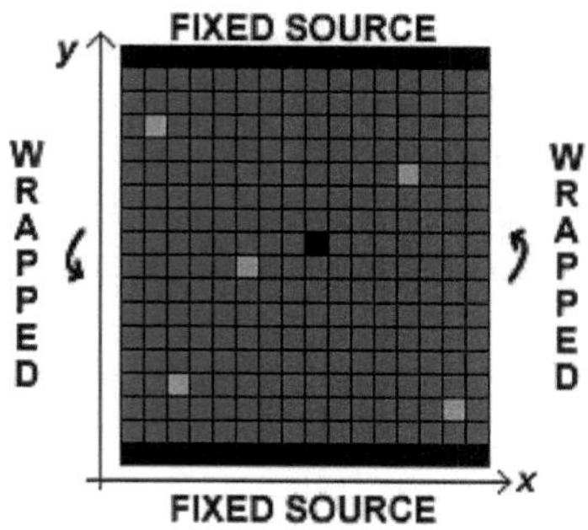

Fig. 9 Schematic of the cellular automata physical domain. The conditions imposed on the four boundaries are indicated. The solid bars (*top* and *bottom*) represent the capillaries while different cell types are shown filling the spaces in the grid

element corresponds in size with the actual biological cells of interest (10–20μm [2, 38]). We simulate the time progression of the system using two main steps. First, we solve the reaction–diffusion equations for the nutrient species, then with dependence upon the new nutrient fields, the actions of the cells (such as migration and proliferation) are carried out. Each iteration therefore corresponds to the period of tumor cell division. That is, a period of approximately 0.5–10 days, depending on the cell type in question [33, 49].

We include two representative nutrient species (such as glucose and oxygen [28, 58])—the first nutrient, N, being a necessary component of the cell division processes, while the second, M, is essential for the cell to survive. The nutrients diffuse throughout the tissue space, and as they do so, they are consumed by the different cells that are resident in tissue.

The nondimensionalized reaction-diffusion equations for the two nutrients are

$$\frac{\partial N}{\partial t} = \nabla^2 N - \alpha(H + I)N - \lambda_N \alpha T N, \tag{9}$$

$$\frac{\partial M}{\partial t} = \nabla^2 M - \alpha(H + I)M - \lambda_M \alpha T M, \tag{10}$$

The cell species are identified by H for host cells (normal tissue), T for tumor cells, and I for immune cells. Also, α[1] is the normal rate of consumption of nutrient by host and immune cells, and $\lambda_N \geq 1$ and $\lambda_M \geq 1$ determine the excess consumption by the tumor cells of the two types of nutrient. We impose Dirichlet boundary conditions at the top and bottom of the domain, to represent the constant nutrient source coming from the blood vessel. The right and left edges of the domain are subject to periodic boundary conditions.

The evolution of the four cell populations proceeds according to a combination of probabilistic and direct rules. A summary of the action and interaction of the cell types follows.

Host cells: The host cells are considered passive: other than their consumption of nutrients, they allow tumor cells to freely divide and migrate.

Tumor cells: The tumor cells in this model can divide, die, or migrate in space. These processes depend upon nutrient levels, the relative abundance of cells of the immune system, and crowding due to the presence of other tumor cells. Tumor cells can die either because of insufficient nutrient levels or from active killing by immune cells.

Immune cells: CTL cells are recruited to the tumor location when natural killer cells lyse tumor cells or when CTLs and tumor cells interact. A single CTL is able to lyse more than one tumor cell [36], a feature reflected in our model.

[1] In references [39] and [27], the authors choose to use α^2 instead of α, since α^2 reflects the squared form of the dimensional terms it replaces. For clarity, we simply use α here.

The immune cell actions in the model are summarized as follows: NK cells are generated at a rate to keep levels in constant proportion to the total number of cells in the domain, CD8+T cells are recruited to the tumor site, both NK and CD8+T cells can move through the computational domain, both NK and CD8+T cells can kill tumor cells, and both NK and CD8+T cells can die, either through deactivation by encounters with tumor cells or through apoptosis.

The cellular automata grid is initiated with a single cancer cell in the domain, along with the normal level, I_0, of natural killer immune cells. The remainder of space available to cells is occupied by non-tumorous host cells.

Spatial Simulations: Tumor Growth, No Immune System

Simulations show that tumor morphology is dictated by relative consumption rates: lower consumption rates of nutrient by tumor cells lead to more compact tumors, while higher consumption rates lead to the papillary morphology.

Figure 10a shows the growth in the total number of tumor cells over time when the tumor is allowed to grow in the absence of any immune response, and tumor cell consumption rates are low relative to normal cells. Note the initially exponential growth phase (iteration 0 through 200), before a phase of linear growth (iteration 200 through 800). These growth characteristics mimic the growth rates of multicell spheroids described experimentally by Folkman [28] and mathematically by Greenspan [30].

Figure 10b displays the state of the system after 800 iterations. A roughly circular tumor with a radius of about 200 cells has developed in the center of the domain and is growing steadily toward the sources of the nutrient. Higher tumor cell densities are seen at the periphery of the tumor where it is surrounded by normal cells comprising

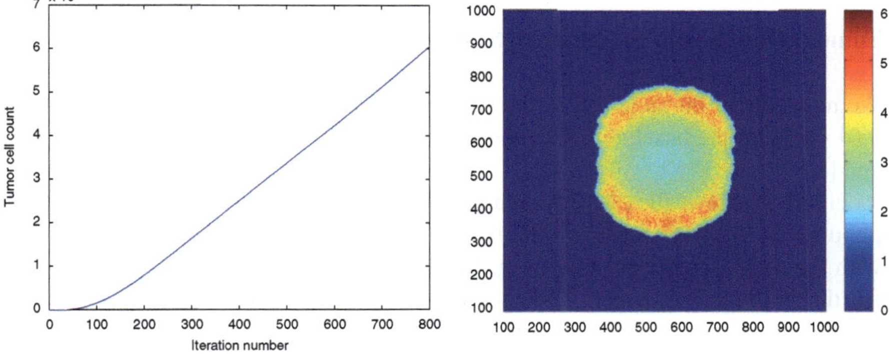

Fig. 10 Compact tumor growth in the absence of immune system interaction. Parameters: Domain size of 1000 elements ≈ 10–20 mm, $t_{end} = 800$ cell division cycles, $\lambda_n = 50$, $\lambda_m = 25$, $\alpha = 1$, $I_0 = 0$. Note the beginning of a necrotic core in Fig. 10b. (**a**)Total tumor cell count over time (**b**) Final tumor cell distribution over the cellular automata grid

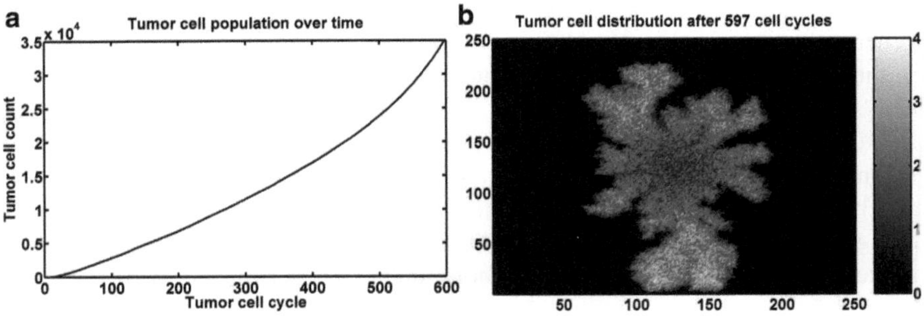

Fig. 11 An example of papillary tumor growth in the absence of immune system interaction. Parameters: Domain size of 250 elements $\approx$ 2.5–4mm, t_{end} = 597 cell cycles, $\lambda_n = 100$, $\lambda_m = 10, \alpha = 2$, $I_0 = 0$. (**a**) Total tumor cell count over time (**b**) Final tumor cell distribution over the cellular automata grid

the host tissue. In the center of the tumor a necrotic core is beginning to form with some necrotic material already appearing. The tumor shown is growing in a domain that is approximately 10–20 mm square.

Figure 11 shows a tumor with relatively high nutrient consumption rates, a "gluttonous" tumor. Figure 11a shows the tumor cell count over time. We observe that the tumor is growing exponentially throughout the time considered without moving to a linear growth rate (as is seen in Fig. 10a). This may be due to the shape of the tumor and the lower requirements of the tumor cells for survival nutrient. Unlike the spherical tumors for which the cell-dense periphery limits the diffusion of nutrients to the tumor center, the papillary tumor grows out quickly from its origin and does not form a cell-dense border. Nutrients diffuse more readily throughout the domain, so a greater percentage of tumor cells is provided with the nutrients to both survive and divide.

Tumor Growth with the Immune System

We now add an immune cell to the simulation. In these experiments we use $I_0 = 0.01$ or 1% of the baseline value, consistent with biological levels measured in [38] and [8].

Figure 12 shows the effect on tumor (left figures) and immune (right figures) cell populations due to changes in CTL recruitment strength. All figures were produced with the same parameters as for Fig. 11, the papillary, "gluttonous," tumor (except for the immune parameters). We see oscillatory population cell counts for both tumor and immune cells. Qualitatively similar results were observed for simulations using the compact tumor parameters (figures not shown).

Figures 12a–b show solutions for the highest level of CTL recruitment and very few oscillations are observed. While it appears in Fig. 12b that little has changed in the immune cell population, the important factor is the *location* of the immune cells. After detection of the tumor, the immune cells are attracted to the location of the tumor mass, thus aiding in its removal.

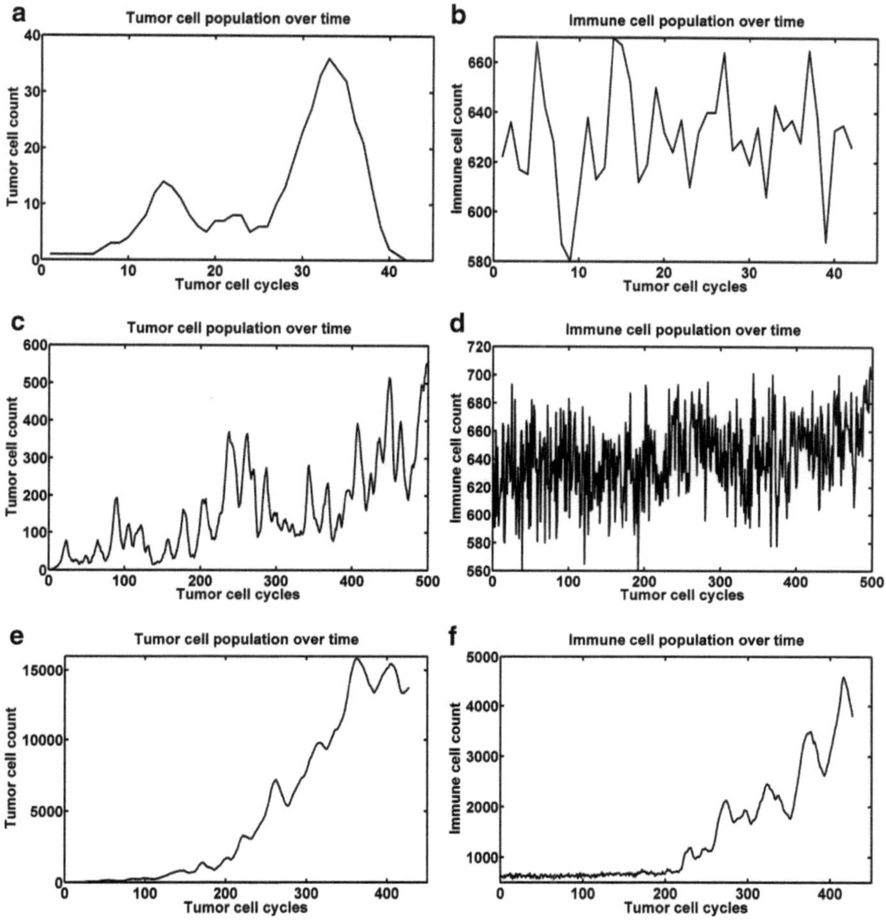

Fig. 12 The effect on tumor and immune cell distributions in compact tumors over time due to changes in CTL recruitment rates, which are inversely related to the parameter θ_L: a larger value of θ_L results in a lower recruitment rate. Parameters: Domain size of 250 elements ≈ 2.5–4mm, $\lambda_n = 100$, $\lambda_m = 10$, $\alpha = 2$, $I_0 = 0.01$. **(a)** $\theta_L = 3$ **(b)**$\theta_L = 3$ **(c)** $\theta_L = 5$ **(d)**$\theta_L = 5$ **(e)** $\theta_L = 7$ **(f)** $\theta_L = 7$

For slightly weaker CTL recruitment, solutions are shown in Figs. 12c–d. The tumor cell population is oscillatory but trending upward. A number of simulations carried out with the same parameter set also exhibited this oscillatory behavior. Experimental evidence for such oscillatory behavior can be found in, for example, Kennedy [32] who studied chronic myelogenous leukemia, and Krikorian et al. [34] who looked at non-Hodgkin's lymphoma.

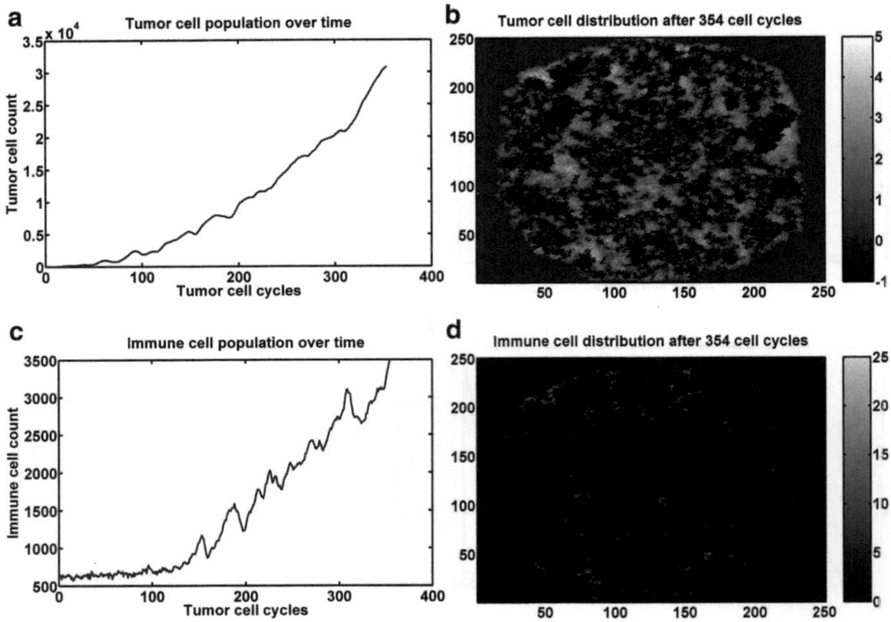

Fig. 13 Immune cell infiltration into a growing tumor. (**a**) and (**b**) show the tumor cell count over time and the final tumor cell distribution over the cellular automata grid, while the same outputs for total immune cells are shown in (**c**) and (**d**). Parameter values are domain size of 250 elements $\approx$ 2.5–4mm, $t_{end} = 354$ cell division cycles, $\lambda_n = 50$, $\lambda_m = 25$,$\alpha = 1$, $I_0 = 1$. Computations were halted when the tumor cells reached the edge of the computational domain

In Figs. 12e–f the tumor and immune cell populations are shown for the case where CTL induction is very low. In this example, the tumor cell population is only slightly oscillatory. In other simulations using this parameter set, the low recruitment of T-cells leads to the tumor undergoing unstable, oscillatory growth.

Lymphocyte Infiltration

Studies have found the relationship between increased survival rates of cancer patients, tumor necrosis and fibrosis, and the presence of intratumoral T-cells, or infiltrated T lymphocytes [51, 54, 60]. The results shown in Fig. 13 simulate the infiltration of immune cells into a growing tumor. These are seen in the darker regions of Fig. 13b where tumor cell necrosis has occurred and in the lighter regions of Fig. 13d where the immune cell numbers are highest. These solution plots are similar to experimental results shown by Schmollinger et al. [51], Soiffer et al. [54], and Zhang et al. [60] where strings of immune cells are moving into the tumor, surrounding individual cells and causing tumor cell necrosis. In Sect. 4.2 we test the effect of injecting immune cells directly into the tumor.

4 Modeling Immunotherapies

We have seen that an individual's immune response to a cancer tumor is a crucial factor in determining the progression of the disease. Cancer immunotherapy is a treatment for cancer that attempts to enhance this immune response. There are a range of different types of immunotherapies, and they can be generally classified into cytokine therapy, cellular transfer, antibody therapy, and vaccines. Newer cytokine treatments called "checkpoint blockades" target specific receptors on T-cells that block or slow their response. These immunotherapy modalities can be broad, targeting the immune system as a whole, or they can be specific, targeting the immune response to a specific cell type. Cancer vaccines are specific: they are intended to enhance the adaptive immune response either by making the tumor-specific immune cells more abundant or more effective or by making the tumor cells more immunogenic, i.e., more recognizable by the cells in the adaptive arm of the immune system. One type of vaccine that has shown some promise in melanoma is a *peptide* vaccine. Peptides are proteins found on cells, and a peptide vaccine targets proteins, or antigens, that are found only on the tumor cells. The idea is to isolate these peptides and to administer them in large doses to the patient in order to stimulate an immune response. The antigen-presenting cells will recognize the peptide, and will initiate the cascade of events roughly depicted in Fig. 6 that will result in the production of tumor-specific CTLs and, ultimately, the destruction of tumor cells. The difficulty is in identifying these peptides, since they must be specific to the patient's own tumor cells but not found, or rarely found, on normal cells. Promising peptides have been found for melanoma and breast cancer [3]. A big advantage that peptide vaccines have over other treatments are their low toxicity, since they promote an immune response targeted only at the tumor cells.

The big questions in the administration of cancer vaccines are: How much? How often? Where? Mathematical models can help suggest answers to these questions. In this section we model three types of cancer immunotherapy and indicate how the model results can inform clinical practice.

4.1 The Kinetics of the Immune Response to Peptide Vaccines: Dose Scheduling

The effectiveness of a peptide vaccine could be measured by the size of the immune response. In the laboratory, the kinetics of this immune response can be measured in mice by injecting them with vaccine and then counting the number of antigen-specific T-cells that result. We would like to know what dosing schedule maximizes this response. The proliferation of T-cells occurs in the lymph organs, so we will model the T-cell populations in the main lymph organ of the mouse: the spleen. We will introduce the helper T-cells into the model, since they are important players in the speed and magnitude of the response.

Once the injected peptides are taken up by antigen-presenting cells, these APCs travel to the spleen where they activate naive T-cells. This activation process takes time to initiate, and we denote this activation time, or *synaptic connection time*, by τ_N. Once activated, the T-cells then begin to proliferate rapidly. After this expansion phase, the activation process is shut down, and the activated T-cells move out of the spleen to find the tumor, become memory T-cells, ready to be activated when the system is next challenged with the same antigen, or become apoptotic, dying off quickly. To describe this process and to capture the dynamics of the T-cell populations when the vaccine is given in repeated doses, the T-cell populations will be divided into five subpopulations: naive, proliferating, active (able to seek and destroy tumor cells), memory cells, and apoptotic cells. The graphic in Fig. 14 shows the five stages, with a dendritic cell as the APC. Note that two synaptic connection times are shown by the dashed lines: τ_N is the time required for activation of the naive T-cells, and τ_M is the activation time required for memory T-cells. Since the memory cells are already trained to recognize the specific antigen, they can be activated more quickly, so that $\tau_M < \tau_N$.

To model the immune activation process mathematically, we use a system of *delay-differential* equations. There is an equation for each state of each cell type

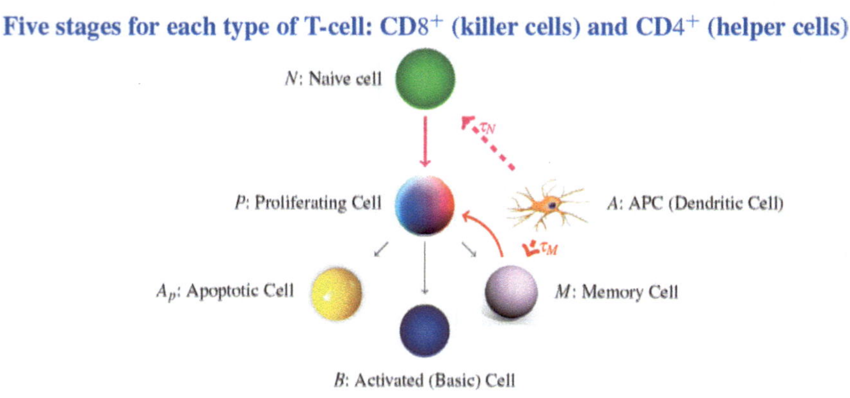

Fig. 14 Schematic diagram representing the model of the T-cell response to antigen. The response takes place in a lymph organ (e.g., the spleen)' and is initiated by an antigen-presenting cell (APC), here represented by a dendritic cell (DC). The T-cells can be in one of five stages, represented here by balls. *Dashed arrows* represent the delays in the model, which show up in the rate of proliferation in response to the presence of the APCs. The two types of T-cells (killer cells and helper cells) respond with different delays. The basic, active immune cells (labeled B) leave the lymph compartment to seek out the antigen

and one for the APCs. This gives a system of 11 equations, with four delays: two synaptic connection times for each of the cell types. For each of the two cell types (killer cells: $CD8^+$ and helper cells: $CD4^+$) we have the five equations 12–16:

APCs arrive from blood and activate naive and memory cells for each type, $CD8^+$ and $CD4^+$:

$$\frac{dA}{dt} = \mu_{BS} D_B - \delta_A A(t) - d_{AT} A(t) \left(N(t) + M(t) + P(t)\right) \tag{11}$$

After synaptic connection times, τ_N and τ_M, activated naive and memory cells begin to proliferate:

$$\frac{dN}{dt} = s_N - d_N N(t) \tag{12}$$

Expansion phase: after contact with antigen, cells reproduce rapidly:

$$\frac{dP}{dt} = g_1 A(t-\tau_N) N(t-\tau_N) + \lambda A(t-\tau_M) M(t-\tau_M) + \rho \frac{A(t-\tau)P(t-\tau)}{\theta + A(t-\tau)}$$

$$-\delta_P P(t) - \frac{1}{T} P(t) \tag{13}$$

Contraction phase: cells become memory (M), die quickly, or go to the blood (B):

$$\frac{dAp}{dt} = \frac{1}{T} P(t) - (r + \alpha + \delta_{Ap}) Ap(t) - \left(\mu_{SB}^* + \frac{\Delta_\mu}{1 + A(t)/\theta_{shut}}\right) Ap(t) \tag{14}$$

$$\frac{dB}{dt} = \left(\mu_{SB}^* + \frac{\Delta_\mu}{1 + A(t)/\theta_{shut}}\right) Ap(t) \tag{15}$$

Memory cells remain after antigen is cleared, with a homeostatic term:

$$\frac{dM}{dt} = rAp(t) + pM(t)\left(1 - \frac{M(t)}{k}\right) \tag{16}$$

Parameters were estimated from the literature and fit to experimental data as described in [47]. All parameter values can be found there. Model simulations show that, after a bolus infusion of peptide vaccine, the CTL population peaks at approximately 6.9 days and the helper T-cell population peaks lower and later, at approximately 8.12 days. See Fig. 15 for a sample simulation. Based on this simulation (which is fit to experimental data with essential, the same kinetic profiles), one might think that administering a second dose of vaccine at approximately Day 6 would result in a high level of T-cells.

To give a more informed answer to the question of when to administer successive doses of the vaccine, we can again formulate and solve an optimization problem, as we did in Sect. 2.1. As in that optimization problem, our goal is to choose the input function that maximizes a desired output. Again, the input function is the dose of the vaccine that produces the source term $D_B(t)$ for the antigen-presenting

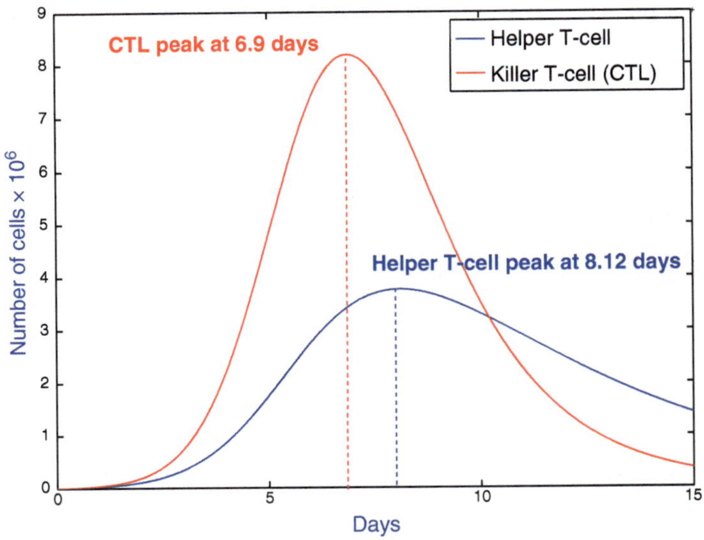

Fig. 15 Model System 11 through 16 simulated for both the CTL and the helper T-cell populations in the spleen. One bolus injection of a peptide antigen was given at time $t = 0$. Total population levels are shown (the sum of naive, proliferating, apoptotic, basic, and memory). The CTL levels peak higher and earlier than the helper T-cell levels. For parameter values and other details see [47]

cells, shown in Equation 11. However, because this system has more equations and four delays, it is too complicated to solve by analytical or direct collocation methods. Instead, we can employ a heuristic optimization scheme to find candidates for optimal solutions. As an example, many runs of a genetic algorithm applied to the system with three different fitness functions yielded the results shown in Fig. 16. In practice, thousands of runs are performed, and those with the highest valued fitness functions are candidates for optimal dosing schedules. In this case, the genetic algorithm suggests that the second dose be given at Day 3—earlier than our first guess of Day 6 based on the kinetics of the response to the first dose. This hypothesis has since been tested in the laboratory on mice confirming the results of the optimization problem [47]. Current vaccine protocols are rigidly set, often at longer intervals than suggested here [9]. We hope that this model and its refinements can serve as a guide for the design of treatment strategies in the future. One possible extension is discussed in Sect. 4.2.

4.2 Dendritic Cell Vaccines

Peptide vaccines, discussed in Sect. 4.1, require the identification of peptides that are expressed on the tumor cells of all patients with that type of cancer (e.g., melanoma), but that are *not* found on normal cells. One of the limitations of this

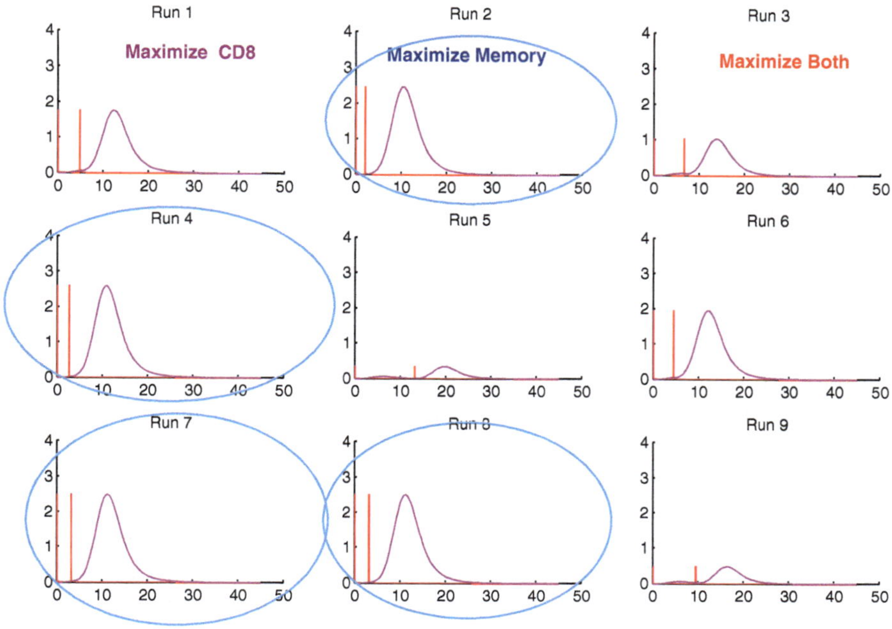

Fig. 16 The solutions with the highest fitness level for 9 different runs of the genetic algorithm. In the first column, the fitness function was the number of $CD8^+$ or CTLs. In the second column, the fitness function was the number of memory T-cells. In the third column, the fitness function was a linear combination of both the CTL population and the memory cell population. Considering all three columns, the runs that yielded the highest fitness values show that the second peptide dose is given at approximately Day 3, (circled runs). *Red bars* indicate the administration times for the two bolus injections of vaccine

type of therapy is that tumor-specific antigens are rare. Another approach is to use vaccines that are developed from the patient's own immune cells, known as "autologous" vaccines. Dendritic cell vaccines are a type of *immune cell-based* vaccine, where immune cells (in this case, dendritic cells, but other APCs are also involved) are removed from the patient, activated by tumor-associated antigen and immune-activating cytokines (such as granulocytes-macrophage colony-stimulating factor, GM-CSF), and cultivated. This expanded colony of activated DCs is then injected into the patient, with the goal of stimulating the tumor-specific immune response modeled in Sect. 4.1. One advantage that autologous DC vaccines have over other, more systemic, treatments is that they have very few toxic side effects: the patient is receiving immune cells of their own making, activated to attack only their own tumor cells. The first such vaccine to be approved by the FDA began to be tested in 2010, with encouraging but not definitive results [9]. During the clinical trial, a fixed dosing regimen was rigidly imposed for all patients: three rounds of vaccines were given every two weeks. Given the sensitivity of the immune kinetics to the timing of the boosting doses, it is likely that outcomes could be improved

Fig. 17 Schematic representation of the DC trafficking model. There are three compartments: spleen, blood and tumor. Different cell populations exist in each department, some of which move from one compartment to another. *Arrows* indicate the flow from one compartment to another, with labels showing which cell populations actually move

by varying the dose and timings. Another variable is the location of the injection: should the vaccine be injected into the tumor site, where the initiation of the immune response normally occurs, or should it be injected into the bloodstream, which is easier and allows the new immune cells to go wherever they are most needed?

In order to answer the three questions: How much? How often? and Where?—a tumor cell population must be added to the model, and the trafficking of the immune cells between the lymph organs and the tumor must be described. We do this in the simplest way possible, by adding a tumor compartment and a blood compartment to the spleen compartment described in Equations 11–16. To reduce the complexity of the model, we focus on the killer T-cells and omit for the time being the helper T-cells. Tumor–immune interactions in the tumor compartment are described by the dePillis–Radunskaya Law, since we are looking at the adaptive response. A schematic is given in Fig. 17.

The dynamics of the trafficking of the immune cells from one compartment to another are complex, with experimentally observed "trapping" effects in both the spleen and the tumor compartments. This results in some rather complicated expressions for the influx and outflow rates in the spleen and tumor compartments: details can be found in [48]. The full set of equations is

Blood compartment:

$$\frac{d}{dt}D_{blood} = -\mu_B D_{blood} + \mu_{TB} D_{tumor} + \nu_{blood}(t) \tag{17}$$

$$\frac{d}{dt}E^a_{blood} = \mu_{SB}(D_{spleen})E^a_{spleen} - \mu_{BB}E^a_{blood} \tag{18}$$

$$\frac{d}{dt}E^m_{blood} = \mu_{SB}(D_{spleen})E^m_{spleen} - \mu_{BB}E^m_{blood} \tag{19}$$

Spleen compartment:

$$\frac{d}{dt}D_{spleen} = D_{\max}\left(1 - e^{\left(\frac{-\mu_{BS}D_{blood}}{MaxD}\right)}\right) - a_D D_{spleen} - b_{DE}E^a_{spleen}D_{spleen} \tag{20}$$

$$\frac{d}{dt}E^a_{spleen} = \mu_{BSE}E^a_{blood} - \mu_{SB}(D_{spleen})E^a_{spleen} + b_a D_{spleen}E^m_{spleen}$$
$$+a_{E_aS}\left(DC_{on}E_{naive} - E^a_{spleen}\right) - r_{am}E^a_{spleen}$$
$$+b_p\frac{D_{spleen}(t-\tau_D)E^a_{spleen}(t-\tau_D)}{\theta_D + D_{spleen}(t-\tau_D)} \tag{21}$$

$$\frac{d}{dt}E^m_{spleen} = r_{am}E^a_{spleen} - \left(a_{E_m} + b_a D_{spleen} + \mu_{SB}(D_{spleen})\right)E^m_{spleen}$$
$$+\mu_{BSE}E^m_{blood}. \tag{22}$$

Tumor compartment:

$$\frac{d}{dt}E^a_{tumor} = \mu_{BTE}(T)E^a_{blood} - a_{E_aT}E^a_{tumor} - cE^a_{tumor}T \tag{23}$$

$$\frac{d}{dt}T = rT\left(1 - \frac{T}{k}\right) - \mathscr{D} \tag{24}$$

$$\frac{d}{dt}D_{tumor} = \frac{mT}{q+T} - (\mu_{TB} + a_D)D_{tumor} + v_{tumor}(t) \tag{25}$$

The "trapping" term which describes the observed phenomenon of activated CTLs being held back in the spleen in the presence of DCs is

$$\mu_{SB}(D_{spleen}) = \mu^*_{SB} + \frac{\Delta\mu}{1 + \frac{D_{Spleen}}{\theta_{shut}}},$$
$$\Delta\mu = \mu^{Normal}_{SB} - \mu^*_{SB}.$$

The functions $v_{blood}(t)$ and $v_{tumor}(t)$ allow us to model injections of DCs into the blood and tumor, respectively. In the tumor compartment, the response rate of the CTLs due to the presence of the tumor has a saturation term

$$\mu_{BTE}(T) = \mu_{BB}(T/(\alpha + T)).$$

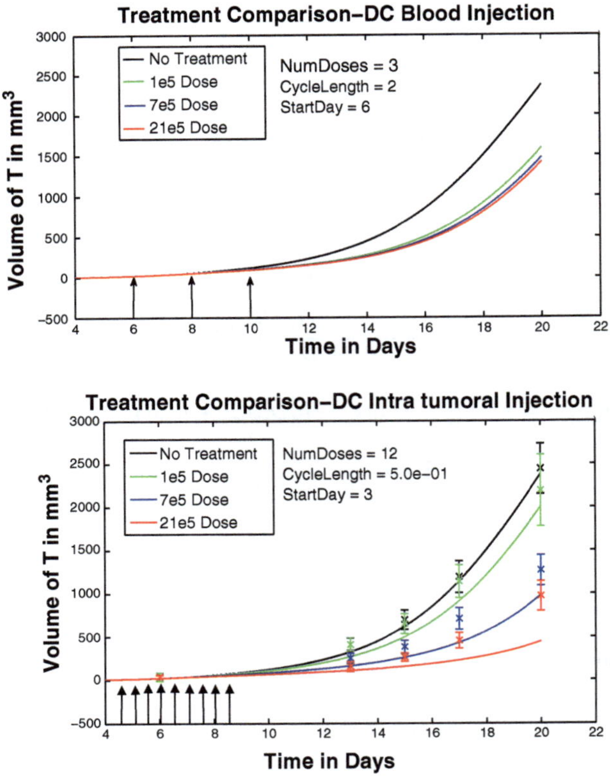

Fig. 18 Comparison of two dosing schedules and injection location. Vaccination treatment starts on Day 6, is injected into the blood, and is given in three doses every two days (*left*), compared to a vaccination regime starting on Day 3, injected directly into the tumor, and given in 12 doses, twice a day (*right*). Times of vaccine doses are shown by arrows (the first two doses are not shown on the *right*)

The ratio-dependent kill rate, $\mathscr{D}$, in Equation 24 is the one described in Sect. 3.1, Equation 8. Note that the effector T-cells denoted by E^a_{tumor} in this model play the role of the CTLs, denoted by L in Equation 8.

This model allows us to experiment with different dosing, timing, and location strategies in order to determine optimal outcomes. Two scenarios are compared in Fig. 18 to give an idea of the flexibility of the model. From these experiments (and others not shown) we can hypothesize that, in the context of DC vaccines, 1) vaccines administered into the blood stream are more effective than those injected directly into the tumor and 2) fractionated dosing schedules are more effective. These hypotheses could be tested in a laboratory or clinical setting, the model refined, and new hypotheses formed. See [48] for more details about the model, parameter settings, and the results of other simulation experiments. Clinical trials of cancer vaccines have had mixed results, where some patients respond well, and others do not [3, 45]. In addition to providing insights into improving treatment protocols, mathematical models can also suggest ways to identify patients who will respond to a particular treatment. Once the model is formulated a sensitivity analysis can indicate which model parameters have the most effect on specific outcomes. A sensitivity analysis using Latin hypercube sampling was performed on this model, indicating that the immune response parameter, d, appearing in the $\mathscr{D}$ term (Equations 24 and 8) has a large impact on the progression of the tumor, and that this effect is enhanced when a patient receives a DC vaccine. A comparison of tumor growth, with and without DC vaccine therapy, with various values of d, is shown in Fig. 19.

4.3 *Monoclonal Antibody Therapy*

In addition to peptide and immune cell vaccines, antibody-mediated therapy has been used either alone or in conjunction with other treatments. In this next model, we capture the dynamics of colorectal cancer growth and its response to monoclonal antibody (mAb) therapy in combination with chemotherapy. We show how the model can be used to simulate clinical trials, a safe and efficient way to lower expenses and speed up the process of treatment design. The work described in this section is extracted from [13], in which further details can be found. Monoclonal antibodies are manufactured to bind to specific proteins. Various protein targets can be used, but epithelial growth factor receptor (EGFR) is a common and useful choice. Circulating epithelial growth factor (EGF) binds to the EGFR and signals a cell proliferation cascade. Many cancerous cells, including colorectal cancer cells, have an EGFR-upregulating mutation, thought to be partly responsible for the high proliferation rate of tumor cells [17, 29, 40, 53]. Monoclonal antibodies can block the EGFR, potentially preventing further tumor cell proliferation.

There are three main pathways for mAb-induced tumor death (see Fig. 20): interactions between mAbs, NK cells, and tumor cells; interactions between mAbs, chemotherapy, and tumor cells; and interactions only between mAbs and tumor cells, resulting in growth rate reduction, complement activation, and possibly other mechanisms for tumor death. In this model, we have chosen to include the following components:

- Cell populations
  - $T(t)$: the total tumor cell population;
  - $N(t)$: the concentration of NK cells per liter of blood (cells/L);
  - $L(t)$: the concentration of CTLs per liter of blood (cells/L);
  - $C(t)$: the concentration of other lymphocytes (cells/L).
- Medications (chemotherapy, cytokines, and monoclonal antibodies); and treatments:
  - $M(t)$: the concentration of chemotherapy per liter of blood (mg/L);
  - $I(t)$: the concentration of interleukin per liter of blood (IU/L);

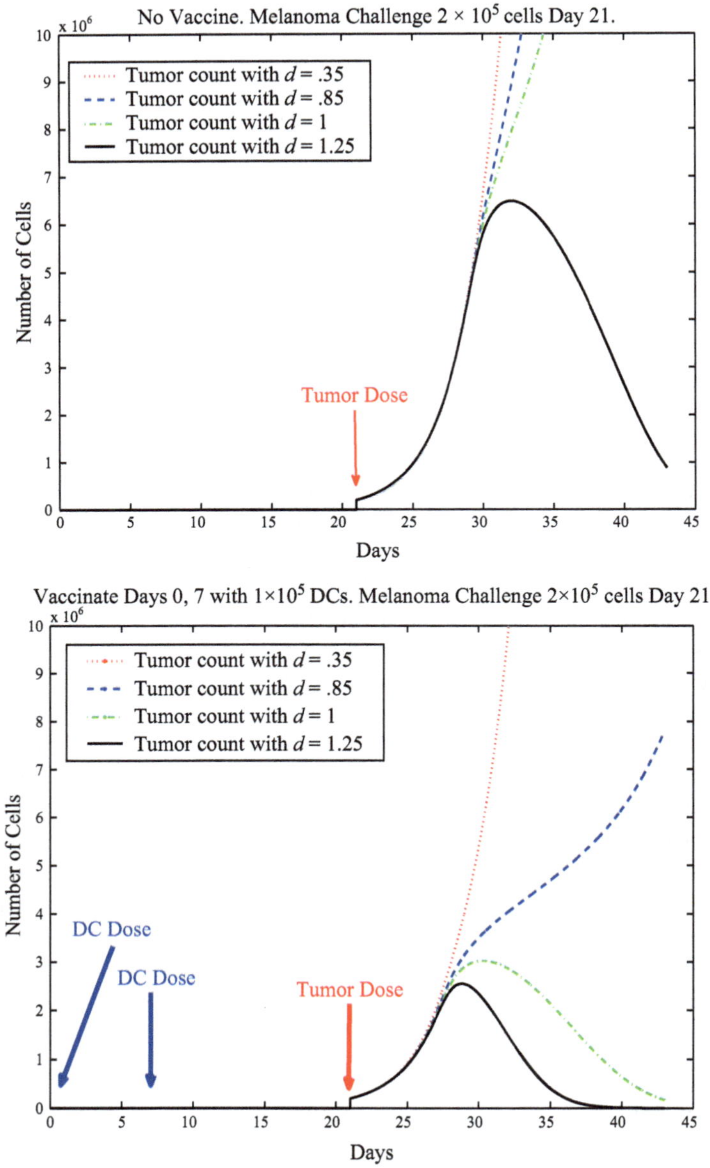

Fig. 19 Prophylactic vaccination and the effect of varying immune response parameter d. *Left*, no vaccine. *Right*, vaccinate with DC treatments on Days 0 and 7 with 1×10^5 DCs per dose. Tumor challenge on Day 21, with 2×10^5 tumor cells. Dosing follows the experiment described in [46]. With $d = 1$, the tumor is controlled as a result of vaccination; with $d = 1.25$ the tumor is controlled independent of vaccination

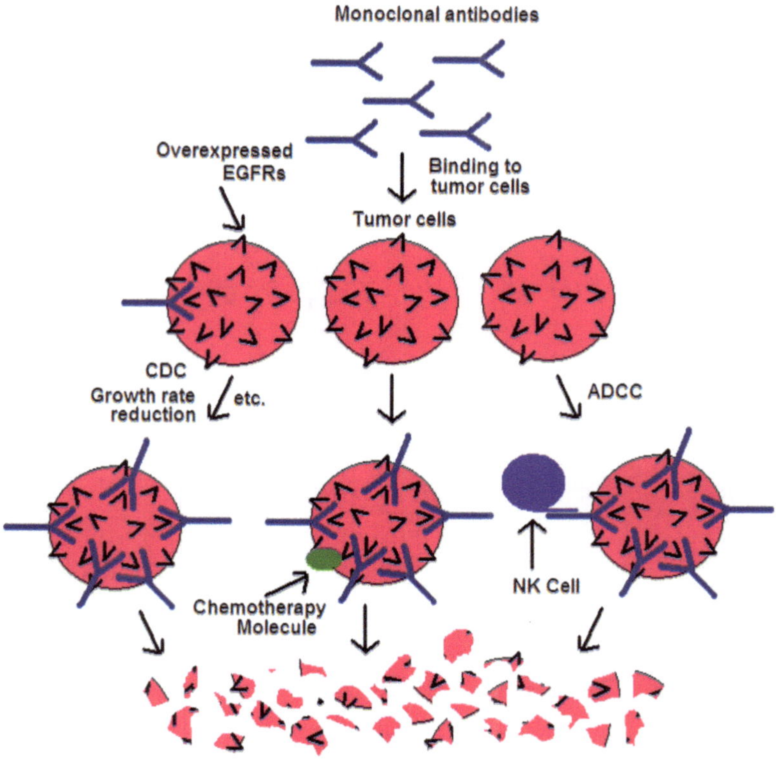

Fig. 20 Three methods of mAb-induced tumor cell death are represented in this model. If an NK cell is present then the cell can undergo ADCC, if a chemotherapy molecule is present then the cell will increase death from the chemotherapy drug, and otherwise, the MAB molecule will cause tumor cell death on its own, through a variety of mechanisms

- $A(t)$: the concentration of monoclonal antibodies per liter of blood (mg/L);
- $v_M(t)$: the amount of irinotecan injected per day per liter of blood (mg/L-day);
- $v_A(t)$: the amount of monoclonal antibodies injected per day per liter of blood (mg/L-day).

Equations (26)–(32) are the system of equations for this model.

$$\frac{dT}{dt} = aT(1-bT) - (c + \xi \frac{A}{h_1 + A})NT - \mathscr{D} - (K_T + K_{AT}A)(1 - e^{-\delta_T M})T - \psi AT \tag{26}$$

(continued)

$$\frac{dN}{dt} = eC - fN - (p + p_A\frac{A}{h_1 + A})NT + \frac{p_N NI}{g_N + I} - K_N(1 - e^{-\delta_N M})N \quad (27)$$

$$\frac{dL}{dt} = \frac{\theta mL}{\theta + I} + j\frac{T}{k + T}L - qLT + (r_1 N + r_2 C)T - \frac{uL^2CI}{\kappa + I} - K_L(1 - e^{-\delta_L M})L + \frac{p_I LI}{g_I + I} \quad (28)$$

$$\frac{dC}{dt} = \alpha - \beta C - K_C(1 - e^{-\delta_C M})C \quad (29)$$

$$\frac{dM}{dt} = -\gamma M + v_M(t) \quad (30)$$

$$\frac{dI}{dt} = -\mu_I I + \phi C + \frac{\omega LI}{\zeta + I} \quad (31)$$

$$\frac{dA}{dt} = -\eta A - \lambda T\frac{A}{h_2 + A} + v_A(t) \quad (32)$$

where the immune response term $\mathscr{D}$ has the familiar form given in Equation 8. The specific treatments that we will explore are the chemotherapeutic drug irinotecan (CPT11), and mAb treatments cetuximab or panitumumab.

4.3.1 Clinical Trial Simulations for mAb Therapy and Chemotherapy

We used the model to explore expected responses to treatment at a population level. In particular, we simulated response to treatment for a group of individuals with a range of immune "strengths." In order to simulate a group of patients having differing immune strengths, we varied the parameters d, λ, and $k_{1/2}$ in Equation 8 for each individual simulated. To reflect the heterogeneity in response to treatment, we also varied the parameters K_T and ψ.

In our clinical trial simulations, we assume that individuals have slightly compromised immune systems after already having been through other immuno-depleting therapies, reflected in a relatively low value of the initial immune cell population. Simulated treatments were administered to each patient, represented by $v_M(t)$ and $v_A(t)$ in model equations (30) and (32). We ran simulations over the set of 64 virtual patients, each identified by a parameter set with different values of the parameters d, $k_{1/2}$, λ, K_T, and ψ, and recorded final tumor size and lymphocyte counts. Lymphocyte count was used as a marker for patient health—if the lymphocyte count dropped low enough for the patient to be considered grade 4 leukopenic, the treatment was considered to be too harsh and not useful.

In order to validate and calibrate our model, we compare the results of the simulated trials to those reported in [11, 17, 29, 31, 37]. Note that the published clinical trial results for cetuximab and panitumumab that we used for comparison reported results as "Response" or "No Response" almost exclusively, so our simulation outcomes reflect this categorization. We carried out monotherapy clinical trial simulations for each of the three drugs used in our model. Monotherapy results can be seen in Fig. 21. Our simulated response to irinotecan was purposefully lower than the clinical trial response, since we assume that our population of patients who may be treated with mAb therapy should be less responsive to chemotherapy than the general population. Our simulated responses to cetuximab and panitumumab were a very close match to the clinical trial outcomes. We also simulated combination therapies, using either irinotecan with cetuximab or irinotecan with panitumumab. These simulations used the common treatments for each drug and gave the two treatments simultaneously. Again, our simulations match the reported clinical trial results fairly closely (see Fig. 22).

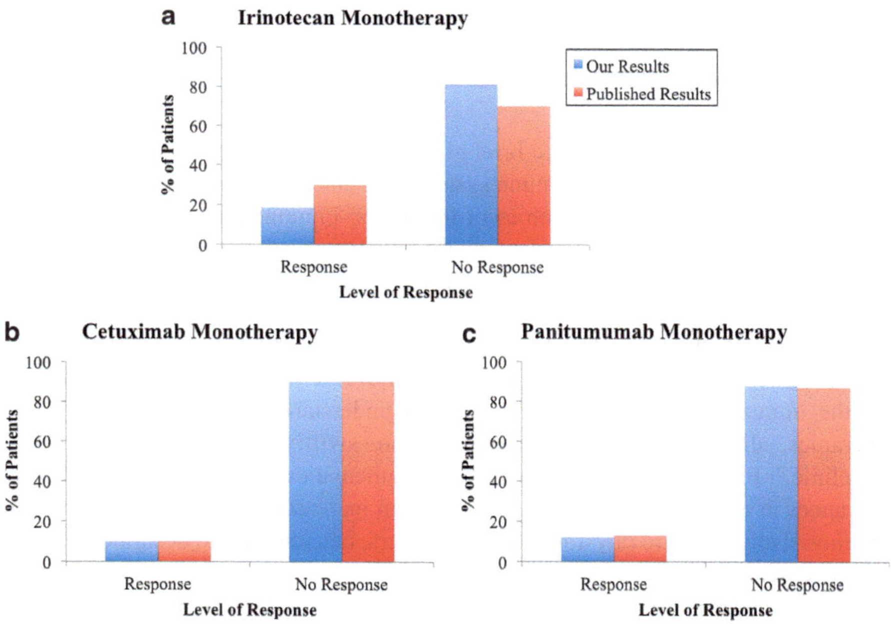

Fig. 21 Our clinical trial simulations compared to reported clinical trial results for irinotecan monotherapy (A), cetuximab monotherapy (B), and panitumumab monotherapy(C). Our simulation results (*left* bars) closely match published results (*right* bars) for both cetuximab and panitumumab monotherapies. For irinotecan monotherapies, the reduced response seen in our simulations is intended, since the patients receiving mAb therapy are often not as responsive as most patients to other treatments

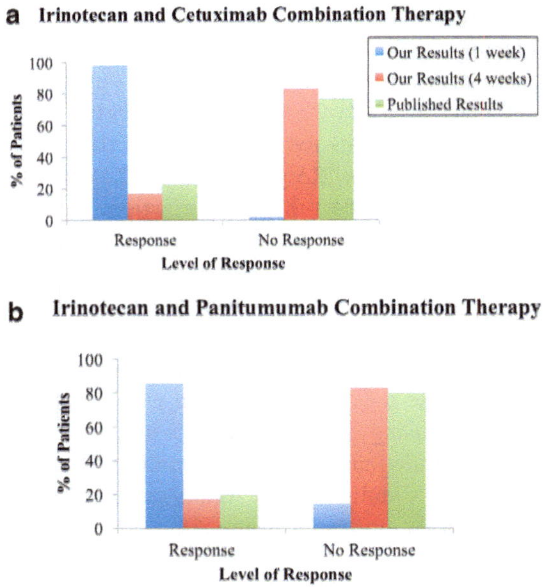

Fig. 22 Our clinical trial simulations compared to reported clinical trial results for irinotecan and cetuximab combination therapy (A) and irinotecan and panitumumab combination therapy (B). When simulation results are measured four weeks post-treatment, our results are very similar to published results

5 Concluding Remarks

In this chapter we have presented brief glimpses of several mathematical models of the interaction between the immune system and cancer tumors. These models can be used to understand the dynamics of the tumor–immune interaction, predict the progression of the disease, and design effective treatment strategies that minimize toxicity. To date, the mathematical models have suggested the advantage of altering traditional protocols, for example, fractionated (or "metronomic") dosing seems to be more effective–and less toxic–than bolus doses. They have helped provide functional forms for cell interactions that can differentiate between the two arms of the immune response. They can help us understand the effects of combined therapies, allowing researchers to move more swiftly and safely from "bench to clinic." Finally, the analysis of these mathematical models, combined with advances in technology that provide us with an increased ability to accurately and noninvasively measure physiological parameters, can lead to the design of patient-specific treatment regimens. Once calibrated to a particular patient, mathematical models can be used to suggest treatment combinations and dosing protocols that are personally optimized.

Much work remains to be done—we hope that the few ideas presented here will motivate mathematicians, clinicians, systems biologists, programmers, and laboratory researchers to continue to collaborate with the goal of understanding and fighting the many diseases known as "cancer".

Acronyms. NK, natural killer Cell; CTL, cytotoxic T lymphocyte, or Killer T-cell, also known as a $CD8^+$ or cytotoxic T-cell because it has a glycoprotein called

CD8 on its surface; $CD4^+$, helper T-cell, expresses the CD4 protein on its surface; ECM, extracellular matrix; IL2, interleukin 2, an immune-stimulating cytokine; APC, antigen-presenting cell; DC, dendritic cell, a type of antigen-presenting cell; EGF, endothelial growth factor; EGFR: endothelial growth factor receptor; mAb, monoclonal antibody

Acknowledgment Earlier, more detailed versions of much of the material in this chapter was published in [13–16,39,47,48]. A. Radunskaya was partially supported by NSF grant DMS-1016136.

References

1. J.A., Adam, N. Bellomo, *A Survey of Models for Tumor Immune Systems Dynamics* (Springer, Newyork, 1997)
2. T. Alarcon, H. Byrne, P. Maini, A cellular automaton model for tumour growth in inhomogeneous environment. J. Theor. Biol. **225**, 257–274 (2003)
3. G. Alatrash, H. Jakher, P.D. Stafford, E.A. Mittendorf, Cancer immunotherapies, their safety and toxicity. Expert Opin. Drug Saf. **12**, 631–645 (2013)
4. N. Bellomo, A. Bellouquid, M. Delitala, Mathematical topics on the modelling complex multicellular systems and tumor immune cells competition. Math. Models Methods Appl. Sci. **14**(11), 1683–1733 (2004)
5. N. Bellomo, L. Preziosi, Modelling and mathematical problems related to tumor evolution and its interaction with the immune system. Math. Comput. Model. **32**(3), 413–452 (2000)
6. A. Cappuccio, M. Elishmereni, , Z. Agur, Cancer immunotherapy by interleukin-21: potential treatment strategies evaluated in a mathematical model. Cancer Res. **66**(14), 7293–7300 (2006)
7. G. Caravagna, A. d'Onofrio, P. Milazzo, R. Barbuti, Tumour suppression by immune system through stochastic oscillations. J. Theor. Biol. **265**(3), 336–345 (2010)
8. A. Cerwenker, L. Lanier, Natural killer cells, viruses and cancer. Nat. Immunol. 41–48 (2001)
9. M.A. Cheever, PROVENGE (Sipuleucel-T) in prostate cancer: the first FDA-approved therapeutic cancer vaccine. Clin. Cancer Res. **17**, 35203,526 (2011)
10. S.E. Clare, F. Nakhlis, J.C. Panetta, Molecular biology of breast cancer metastasis: The use of mathematical models to determine relapse and to predict response to chemotherapy in breast cancer. Breast Cancer Res. **2**, 430–435 (2000)
11. D. Cunningham, Y. Humblet, S. Siena, Cetuximab monotherapy and cetuximab plus irinotecan in irinotecan-refractory metastatic colorectal cancer. N. Engl. J. Med. **351**, 337–45 (2004)
12. L. de Pillis, D. Mallet, A. Radunskaya, Spatial tumor-immune modeling. Comput. Math. Methods Med. **7**(2–3), 159–176 (2006)
13. L. de Pillis, A. Radunskaya, H. Savage, Mathematical model of colorectal cancer with monoclonal antibody treatments. URL http://arxiv.org/abs/1312.3023. Preprint
14. L. de Pillis, A.E. Radunskaya, The dynamics of an optimally controlled tumor model: A case study. Math. Comput. Model. (Special Issues) **37**(11), 12211,244 (2003)
15. L.G. de Pillis, A.E. Radunskaya, A mathematical tumor model with immune resistance and drug therapy: An optimal control approach. J. Theor. Med. **3**, 79–100 (2001)
16. L.G., de Pillis, A.E. Radunskaya, C.L. Wiseman, A validated mathematical model of cell-mediated immune response to tumor growth. Cancer Res. **65**(17), 7950–7958 (2005)
17. V.J. De Vita, S. Hellman, S. Rosenberg, *Cancer: Principles and Practice of Oncology*, 7 edn. (Lippincott Wiliams & Wilkins, Sydney 2000)
18. A. Diefenbach, E.R. Jensen, A.M. Jamieson, D.H.: Raulet, Rae1 and h60 ligands of the nkg2d receptor stimulate tumour immunity. Nature **413**(6852), 165–171 (2001)

19. A. d'Onofrio, A general framework for modeling tumor-immune system competition and immunotherapy: Mathematical analysis and biomedical inferences. Phys. D **208**(3), 220–235 (2005)
20. A. d'Onofrio, Metamodeling tumor–immune system interaction, tumor evasion and immunotherapy. Math. Comput. Model. **47**(5), 614–637 (2008)
21. A. d'Onofrio, A. Gandolfi, Resistance to antitumor chemotherapy due to bounded-noise-induced transitions. Phys. Rev. E **82**, 061,901 (2010)
22. C. DuBois, J. Farnham, E. Aaron, A. Radunskaya, A multiple time-scale computational model of a tumor and its micro environment. MBE **10**(1), 121–150 (2013)
23. M.E. Dudley, J.R. Wunderlich, P.F. Robbins, J.C. Yang, P. Hwu, D.J. Schwartzentruber, S.L. Topalian, R. Sherry, N.P. Restifo, A.M. Hubicki, M.R. Robinson, M. Raffeld, P. Duray, C.A. Seipp, L. Rogers-Freezer, K.E. Morton, S.A. Mavroukakis, D.E. White, S.A. Rosenberg, Cancer regression and autoimmunity in patients after clonal repopulation with antitumor lymphocytes. Science **298**(5594), 850–854 (2002)
24. O.K. Dzivenu, J. O'Donnell-Tormey, Cancer and the immune system: the vital connection. Online (2003). URL http://www.cancerresearch.org/cancer-immunotherapy/resources/cancer-and-the-immune-system
25. P. Ehrlich, Über den jetzigen stand der karzinomforschung. Ned. Tijdschr. Geneeskd. **5**, 273–290 (1909)
26. A. Farrell, Milestone 3, (1909) Immune surveillance, hide and seek. Nat. Med. (2006)
27. S.C. Ferreira, M.L. Martins, M.J. Vilela, Reaction-diffusion model for the growth of avascular tumor. Phys. Rev. E **65**, 021,907 (2002)
28. J. Folkman, M. Hochberg, Self-regulation of growth in three dimensions. J. Exp. Med. **138**, 745–753 (1973)
29. C. Gravalos, J. Cassinello, P. Garcia-Alfonso, A. Jimeno, Integration of panitumumab into the treatment of colorectal cancer. Crit. Rev. Oncol. Hematol. **74**(1), 16–26 (2010)
30. H. Greenspan, Models for the growth of a solid tumor by diffusion. Stud. Appl. Math. **51**, 317–338 (1972)
31. A.M. Grothey, Defining the role of panitumumab in colorectal cancer. Community Oncology **3**, 10–16 (2006)
32. B.J. Kennedy, Cyclic leukocyte oscillations in chronic myelogenous leukemia during hydroxyurea therapy. Blood **35**(6), 751–760 (1970)
33. D.D. Kirschner, J.C. Panetta, Modeling immunotherapy of the tumor - immune interaction. J. Math. Biol. **37**(3), 235–252 (1998)
34. Krikorian, J., Portlock, C., Cooney, D., Rosenberg, S.: Spontaneous regression of non-hodgkin's lymphoma: A report of nine cases. Cancer **46**, 2093–2099 (1980)
35. N.N. Kronik, Y. Kogan, V. Vainstein, Z. Agur, Improving alloreactive ctl immunotherapy for malignant gliomas using a simulation model of their interactive dynamics. Cancer Immunol. Immunother. **57**(3), 425–439 (2008)
36. V. Kuznetsov, in *A Survey of Models for Tumor-Immune System Dynamics*, eds. by J. Adam, N. Bellomo Basic Models of Tumor-Immune System Interactions- Identification, Analysis and Predictions (Birkhauser, Basel 1997)
37. H.J. Lenz, Cetuximab in the management of colorectal cancer. Biologics **2**, 77–91 (2007)
38. A. Lin, A model of tumor and lymphocyte interactions. Discrete Contin. Dyn. Syst. Ser. B **4**(1), 241–266 (2004)
39. Mallett, D., de Pillis, L.: A cellular automata model of tumor-immunesystem interactions. J. Theor. Biol. **239**, 334–350 (2006)
40. E. Martinelli, R. De Palma, M. Orditura, F. De Vita, F. Ciardiello, Anti-epidermal growth factor receptor monoclonal antibodies in cancer therapy. Clin. Exp. Immunol. **158**, 1–9 (2009)
41. A. Matzavinos, M.A. Chaplain, V.A. Kuznetsov, Mathematical modelling of the spatio-temporal response of cytotoxic t-lymphocytes to a solid tumour. Math. Med. Biol. **21**(1), 1–34 (2004)
42. L. Norton, R. Simon, H. Brereton, A. Bogden, Predicting the course of gompertzian growth. Nature **264**, 542–545 (1976)

43. W. Paul, Fundamental Immunology, 5 edn. Lippincott, Williams and Wilkins Publishers, Sydney (2003)
44. A.S. Perelson, G. Weisbuch, Immunology for physicists. Rev. Mod. Phys. **69**, 1219–1268 (1997)
45. S. Pilon-Thomas, M., Verhaegen, J. Mulé, Dendritic cell-based therapeutics for breast cancer. Heart Disease **20**, 65–71 (2004)
46. O. Preynat-Seauve, E. Contassot, P. Schuler, L.E. French, B. Huard, Melanoma-infiltrating dendritic cells induce protective antitumor responses mediated by t cells. Melanoma Res. **17**, 169–176 (2007)
47. A. Radunskaya, S. Hook, in: *New Challenges for Cancer Systems Biomedicine* eds. by A. d'Onofrio, P. Cerrai, A. Gandolfi, Modeling the kinetics of the Immune Response (Springer, Newyork, 2012), pp. 267–282.
48. A. Radunskaya, L. de Pillis, A. Gallegos, A model of dendritic cell therapy for melanoma. Front. Oncology **3**(56), 223–228 (2013)
49. H. Riedel, in *The Cancer Handbook Wiley*, ed. by M. Alison Models for tumour growth and differentiations (New Jersey Institute of Technology, New Jersey 2004)
50. S.A. Rosenberg, J. Yang, S.L.E.A. Topalian, Treatment of 283 consecutive patients with metastatic melanoma or renal cell cancer using high-dose bolus interleukin 2. JAMA **271**, 907–913 (1994)
51. J. Schmollinger, R. Vonderhelde, K. Hoar, R. Vonderheide, K. Hoar, Maecker, B., J., F.S., H. Schultze, R. Soiffer, K. Jung, M. Kuroda, N. Letvin, E. Greenfield, M. Mihm, J. Kutok, G.Dranoff, Melanoma inhibitor of apoptosis protein (ml-iap) is a target for immune-mediated tumor destruction. Proc. Natl. Acad. Sci. USA **100**(6), 3398–3403 (2003)
52. J. Sherratt, M. Nowak, Oncogenes, anti-oncogenes and the immune response to cancer. Proc. R. Soc. Lond. B **248**, 261–271 (1992)
53. S. Siena, A. Sartore-Bianchi, F. Di Nicolantonio, J. Balfour, A. Bardelli, Biomarkers predicting clinical outcome of epidermal growth factor receptor-targeted therapy in metastatic colorectal cancer. J. Natl. Cancer Inst. **101**, 1–17 (2009)
54. R. Soiffer, T. Lynch, M. Mihm, K. Jung, C. Rhuda, J. Schmollinger, F. Hodi, L. Liebster, P. Lam, S. Mentzer, S. Singer, K. Tanabe, A. Cosimi, R. Duda, A. Sober, A. Bhan, J. Daley, D. Neuberg, G. Parry, J. Rokovich, L. Richards, J. Drayer, A. Berns, S. Clift, L. Cohen, R. Mulligan, G. Dranoff, Vaccination with irradiated autologous melanoma cells engineered to secrete human granulocyte macrophage colony-stimulating factor generates potent antitumor immunity in patients with metastatic melanoma. Proc. Natl. Acad. Sci. USA **95**, 13,141–13,146 (1998)
55. J. Speer, V. Petrosky, M. Retsky, R. Wardwell, A stochastic numerical model of breast cancer growth that simulates clinical data. Cancer Res. **44**, 41244,130 (1984)
56. C.O. Starnes, Coley's toxins in perspective. Nature **357**, 11–12 (1992)
57. O. von Stryk, User's guide for DIRCOL: A direct collocation method for the numerical solution of optimal control problems. Lehrstuhl M2 Numerische Mathematik, Technische Universitaet Muenchen (1999)
58. R.M. Sutherland, Cell and environment interactions in tumor microregions: the multicell spheroid model. Science **240**, 177–184 (1988)
59. R. Thomlinson, Measurement and management of carcinoma of the breast. Clin. Radiol. **33**(5), 481–493 (1982)
60. L. Zhang, J. Conejo-Garcia, D. Katsaros, P. Gimotty, M. Massobrio, G. Regnani, A. Makrigiannakis, H. Gray, K. Schlienger, M. Liebman, S. Rubin, G. Coukos, Intratumoral t cells, recurrence, and survival in epithelial ovarian cancer. N. Engl. J. Med. **348**(3), 203–213 (2003)

A Hybrid Multiscale Approach in Cancer Modelling and Treatment Prediction

Gibin Powathil and Mark A.J. Chaplain

Abstract Cancer is a complex multiscale disease involving inter-related processes across a wide range of temporal and spatial scales. Multiscale mathematical models can help in studying cancer progression and serve as an in silico test base for comparing and optimizing various multi-modality anticancer treatment protocols. Here, we discuss one such hybrid multiscale approach, interlinking individual cell behavior with the macroscopic tissue scale. Using this technique, we study the spatio-temporal dynamics of individual cells and their interactions with the tumor microenvironment. At the intracellular level, the internal cell-cycle mechanism is modelled using a system of coupled ordinary differential equations, which determine cellular growth dynamics for each individual cell. The evolution of these individual cancer cells are modelled using a cellular automaton approach. Moreover, we have also incorporated the effects of oxygen distribution into this multiscale model as it has been shown to affect the internal cell-cycle dynamics of the cancer cells. The hybrid multiscale model is then used to study the effects of cell-cycle-specific chemotherapeutic drugs, alone and in combination with radiotherapy, with a long-term goal of predicting an optimal multimodality treatment plan for individual patients.

Keywords Cancer • Hybrid multiscale model • Cell cycle • Hypoxia • Chemotherapy • Radiotherapy

1 Introduction

Cancer has been and still is one of the most devastating diseases known to the developed world. Even with numerous technological, medical and pharmaceutical developments in detecting and treating different types of cancer, the median survival rates of most types of cancer remain unchanged for the last 20–30 years. One way to

G. Powathil • M.A.J. Chaplain (✉)
Division of Mathematics, University of Dundee,
Dundee, Scotland, DD1 4HN, UK
e-mail: gibin@maths.dundee.ac.uk; chaplain@maths.dundee.ac.uk

© Springer Science+Business Media New York 2014

A. d'Onofrio, A. Gandolfi (eds.), *Mathematical Oncology 2013*,
Modeling and Simulation in Science, Engineering and Technology,
DOI 10.1007/978-1-4939-0458-7_8

improve this situation is to increase the survival chances of cancer patients through the intelligent planning and optimum delivery of multimodality therapies.

The most common treatments for cancer management are surgery, chemotherapy, radiation therapy and their various combinations. Chemotherapeutic drugs function by killing the cancer cells through interfering with the cell-cycle mechanism, which regulates complex intracellular processes such as proliferation, cell division and DNA replication [66]. The cell-cycle mechanism is very much dynamic in nature and is influenced by numerous intracellular pathways, extracellular interactions and the tumor microenvironment, in particular oxygen [10, 66]. Two important pathways out of many that affect the regulation of the cell cycle are the HIF-1 pathway, which is upregulated by the presence of hypoxia, and the wee1 pathway, which is influenced by circadian rhythms [37, 38]. These intracellular and extracellular heterogeneities as well as dynamical changes in the tissue microenvironment directly or indirectly contribute towards cell-cycle-mediated drug resistance and poor treatment outcome [10]. An effective way to address this is by using an appropriate combination of cell-cycle-specific chemotherapeutic drugs that targets a cancer cell at its various phases of cell division.

In most cases, chemotherapy is administered in combination with the radiation therapy, although in some cases either one of them is given alone. Just as with chemotherapy, cell-cycle dynamics also play a vital role in mediating a cell's sensitivity towards radiation therapy since the cell-cycle phase determines the cell's relative radiosensitivity [42, 53]. Previous studies have shown that cells that are in G2-M phase are more sensitive to the radiation than those in G1 phase [53]. Moreover, irradiation can also alter a cell's cell-cycle dynamics through the activation of various intracellular pathways and thus can delay the rate of progression of a cell's cell cycle, causing a group of cancer cells to accumulate in a particular phase of the cell cycle [42, 53]. The treatment-dependent perturbations of cell-cycle dynamics together with cell-cycle-dependent therapeutic sensitivity are one of many rationales behind the combination treatment protocols of chemotherapy and radiation therapy [25, 33, 53].

Clinically driven mathematical models can be used as powerful tools to understand, study, and provide useful predictions related to the outcome of various treatment protocols used to treat human malignancies. The multiscale complexity of cancer progression warrants a multiscale modelling approach to produce truly predictive mathematical models. In order to capture all the dynamics of tumor progression, we need to couple processes that are occurring at various spatial and temporal scales. In this chapter, we discuss one such hybrid multiscale model that incorporates some of the relevant intracellular, cellular and macroscopic dynamics.

2 Multiscale Mathematical Model: Growth and Progression

We consider a hybrid cellular automaton model for cancer growth and progression, which models the spatio-temporal dynamics at the cell level, interlinking cell-level dynamics to the molecular variations of intercellular signalling and macroscopic behavior of tissue oxygen dynamics [57]. The evolution of cancer cells is modelled using a cellular automaton (CA) model where each cell has its own cell-cycle dynamics incorporated using a set of ordinary differential equations. The cells are located spatially in a dynamic microenvironment due to the variations in oxygen concentration, and the changes in oxygen distribution and drug concentration are modelled using partial differential equations. A cell-based model is considered to study the multiple effects of radiation therapy [56]. Finally, this hybrid multiscale CA modelling approach is used to analyze and study the therapeutic outcome when chemotherapy and radiation therapy are given alone and in combination with each other.

The computational simulations are performed on a two-dimensional spatial grid, where each grid represents either a cancer cell, the cross section of a blood vessel or the extracellular matrix. The spatial size of this computational grid has been chosen to approximately match the size of a single cell. To simulate the spatio-temporal progression of the cancer cells and their response to chemotherapeutic drugs and radiation, each automaton cell is associated with four major components – cells, the local oxygen (and hence HIF-1 α) concentration, initial blood vessel distribution and drug concentrations. A schematic overview of the model with the scales involved is given in Fig. 1 [56].

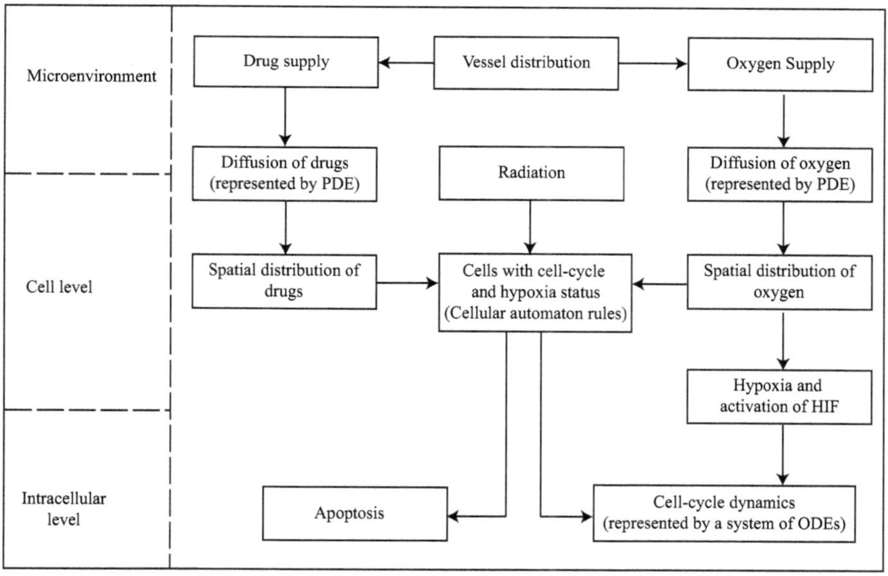

Fig. 1 Schematic diagram of the model showing the appropriate scales involved. Adapted from [56]

2.1 Intracellular Dynamics: Cell-Cycle Model

There are many different models of the cell-cycle dynamics incorporating varying underlying complexities, and details of these can be found in the key papers [7, 26, 30, 48, 74]. For our multiscale model, to model the cell-cycle dynamics within each cell, we use a simple model which was originally developed by Tyson and Novak [47, 74] that includes various relevant interactions for cell-cycle regulation and control. Using these kinetic relations, Tyson and Novak [47, 74] explain the transitions between two main steady states, G1 and S-G2-M, of the cell cycle through the changes in cell mass. Although, we use this simple model in our simulations, one could easily replace this with more complex models that describe the further complicated processes involved in the mammalian cell cycle [48].

To make the current six-variable model more relevant to the mammalian cell, we have used the equivalent mammalian proteins stated in Tyson and Novak's paper, namely the Cdk-cyclin B complex [CycB], the APC-Cdh1 complex [Cdh1], the active form of the p55cdc-APC complex [p55cdc_A], the total p55cdc-APC complex [p55cdc_T], the active form of Plk1 protein [Plk1] and the mass of the cell [mass] [57]. Following Tyson and Novak's model, the evolution of the concentrations of these components is modelled using the following system of six ODEs:

$$\frac{d[\text{CycB}]}{dt} = k_1 - (k_2' + k_2''[\text{Cdh1}])[\text{CycB}], \tag{1}$$

$$\frac{d[\text{Cdh1}]}{dt} = \frac{(k_3' + k_3''[\text{p55cdc}_\text{A}])(1-[\text{Cdh1}])}{J_3 + 1 - [\text{Cdh1}]} - \frac{k_4[\text{mass}][\text{CycB}][\text{Cdh1}]}{J_4 + [\text{Cdh1}]}, \tag{2}$$

$$\frac{d\,[\text{p55cdc}_\text{T}]}{dt} = k_5' + k_5''\frac{([\text{CycB}][\text{mass}])^n}{J_5^n + ([\text{CycB}][\text{mass}])^n} - k_6\,[\text{p55cdc}_\text{T}], \tag{3}$$

$$\frac{d\,[\text{p55cdc}_\text{A}]}{dt} = \frac{k_7[\text{Plk1}](\,[\text{p55cdc}_\text{T}] - [\text{p55cdc}_\text{A}])}{J_7 + [\text{p55cdc}_\text{T}] - [\text{p55cdc}_\text{A}]} - \frac{k_8[Mad]\,[\text{p55cdc}_\text{A}]}{J_8 + [\text{p55cdc}_\text{A}]} - k_6\,[\text{p55cdc}_\text{A}], \tag{4}$$

$$\frac{d[\text{Plk1}]}{dt} = k_9[\text{mass}][\text{CycB}](1\text{-}[\text{Plk1}]) - k_{10}[\text{Plk1}], \tag{5}$$

$$\frac{d[\text{mass}]}{dt} = \mu[\text{mass}]\left(1 - \frac{[\text{mass}]}{m_*}\right), \tag{6}$$

where k_i are the rate constants and the values are chosen in proportional to those in Tyson and Novak so that the time scale is relevant to mammalian cell cycle [47,74].

Here, the equation governing the change of mass accounts for the growth of each cell. A cell is assumed to divide when the concentration of Cdk-cyclin B complex [CycB] crosses a specific threshold value $[\text{CycB}]_{th}$ which is assumed to be 0.1, from above and then the mass, [mass] is halved. To introduce a random growth rate for individual cells which in turn introduces cell-cycle heterogeneity in the population, we consider a varying growth rate μ:

$$\mu = \mu^{+} + \varepsilon\hat{\mu}, \tag{7}$$

where $\mu^{+} = 0.03$, $\varepsilon = 0.006$ and $\hat{\mu}$ is a probability density function with uniform distribution between -1 and 1. The rest of the parameter values of the cell-cycle model can be found in Powathil et al. [57].

Figure 2 shows the changes in various protein concentrations that have been included in the current cell-cycle model for one single automaton cell. Every cell in this multiscale model has similar cell-cycle dynamics built-in which further control the division and cell-cycle phases of the respective cells. In this representative figure (Fig. 2), a cell undergoes division constantly as long as there is enough space to divide and the surrounding microenvironment is favorable for its division. However, as soon as its neighboring spaces are occupied, the cell moves to a resting phase where the concentrations are maintained in a constant level and in Fig. 2, this happens at around 190 h.

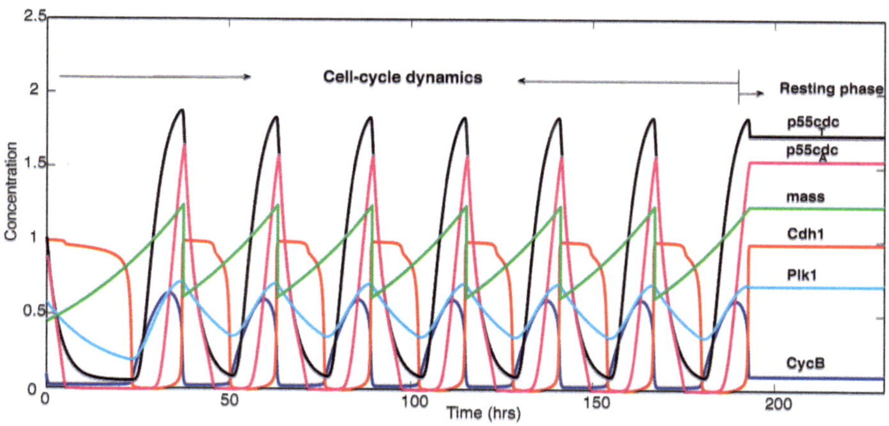

Fig. 2 Plot of the concentration profiles of the various intracellular proteins and the cell mass over a period of 200 h for one automaton cell in the model. This is obtained by solving the system of equations, Equations 1 to 6, with the relevant parameter values from Table 1. Adapted from [57]

Table 1 cell-cycle model parameters from [55]

Component	Rate constants (hr^{-1})	Dimensionless constants
[CycB]	$k_1 = 0.12$, $k_2' = 0.12$, $k_2'' = 4.5$, [p27/p21]=1.05	$[\text{CycB}]_{th} = 0.1$
[Cdh1]	$k_3' = 3$,$k_3'' = 30$, $k_4 = 105$	$J_3 = 0.04$,$J_4 = 0.04$
[p55cdc$_T$]	$k_5' = 0.015$, $k_5'' = 0.6$, $k_6 = 0.3$	$J_5 = 0.3$, $n = 4$
[p55cdc$_A$]	$k_7 = 3$,$k_8 = 1.5$	$J_7 = 0.001$,$J_8 = 0.001$,[Mad]=1
[Plk1]	$k_9 = 0.3$, $k_{10} = 0.06$	
[mass]	$\mu^{+} = 0.03$	$m_{*} = 10$

2.2 Macroscopic Effects: Oxygen Dynamics and Hypoxia

At the macroscopic level, the effects of the changing tissue microenvironment are incorporated into the model by introducing oxygen dynamics which are modelled using a suitable partial differential equation incorporating vessels as sources of oxygen. Computationally, there are several ways of introducing vascular dynamics into the model depending on various temporal and spatial scales of interest [9, 15, 67]. Here, we have considered blood vessel cross sections distributed randomly throughout the two-dimensional domain with density $\phi_d = N_v/N^2$, where N_v is the number of vessel cross sections [18,36]. If $K(x,t)$ denotes the oxygen concentration at position x at time t, then its spatio-temporal evolution can be expressed as,

$$\frac{\partial K(x,t)}{\partial t} = \nabla.(D_K(x)\nabla K(x,t)) + r(x)m(x) - \phi K(x,t)\text{cell}(x,t), \tag{8}$$

where $D_K(x)$ is the diffusion coefficient and ϕ is the rate of oxygen consumption by a cell at position x at time t ($\text{cell}(x,t) = 1$ if position x is occupied by a cancer cell at time t and zero otherwise). Here, $m(x)$ denotes the vessel cross section at position x ($m(x) = 1$ for the presence of blood vessel at position x, and zero otherwise); thus the term $r(x)m(x)$ describes the production of oxygen at rate $r(x)$ [57]. The diffusion coefficient and the supply rate of the oxygen vary depending on the location of the cancer cells and blood vessels [57]. Since it has been observed that when a vessel is surrounded by a mass of densely packed cancer cells its perfusion and diffusion capabilities are seriously impaired, this is incorporated in our model by considering a lower diffusion rate and a lower supply rate in the tumor as compared to the normal vessels [57]. This equation is solved using no-flux boundary conditions and an appropriate initial condition [55]. Figure 3 shows a representative profile of the spatial distribution of oxygen concentration after solving Equation (8) with relevant parameters as given below.

The oxygen diffusion length scale L can be considered to be approximately equal to 100 μm and the value of the diffusion constant is taken as $2 * 10^{-5}$ cm^2/s [50]. Using these and the relation $L = \sqrt{D/\phi}$, the mean oxygen uptake can be approximately estimated as 0.2 s^{-1}. The oxygen supply through the blood vessel is approximately taken as $8.2 * 10^{-3}$mols s^{-1} [44]. The appropriate nondimensionalization will yield a time scale of $T = 0.001$hr and hence each time step is set to be 0.001hr for both CA time step and oxygen dynamics. The length scale of 100μm will give a square grid of length $\Delta x * L$=20 μm, approximate diameter of a cell and thus a tumor of radius 1 mm [57].

The changes in the oxygen concentration, especially hypoxia, may affect various intra and intercellular processes of the cells that constitute the tumor mass. In the present model, the effects of hypoxia are included through the activation and inactivation of HIF-1 α which further results in changes in intracellular cell-cycle dynamics. When oxygen concentration at a specific position x falls below 10% (hypoxic cell), HIF-1α is assumed to become active from an inactive phase, which

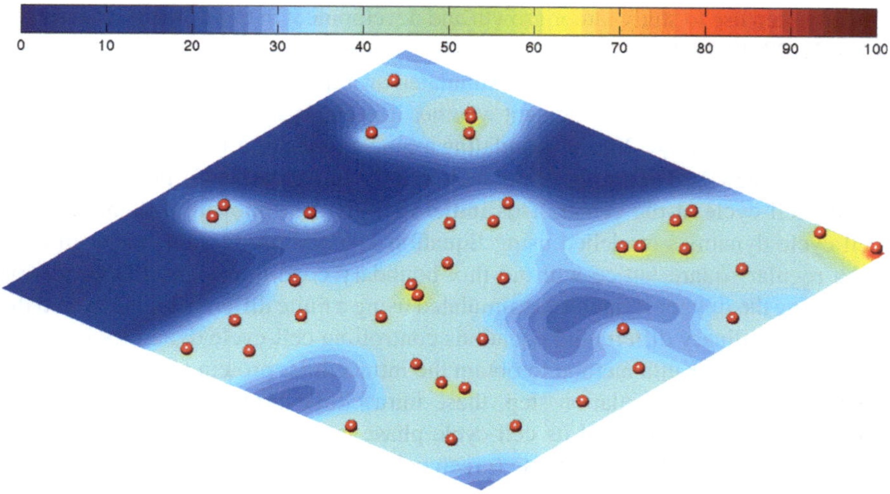

Fig. 3 Plot showing the concentration profile of oxygen supplied from the vasculature. The *red-colored spheres* represent the blood vessel cross sections and the color map shows the percentage of oxygen concentration. Adapted from [56]

further delays the cell-cycle dynamics through the upregulation of p27/p21 pathway [29, 57]. This is incorporated into Equation (1) using an additional decay term proportional to the concentration of p27/p21 (which is considered here as constant) [4,57]. Thus the modified equation for cyclic-CDK dynamics can be written as

$$\frac{d[\text{CycB}]}{dt} = k_1 - (k_2' + k_2''[\text{Cdh1}] + [\text{p27/p21}][HIF])[\text{CycB}], \tag{9}$$

where

$$[HIF] = \begin{cases} 1: & K(x,t) \leq 10\% \\ 0: & K(x,t) > 10\%. \end{cases} \tag{10}$$

2.3 *Cancer Growth and Progression: Cellular Automaton Model*

Here, we use a cellular automaton (CA) approach to study the spatial and temporal dynamics at the cell level incorporating the intracellular and macroscopic details described above. Previously, cellular automaton modelling approaches have been used to model various aspects of tumor development and progression, including the formation of multicellular spheroids [35, 52], tumor-induced angiogenesis [9], cancer cell adhesion and invasion [73]. Moreover, several hybrid CA models have already seen use to study tumor growth and development in multiple scales and time [3,5,21,28,49,52,54,60].

Here, the computational model is simulated on a spatial grid of size 100×100 grid points and each automaton element whether it is empty or occupied has a physical size of $l\times l$, where $l = 20\ \mu m$, simulating a cancer tissue of $2\times2\ mm^2$ area. The CA begins as a new grid of empty points with a single initial cell in G1 phase (blue cell) of the cell cycle at the center. This initial cell divides following the intracellular cell-cycle dynamics modelled using Equation 1–6 and produces a cluster of cells on a regular square lattice with no-flux boundary conditions. The PDE equation governing the oxygen dynamics is simulated using a finite difference scheme and the system of ordinary differential equations controlling cell-cycle dynamics is solved using the Runge-Kutta method to obtain the intracellular protein concentrations for each cell. At each simulation step, these intracellular protein levels are checked for each individual cell and its cell-cycle phase is updated accordingly. A cell is considered to be in S-G2-M phase (green cell) if its [CycB] level is greater than a specific threshold and if it is lower than this value, the cell is in G1 phase. If the cyclin B-cdk complex concentration [CycB] crosses this threshold from above, the cell is considered ready for division. Once the cell is marked for division, its neighborhood of order 3 is checked for an empty space with highest oxygen gradient and the cell undergoes cell division and its mass [mass] is halved. If the cell's neighborhood has "no space" then its growth rate μ is set to zero and it enters a resting state (magenta cell). After the division, the new cell is placed with a G1 state of cell cycle and is assigned a value for μ randomly from the range of values of μ. If there are more than one empty space with same oxygen gradient, a position is chosen randomly. The position of the new daughter cells is determined by Moore and von Neumann neighborhood alternatively to avoid the associated cell distribution patterns specific to each method [56,57].

As this multiscale model evolves over time, the number of cells increases and the oxygen consumption also increases accordingly, eventually leading to tissue hypoxia. Here, the hypoxic cells that are in G1 phase are represented by rose color-coded cells while hypoxic S-G2-M cells and hypoxic resting cells are denoted by the colors yellow and silver, respectively. Figure 4 shows a snapshot of the multiscale hybrid CA model after 700 h where Fig. 4a illustrates the distribution of oxygen in a percentage scale, Figure 4(b) gives the HIF-1α map and Fig. 4c shows the distribution of cells in various cell cycle phases. Proliferating cells which are active in cell-cycle are mainly seen near the high oxygen concentration regions of the tumor boundary, creating fingerlike growth pattern in the tumor cell distribution. This is mainly due the to effect of surrounding microenvironment as the tumor tends to advance towards the most favorable microenvironment that supports its growth and invasion. In Fig. 5 we plot the corresponding total number of cells along with the number of cells in each of the various cell-cycle phases against time for this asynchronous population.

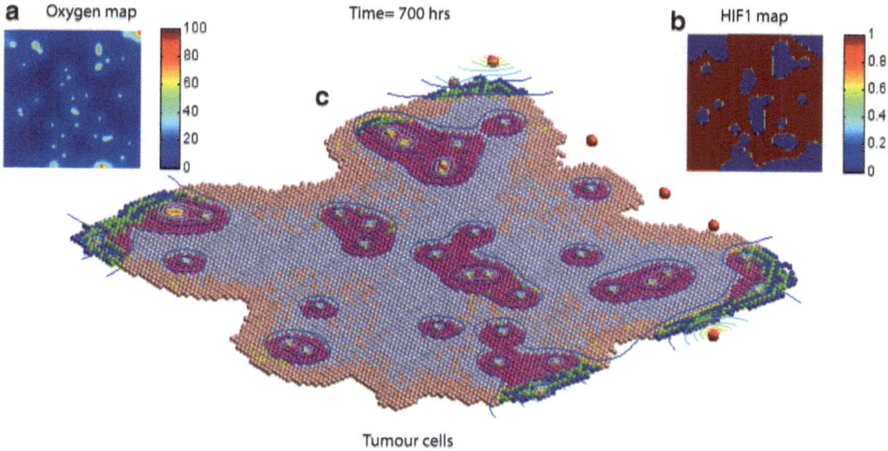

Fig. 4 Plots showing a snapshot of the simulation results of the model at time= 700 h. (**a**) plot of the oxygen concentration distribution within the spatial domain, (**b**) plot of the HIF-1 α concentration within the spatial domain and (**c**) plot of the spatial distribution of the cells in different stages of the cell cycle which are G1 (*blue*), S-G2-M (*green*), resting (*magenta*), hypoxic cells in G1 (*rose*), hypoxic cells in S-G2-M (*yellow*) and hypoxic cells in resting (*silver*). The contours represent the oxygen concentration profiles. Adapted from [57]

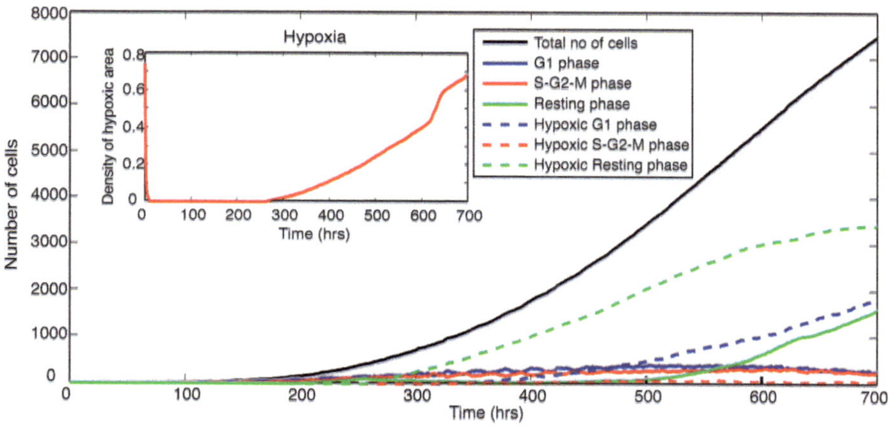

Fig. 5 Plot showing the total number of cells in the different phases of the cell cycle over the course of one simulation run representing 700 h. The subplot shows the density of hypoxic area as a function of time. Adapted from [57]

3 Mathematical Model: Effects of Anticancer Treatments

Chemotherapy and radiotherapy play important roles in the primary treatment of many cancers and in improving the survival after cancer surgery. Currently, numerous chemotherapeutic drugs and irradiation techniques are employed, which have evolved over several decades through empirical clinical usage. The effectiveness of these treatment protocols is also considerably affected by internal tumor heterogeneities caused by perturbations in the intracellular pathways as well as by the dynamical changes in the tissue microenvironment, in particular oxygen concentration [10]. Hence, it is also important to consider such heterogeneity when studying various optimization protocols, as this can help in improving the delivery of multi-modality treatments. Mathematical modelling of such complex dynamic situations might provide one solution to this problem, and speed up the delivery of efficacious treatments to patients while preventing the use of potentially sub-optimal treatment combinations. Mathematical and computational models can also be very helpful in gaining valuable insights into the mechanisms and consequences of various complex intra-cellular and intercellular changes during and after the therapy. Here, we use the hybrid multiscale cellular automaton model described previously, incorporating the effects of oxygen heterogeneity and cell-cycle dynamics to study the multiple effects of cell-cycle-dependent chemotherapy and radiotherapy.

3.1 Modelling the Effects of Cell-Cycle-Specific Chemotherapy

Several previous mathematical models have shed light on understanding the effects of chemotherapy and its optimal delivery [1, 17, 60, 68]. Chemotherapy is a commonly used treatment for cancer. Chemotherapeutic drugs act on rapidly proliferating cells, such as cancer cells, by interfering with the cell-cycle and other cell-cycle specific targets. Cell-cycle-specific anticancer drugs are more effective on dividing cells by interfering with the cell-cycle and other cell-cycle-specific targets. Hence, it might be more efficient to use a combination of multiple cell-cycle phase-specific drugs that target the cells in different phases of the cell cycle. Here, the concentration of chemotherapeutic drug type i, $C_i(x,t)$ is governed by a similar equation as that of oxygen distribution (8), given by;

$$\frac{\partial C_i(x,t)}{\partial t}=\nabla.(D_{ci}(x)\nabla C_i(x,t))+r_{ci}(x)m(x)-\phi_{ci}C_i(x,t)\text{cell}(x,t)-\eta_{ci}C_i(x,t), \tag{11}$$

where $D_{ci}(x)$ is the diffusion coefficient of the drug, ϕ_{ci} is the rate by which the drug is taken in by a cell (assumed to be zero as it is very negligible when compared to oxygen uptake), r_{ci} is the drug supply rate by the pre-existing vascular network and η_{ci} is the drug decay rate [55].

Drug molecules are much bigger in size than oxygen molecules and their supply through the normal blood vessels is minimal. Additionally, the drug diffusion rate varies depending on the vessel location within the domain and with other factors such as pressure. Hence, similar to our assumptions for oxygen, these effects of pressure are taken into account through the following assumptions related to the transport properties and delivery rate of the drug [57]. The parameters that are used in the equations governing the dynamics of chemotherapy are chosen in comparison with oxygen molecules and other known compounds of similar molecular mass [57]. In the model, chemotherapeutic drugs are assumed to be effective in killing a cell with a probability p_i, if its concentration at that location of the cell is above a fixed threshold value θ_{ci} and below which the drug has no effect on any cells. Using this model, we have further analyzed the effects of these drugs which are delivered at a same rate in two different combinations targeting either the same phase or a different phase of the cell-cycle.

The temporal evolution of the tumor dynamics when the time delay between two drug doses is 84 h (chosen as a representative illustration) is given in Figs. 6, 7 and 8. Figs. 6a–d and 7a–d show the changes in the total number of cells as well as the number of cells that are in various phases of the cell-cycle when the phase-specific drugs are delivered in multiple combinations. Figure 8 illustrates the spatial evolution of the cancer cell distribution during the administration of the chemotherapeutic drugs. These dynamical changes in the spatial evolution of cancer cell distribution will have a significant impact on the final cell distribution when more than two doses of cell-cycle-specific chemotherapy are administered subsequently. Hence, it is very important to consider the cell-cycle heterogeneity together with the oxygen heterogeneity while studying the effects of these drugs to obtain an accurate prediction.

The graph in Fig. 6a shows that when the delay is 84 h, the combination of two S-G2-M state-specific drugs gives a slightly better cell kill at the final simulation time as compared to the other three possible combinations of these two drug doses. The worst combination is the delivery of two G1-specific drug doses while other two combinations give a similar outcome as that of the application of two S-G2-M drugs. This can be explained by the presence of internal heterogeneity due to the cell-cycle dynamics which oscillates between the G1 and S-G2-M states. When the first dose is given, the majority of the cells that are in the vicinity of the diffused drug concentration might happen to be in the G1 state, leaving the cells in the other cycling phase, hence synchronizing the cell population. When a fraction of the cells is killed by the delivered drug, the remaining cells undergo a spatial redistribution that further changes the dynamics of oxygen and drug concentrations, which will eventually affect the future drug delivery. Interestingly, although the first or second doses of G1-specific drug kill more cells than that of S-G2-M-phase drug, some combination involving a G1-phase drug results in a higher number of surviving cells at the final time, 700 h. This is because the higher rate of cell kill eliminates most of the cells in the resting phase, increasing the number of available free space and decreasing the hypoxic area (Figs. 7, 8). As free space is created by the cell kill, the resting cells will return to their active cell cycle and these viable cells utilize this favorable microenvironment to proliferate rapidly and reach a maximum size in the absence of further doses of chemotherapy. An extensive analysis and results are given in [57].

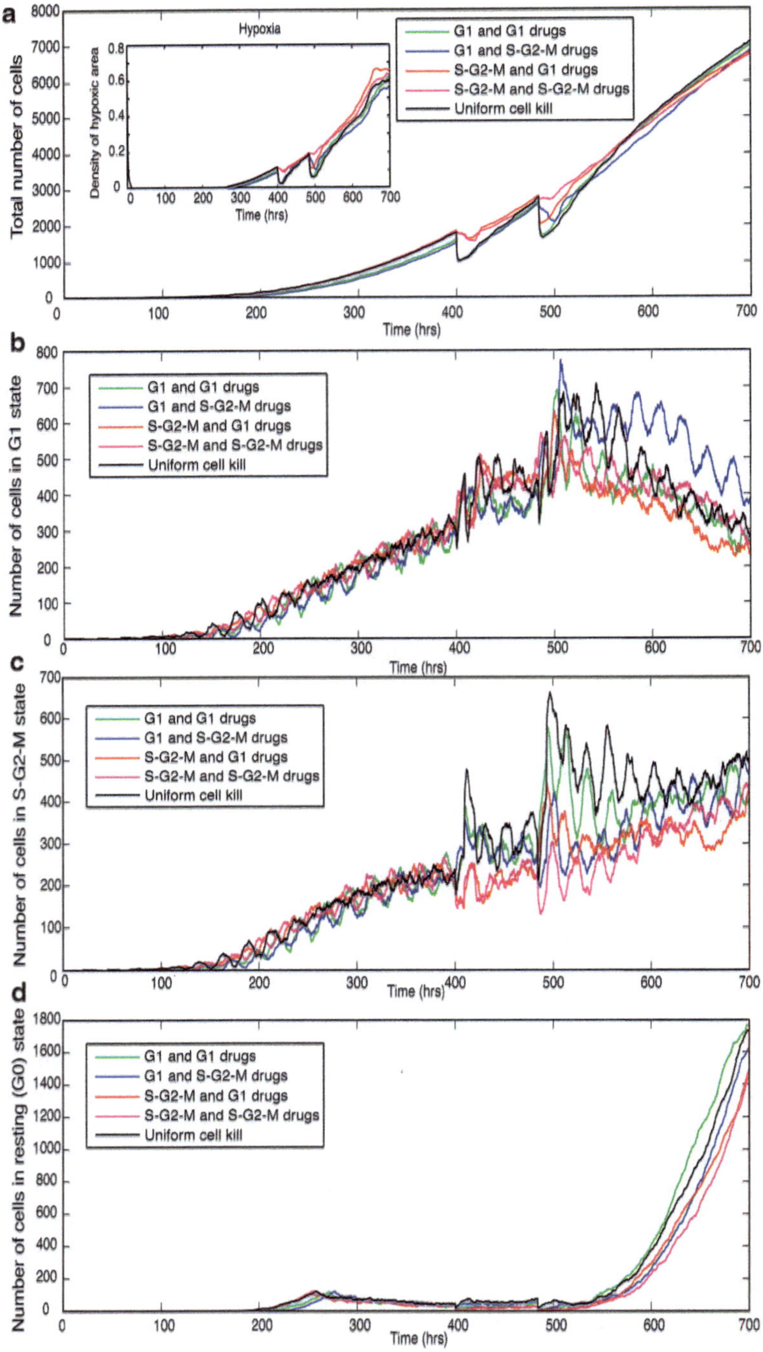

Fig. 6 Plots showing the total number of cells as well as the number of cells in different states as a function of time when two drug doses are given with a time delay of 84 h. (**a**) plot showing the total number of cells and subplot showing the density of hypoxic area over time, (**b**) Plot showing the number of cells in G1 state, (**c**) plot showing the number of cells in S-G2-M state and (**d**) Plot showing the number of cells in G0 state. Adapted from [57]

a

b

c

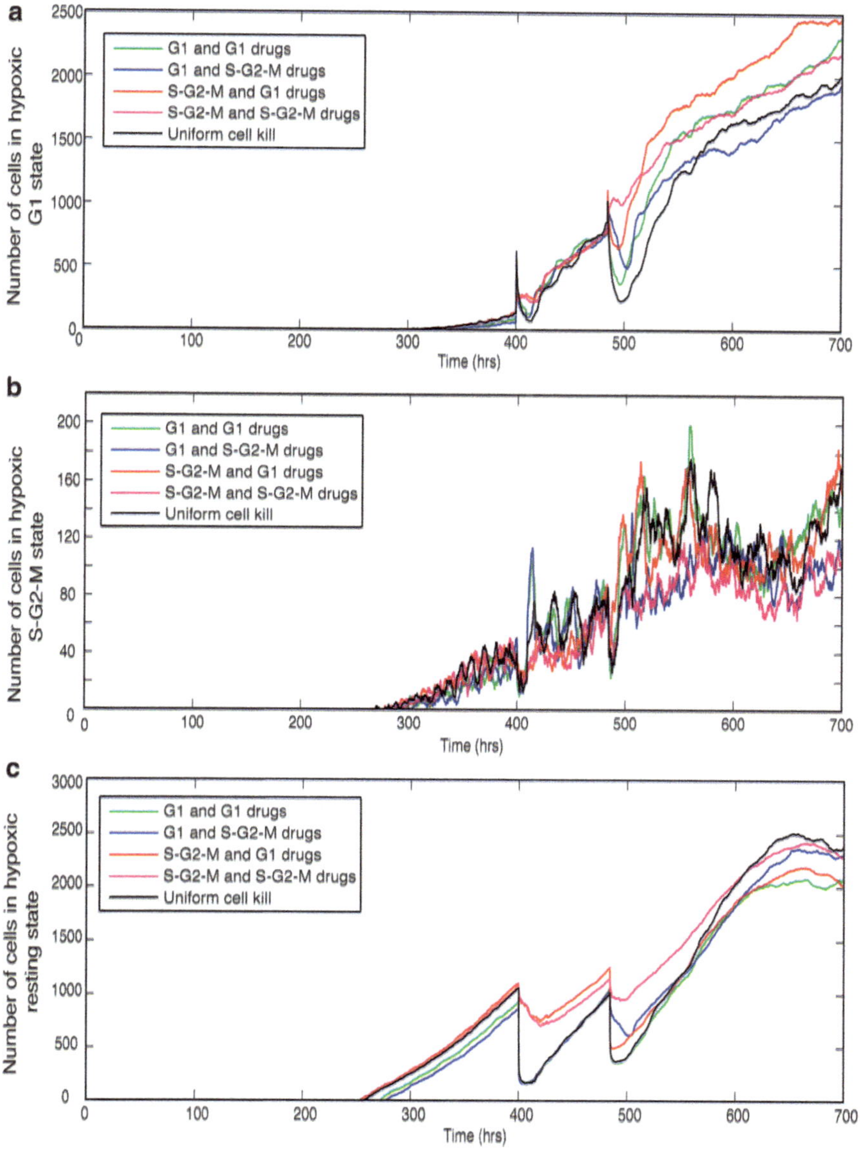

Fig. 7 Plots showing the number of cells in different states as a function of time when two drug doses are given with a time delay of 84 h. (**a**) Plot showing the number of cells in G1 state that are in hypoxic region, (**b**) plot showing the number of cells in S-G2-M state that are in hypoxic region and (**d**) plot showing the number of cells in G0 state that are in hypoxic region. Adapted from [57]

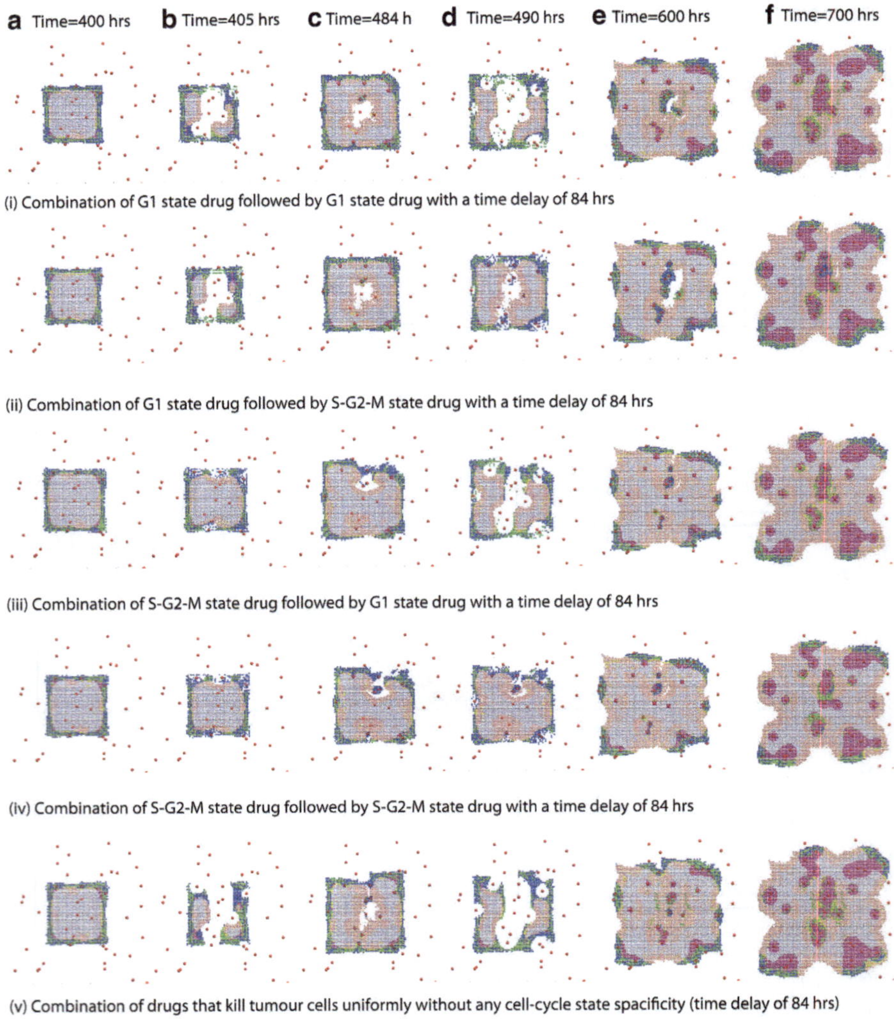

Fig. 8 Plots showing the spatial distributions of cells within a growing solid tumor at time = 400 h (delivery of 1st drug, column (**a**)), time=405 h (column (**b**)), time=484 h (2nd drug delivery, column (**c**)), time=490 h (column (**d**)), time=600 h (column (**e**)) and time=700 h (final time, column (**f**)), for various combinations of cell-cycle-specific drugs (i to v). These figures show that consideration of spatial distribution of cells may highly benefit in planning and optimizing the delivery of cell-cycle-dependent drugs. Adapted from [57]

3.2 Modelling the Effects of Radiation Therapy

In a similar manner to chemotherapy, the key intracellular processes such as cell-cycle dynamics and external factors including oxygen distribution also play a vital role in determining the radiosensitivity of cells that are irradiated [53, 71].

In addition, the delivered radiation fractions (treatments) further dynamically change this radiosensitivity over time by redistributing the cancer cells within the cell cycle by inducing repopulation of the tumor mass, by allowing reoxygenation of the tumor and by activating DNA repair mechanisms [42, 53, 71]. Traditionally, the survival fraction of cells after they are irradiated is calculated using the linear quadratic (LQ) model [65]. The LQ model considers the effects of both irreparable damage and repairable damage susceptible to misrepair. One of the basic assumptions of the LQ model is that a cell is damaged through double-strand breaks (DSBs) of its DNA, leading to reproductive cell death. This damage to DNA can happen in two different ways, which are captured in the LQ model through its linear and quadratic terms. The linear term accounts for the DSBs due to one single hit of radiation whereas the quadratic term represents the effects of two separate ionizing events that eventually cause DNA DSBs [65]. Assuming that the probability of one interaction causing a DSB is linearly proportional to the dose d, the survival fraction in the LQ model can be written as

$$S(d) = \exp(-\alpha d - \beta d^2), \tag{12}$$

where d is the radiation dose and α and β are sensitivity parameters, taken to be $\alpha = 0.3\ Gy^{-1}$ and $\beta = 0.03\ Gy^{-2}$ [55].

As discussed above, experimental observations indicate that the relative radiation sensitivity of a cancer cell depends on multiple factors, including its oxygenation status and its cell-cycle phase [6, 53, 75]. While the cells are found to be more sensitive when in the S-G2-M phase as compared with the G1 phase, their relative radiation sensitivity is minimal at low oxygen levels. The effect of changing tissue oxygen levels within the spatial domain on the radiation sensitivity can be incorporated into the LQ model (Eq. 12) by using the concepts of an oxygen enhancement ratio (OER) or oxygen modification factor (OMF) [55], defined as

$$\text{OMF} = \frac{\text{OER}(pO_2)}{\text{OER}_m} = \frac{1}{\text{OER}_m}\frac{\text{OER}_m . pO_2(x) + K_m}{pO_2(x) + K_m}, \tag{13}$$

where $pO2(x)$ is the oxygen concentration at position x, OER is the ratio of the radiation doses needed for the same cell kill under anoxic and oxic conditions, $\text{OER}_m = 3$ is the maximum ratio and $K_m = 3$ mm Hg is the $pO2$ at half the increase from 1 to OER_m [55]. Furthermore, a varying radiation sensitivity based on the cell's cell-cycle status is incorporated into the LQ model using a parameter γ, which varies from 0 to 1, depending on the individual cell's position at the time of the irradiation. Here, we assumed that the cells in S-G2-M phase have maximum sensitivity with $\gamma = 1$, while the cells in G1 phase and the resting phase have relative sensitivities of $\gamma = 0.5$ and $\gamma = 0.25$, respectively.

The equation for survival probability (LQ model, Eq. 12) is then modified by incorporating this varying sensitivity due to the changes in cell-cycle phase and oxygen levels by additional terms γ and OMF [56], as follows:

$$S(d) = \exp\left[\gamma\left(-\alpha.\text{OMF}.d - \beta(\text{OMF}.d)^2\right)\right]. \tag{14}$$

The effects of cellular repair are included in the model by assuming that 98% of damage caused by the radiation is likely to be repaired within few hours of radiation, if they are treated with low-dose radiation (< 5 Gy) [31, 72] and allowing the cells to stay in the same phase (divisional delay) for an extra time duration of up to 9 h for repairing the damage [42]. This survival probability is then used to calculate the survival chances of each cell when they are irradiated with the radiation rays. To study this survival chance of an individual cell, a random number is drawn for each cell at every time when it is irradiated and compared against the calculated survival probability. The irradiated cell survives if the random number is smaller than the survival probability and dies otherwise.

This single-cell-based radiation model is used to study multiple factors involving radiation therapy. To study cell-cycle dynamics when cells are treated with multiple fractions of radiation, we have simulated the radiation therapy of the cancer cells with 2.5 Gy/day for 5 days, up to 12.5 Gy starting at time = 400 h. This is compared against the results of irradiation with a single dose of 12.5 Gy given at time = 400 h and the control cell distribution. The total number of cells and the number of cells that are in G1, G2 and resting phases (b) when cells are treated with a single dose of radiation and (c) when the cells are treated with fractional doses of radiations are plotted against time in Figs. 9b and c, respectively [56]. The subplot in these Figures show the percentage of hypoxic area with respect to the time. Figure 9b shows that when the cells are irradiated with a single dose of radiation, the majority of the cells stay in G2 phase of the cell-cycle after the radiation for a short period and then the majority move into G1 and stay in that distribution for about 60–70 h (partial synchronization) before eventually recovering and following the cycling pattern seen in the case of the control cell population. However, when the total dose of 12.5 Gy is given in 5 small doses, the synchronization is observed only during the treatment time and is lost as soon as the radiation is stopped (Fig. 9c). In both cases, the number of cells in resting phase decreases as the resting cells re-enter the active phase of the cell cycle (into G1 phase) with the creation of empty spaces and a favorable microenvironment.

Figure 10 shows the analysis of various factors affecting a cell's radiation sensitivity. Figure 10a shows the temporal changes in the total number of cells and the cells that are in various phases of the cell cycle, when cells are irradiated with 2.5 Gy/day for 5 days in a 100 % oxygenated microenvironment. In Fig. 10b, we plot the number of cells if no cell-cycle delay after the radiation is assumed. The effects of radiation therapy on the number of cells when no cell-cycle specificity for the radiation sensitivity is assumed are given in Fig. 10c. The number of cells for the case where no DNA repair is assumed is plotted in Fig. 10d and shows an increase in the number of cells killed as compared to the normal case as expected.

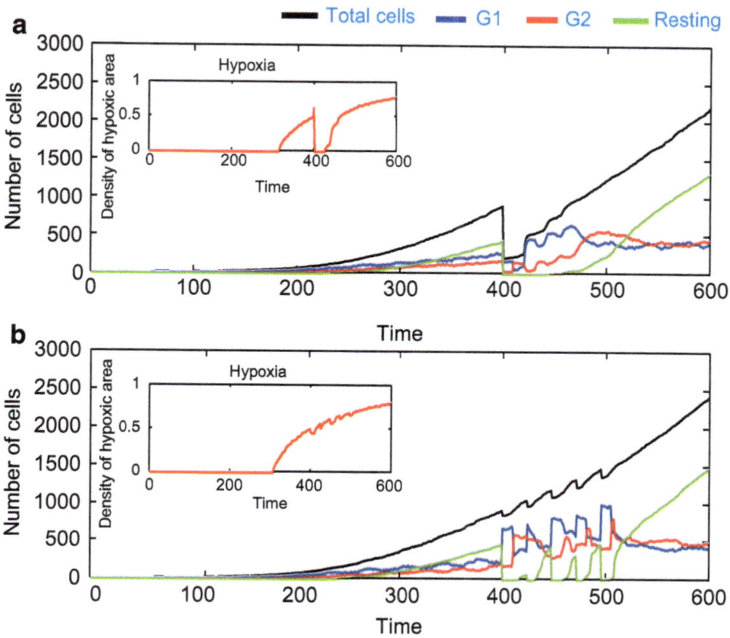

Fig. 9 Number of cells in G1, G2 and resting phases for a heterogeneous environment. The number of cells (**a**) under single dose of radiation with 12.5 Gy and (**b**) under fractional radiation starting at time =400 h (5 fractions of 2.5 Gy). Adapted from [56]

In all four cases (Fig. 10), after the irradiation, the presence of a favorable tissue microenvironment increases the number of cells in active cell cycle. It can be seen from Fig. 10 that out of four factors that we have considered, the cell-cycle phase-specific radiation sensitivity of the individual cells and the activation of the repair mechanisms within the cell significantly affect the relative radio-sensitivity of a cell.

We have also analyzed the usefulness of the developed model by comparing the simulation results of the model with the experimental results when the cells are irradiated with a radiation dose of 3 Gy. The comparison results are given in Fig. 11 [16, 56]. The results show that the controls have cells predominantly distributed in G1 as compared to G2. After irradiation, the majority of the cells started to accumulate in G2 phase, about 12 h after irradiation, and stayed in G2 phase before going back to a G1-phase-dominant cell distribution by 22–24 h after irradiation. The results show a qualitative agreement with the experimental results. More analyses about the model and additional results can be found in [56].

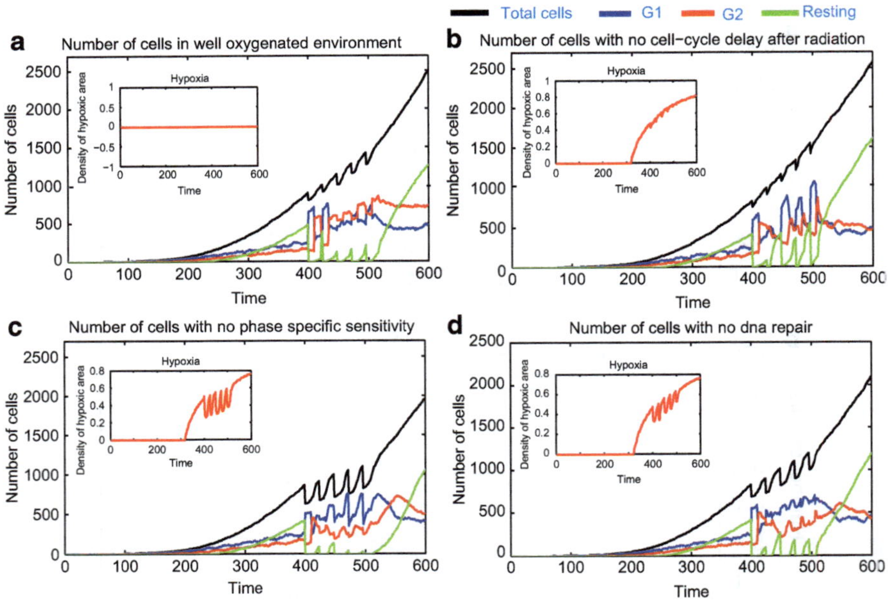

Fig. 10 Number of cells under various conditions that influence the radiation damage after the irradiation. (**a**) Plots for a well-oxygenated microenvironment, (**b**) plot assuming no cell-cycle delay for repair after the radiation, (**c**) plots assuming there is no cell-cycle phase-specific sensitivity for repair after the radiation and (**d**) plots when there is no DNA repair after the radiation. Adapted from [56]

3.3 Modelling the Effects of Combination Therapy

Clinically, a kinetically based administration of chemotherapy and radiation therapy is often observed to give a better outcome than that of chemotherapy or radiation therapy alone [25, 33, 53]. Studies have shown that both radiation therapy and chemotherapeutic drugs can induce a cell-cycle synchrony and arrest cells at a particular cell-cycle phase which improves the effectiveness of the next dose of radiation/chemotherapy [25,53]. However, most of these interactions are dependent on the type of drugs given and their combinations among their doses and radiation fractions. Hence an appropriate combination of these therapeutic modalities is an essential requirement to achieve a maximum survival [32, 34]. Here, we use our multiscale model to study two different combination regimes of radiation therapy and chemotherapy and some of the representative results are given in Figs. 12 and 13. Note that in these combinations, we keep the total treatment time constant to compare their effects on tumor control. We have also used the same set of parameter values and doses for each phase-specific chemotherapy [57].

Figure 12 shows the effects of combination treatments when two doses of the chemotherapy drugs are given after radiation therapy at times = 466 h and 496 h.

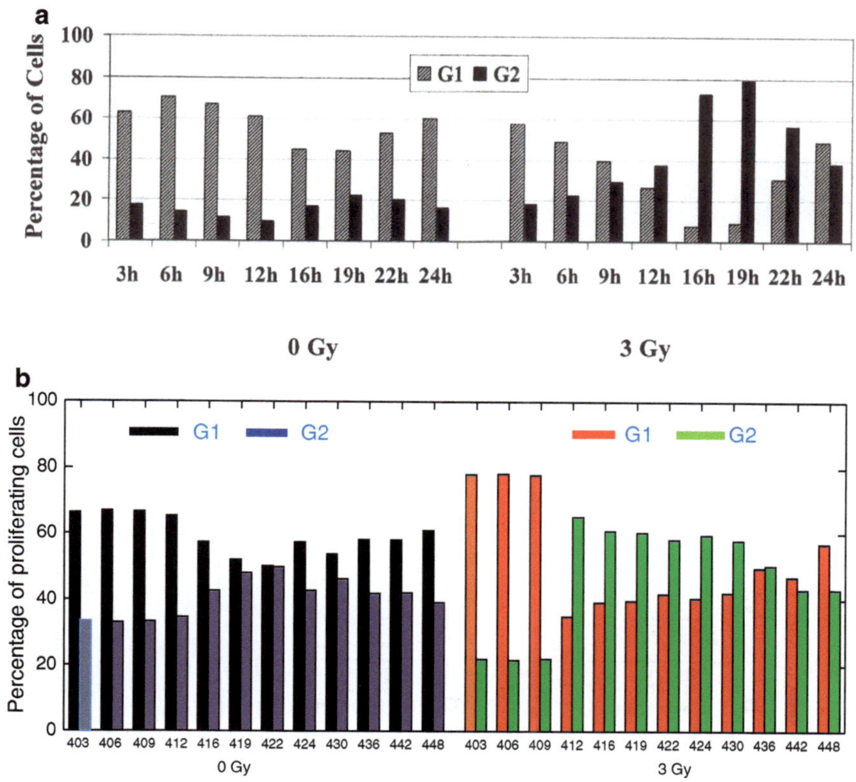

Fig. 11 Percentage of cells in G1 and G2 states with and without radiation. (**a**) Experimental results from [16]. Cells were irradiated with 3 Gy and samples were taken after 3 h, 6 h, 9 h, 12 h, 16 h, 22 h and 24 h for flow cytometry analysis. The cell-cycle distribution of irradiated cells was compared to unirradiated cells collected at the same time points. The left peak in each case represents cells in G1 and the right peek represents cells in G2 phase. From Chaudhry [16] (BioMed Central OpenAccess). and (**b**) Simulation results (percentage of proliferating cells). The percentage of proliferating cells in G1 and G2 phases when the cell is irradiated with dose=3 Gy at time=400, 403, 406, 409, 412, 416, 419, 422, 424, 430, 436, 442 and 448 h. The cell-cycle distribution of irradiated cells was compared to unirradiated cells collected at the same time points. The results show a qualitative agreement with the experimental results. Adapted from [56]

The radiation starts at time= 340 h with a fractional daily dosage of 2 Gy for 5 days. Here, the radiation given before two doses of chemotherapy introduces a partial cell-cycle synchrony of cell distribution that remains until the end of the therapy. The cell-phase distribution for the case when the doses of chemotherapeutic drug are given before and after the radiation is plotted in Fig. 13. The administration of a G1-phase-specific chemotherapeutic drug redistributes the cells so that the majority of the cells are in G1 phase throughout the radiation period. This is the consequence of an increased cell kill, that promotes an increased proliferation as cells will re-enter the active phase of the cell cycle (G1) when conditions become

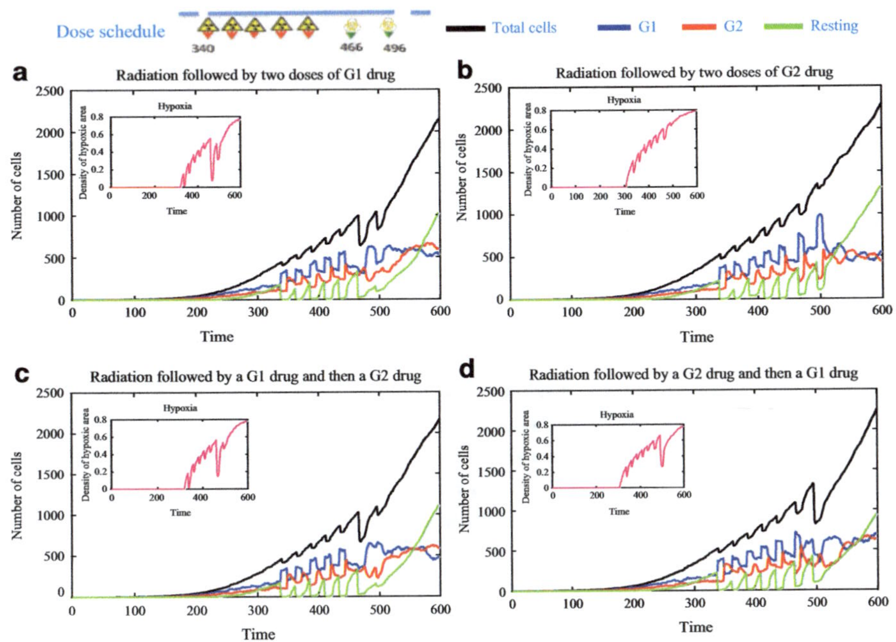

Fig. 12 Number of cells when chemotherapy is given after radiation therapy. Two doses of G1 and/or G2 drugs are given at time=466 h and 496 h, after 5 fractions of radiation therapy (1 week) with a daily dose of 2.5 Gy starting at time=340 h. (**a**) Plots when two G1-phase-specific drugs are given after radiation, (**b**) plots when two G2-phase-specific drugs are given after radiation, (**c**) plots when a G1-phase-specific drug followed by a G2 specific drug are given after radiation and (**d**) plots when a G2-phase-specific drug followed by a G2-specific drug are given after radiation. Adapted from [56]

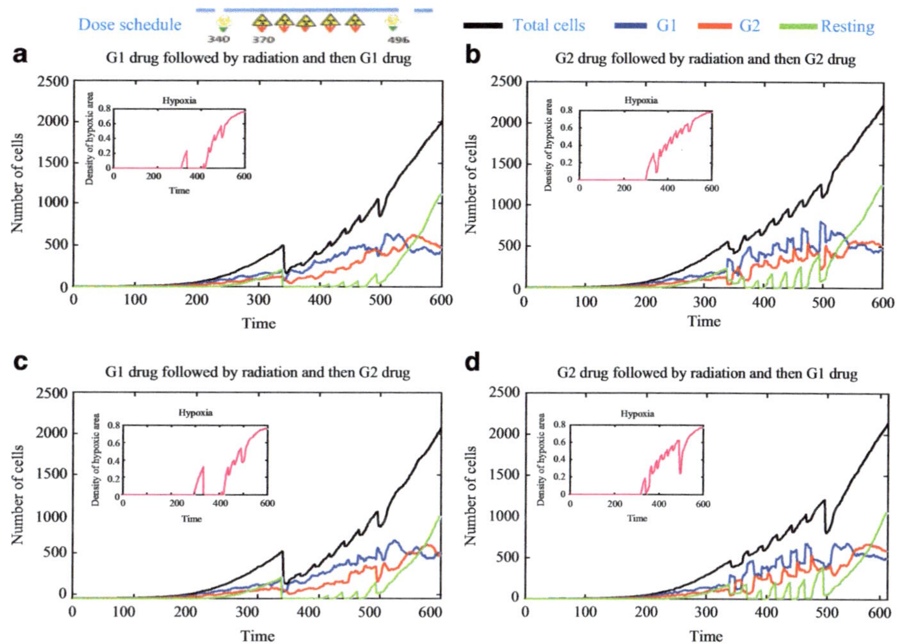

Fig. 13 Number of cells when a chemotherapy (one dose) is given before and after radiation therapy. Two doses of G1 and/or G2 drugs are given at time=340 h and 496 h and 5 fractions of radiation therapy (1 week) with a daily dose of 2.5 Gy are given in between the chemotherapy doses, starting at time=370 h. (**a**) Plots when two G1-phase-specific drugs is given, each before and after radiation, (**b**) plots when two G2-phase-specific drugs are given, each before and after radiation, (**c**) plots when a G1-phase-specific drug is given before the radiation followed by a G2-specific drug and (**d**) plots when a G2-phase-specific drug is given before radiation followed by a G2 specific drug. Adapted from [56]

favorable. Alternatively, the administration of a G2-specific drug, which kills fewer cells compared to the G1-phase-specific drug, helps to keep the cells in synchrony throughout the treatment time. However, in both cases shown, in the absence of additional fractions of radiation and further doses of chemotherapy, all the schedules perform in a similar fashion, although some give a better cell kill.

4 Conclusions

Along with the rapid growth in acquisition of genetic, proteomic and other biochemical and biological data, there has been a parallel development from the theoretical side in terms of modelling. In particular, systems biology has emerged as a field of research over the past decade applied to a wide range of problems in the biomedical sciences. Systems biology seeks to bring to bear a range of interdisciplinary skills and tools on complex biomedical problems. By adopting a holistic or integrative approach (as opposed to the more traditional reductionist logic), systems biology aims to predict emergent behavior that will arise from complex biomedical systems i.e., behavior that appears over time due to the interactions between genes, proteins, cells and tissues across a range of spatial and temporal scales. Given the complexity of most biomedical systems and the inherent nonlinearities in such systems, without adopting some kind of systems approach it is not possible to make accurate predictions. Indeed, in the last few years, systems biology itself has evolved and further developed seeking not just to understand events at the separate biological scales in a qualitative manner, but there are now mathematical models which are truly multiscale, leading to the emergence of quantitative systems biology or quantitative integrative biology. This novel systems approach is now being brought to bear on cancer modelling and a related discipline of what may be termed systems oncology now exists in its own right to develop predictive multi-scale models of cancer growth and spread.

Established mathematical models now exist for all the key phases of solid tumor growth, i.e., avascular growth [13, 14], tumor-induced angiogenesis [9, 45], the immune response to cancer [2, 43], invasion and metastasis [8, 19, 27, 58, 59] and vascular growth [41, 76]. New areas are also now being investigated concerning the spatio-temporal modelling of intracellular pathways associated with cancer such as p53-Mdmd2 [69, 70]. A comprehensive overview of the field may be found in the review article [40]. In the past few years especially, multiscale models of solid tumor growth have been developed in order to account for the different spatial and temporal scales (from genes to tissues) that occur not only in cancer but in all biological systems [3, 5, 63, 77]. A review of recent models in this area may be found in the paper of Deisboeck et al. [20]. There has also been a concerted effort to integrate mathematical models of cancer with real data in a genuine attempt to develop quantitative, predictive models [12]. Alongside these developments in the field in general, it is natural that models of cancer treatment have begun to be formulated both for chemotherapy [1, 23, 24, 39, 46, 51, 62] and radiotherapy [22, 61, 64]. More recently models adopting a multiscale approach to treatment modelling have been developed [11].

In this chapter we have adopted a multiscale approach to modelling cancer treatment. Specifically, we have presented a hybrid multiscale cellular automaton model to study the effects of cell-cycle-phase specific chemotherapy and radiation therapy, alone and in combinations [56, 57]. The clinical and experimental observations indicated the internal and external heterogeneities within a cancer cell play important roles in prescribing the effects of chemotherapeutic drug as well as radiation therapy [53]. For this reason we incorporated the effects of cell-cycle-mediated chemotherapeutic and radiation sensitivity as well as the effects of changing oxygen and tissue dynamics within this hybrid cell-based modelling framework. The results obtained from the model are in qualitative agreement with experimental results [56] and thus show its potential usefulness in studying and understanding a kinetic administration of cell-cycle phase-specific chemotherapeutic drugs in combination with radiation therapy. Future work will consider other factors that may interfere with the cell-cycle dynamics such as circadian rhythms and wee1 dynamics and explore the therapeutic benefits of "chronotherapy," alone and in combination with the radiation therapy. Furthermore, we will also incorporate the interplay between normal cells and cancer cells, as their interactions are equally important in studying the therapeutic benefit with a minimum damage to the normal cells.

The long-term goal of such interdisciplinary, multiscale "systems oncology" modelling is to build a virtual cancer made up of different but connected mathematical models at the different biological scales (from genes to tissue to organ). The development of quantitative, predictive models (based on sound biological evidence and underpinned and parameterized by biological data) will no doubt have a positive impact on patients suffering from diseases such as cancer through improved clinical treatment and is a real motivation for becoming involved in such modelling.

Acknowledgements The authors gratefully acknowledge the support of the ERC Advanced Investigator Grant 227619, M5CGS - From Mutations to Metastases: Multiscale Mathematical Modelling of Cancer Growth and Spread.

References

1. Z. Agur, R. Hassin, S. Levy, Optimizing chemotherapy scheduling using local search heuristics. Operat. Res. **54**(5), 829–846 (2006)
2. M. Al-Tameemi, M. Chaplain, A. d'Onofrio, Evasion of tumours from the control of the immune system: consequences of brief encounters. Biol. Direct **7**, 31 (2012)
3. T. Alarcon, H.M. Byrne, P.K. Maini, A cellular automaton model for tumour growth in inhomogeneous environment. J. Theor. Biol. **225**, 257–274 (2003)
4. T. Alarcon, H.M. Byrne, P.K. Maini, A mathematical model of the effects of hypoxia on the cell-cycle of normal and cancer cells. J. Theor. Biol. **229**, 395–411 (2004)
5. T. Alarcon, H.M. Byrne,, Maini, .: A multiple scale model for tumour growth. Multiscale Model. Sim. **3**, 440–475 (2005)

6. Alper, T., Howard-Flanders, P.: Role of oxygen in modifying the radiosensitivity of E. coli B. Nature **178**(4540), 978–979 (1956)
7. A. Altinok, F. Levi, A. Goldbeter, A cell cycle automaton model for probing circadian patterns of anticancer drug delivery. Adv. Drug Deliv. Rev. **59**, 1036–1053 (2007)
8. V. Andasari, A. Gerisch, G. Lolas, A.P. South, M.A. Chaplain, Mathematical modeling of cancer cell invasion of tissue: biological insight from mathematical analysis and computational simulation. J. Math. Biol. **63**(1), 141–171 (2011)
9. A.R. Anderson, M.A. Chaplain, Continuous and discrete mathematical models of tumor-induced angiogenesis. Bull. Math. Biol. **60**, 857–899 (1998)
10. J.C. Bailar, H.L. Gornik, Cancer undefeated. N. Engl. J. Med. **336**, 1569–1574 (1997)
11. F. Billy, B. Ribba, O. Saut, H. Morre-Trouilhet, T. Colin, D. Bresch, J.P. Boissel, E. Grenier, J.P. Flandrois, A pharmacologically based multiscale mathematical model of angiogenesis and its use in investigating the efficacy of a new cancer treatment strategy. J. Theor. Biol. **260**(4), 545–562 (2009)
12. H.M. Byrne, Dissecting cancer through mathematics: from the cell to the animal model. Nat. Rev. Cancer **10**, 221–230 (2010)
13. H.M. Byrne, M.A. Chaplain, Growth of nonnecrotic tumors in the presence and absence of inhibitors. Math. Biosci. **130**(2), 151–181 (1995)
14. H.M. Byrne, M.A. Chaplain, Growth of necrotic tumors in the presence and absence of inhibitors. Math. Biosci. **135**(2), 187–216 (1996)
15. M. Chaplain, A. Anderson, Mathematical modelling of tumour-induced angiogenesis: network growth and structure. Cancer Treat. Res. **117**, 51–75 (2004)
16. M.A. Chaudhry, Base excision repair of ionizing radiation-induced DNA damage in G1 and G2 cell cycle phases. Cancer Cell. Int. **7**, 15 (2007)
17. J. Clairambault, A step toward optimization of cancer therapeutics. Physiologically based modeling of circadian control on cell proliferation. IEEE Eng. Med. Biol. Mag. **27**, 20–24 (2008)
18. A. Dasu, I. Toma-Dasu, M. Karlsson, Theoretical simulation of tumour oxygenation and results from acute and chronic hypoxia. Phys. Med. Biol. **48**, 2829–2842 (2003)
19. N.E. Deakin, M.A. Chaplain, Mathematical modeling of cancer invasion: the role of membrane-bound matrix metalloproteinases. Front. Oncol. **3**, 70 (2013)
20. T.S. Deisboeck, Z. Wang, P. Macklin, V. Cristini, Multiscale cancer modeling. Annu. Rev. Biomed. Eng. **13**, 127–155 (2011)
21. S. Dormann, A. Deutsch, Modeling of self-organized avascular tumor growth with a hybrid cellular automaton. In Silico Biol. (Gedrukt) **2**, 393–406 (2002)
22. H. Enderling, A.R. Anderson, M.A. Chaplain, A.J. Munro, J.S. Vaidya, Mathematical modelling of radiotherapy strategies for early breast cancer. J. Theor. Biol. **241**(1), 158–171 (2006)
23. K. Fister, J. Panetta, Optimal control applied to cell-cycle-specific cancer chemotherapy. SIAM J. Appl. Math. **60**, 1059–1072 (2000)
24. H.B Frieboes, M.E. Edgerton, J.P. Fruehauf, F.R. Rose, L.K. Worrall, R.A. Gatenby, M. Ferrari, V. Cristini, Prediction of drug response in breast cancer using integrative experimental/computational modeling. Cancer Res. **69**, 4484–4492 (2009)
25. K. Fu, Biological basis for the interaction of chemotherapeutic agents and radiation therapy. Cancer **55**(S9), 2123–2130 (1985)
26. C. Gerard, A. Goldbeter, Temporal self-organization of the cyclin/Cdk network driving the mammalian cell cycle. Proc. Natl. Acad. Sci. U.S.A. **106**, 21,643–21,648 (2009)
27. A. Gerisch, M.A. Chaplain, Mathematical modelling of cancer cell invasion of tissue: local and non-local models and the effect of adhesion. J. Theor. Biol. **250**(4), 684–704 (2008)
28. P. Gerlee, A.R. Anderson, An evolutionary hybrid cellular automaton model of solid tumour growth. J. Theor. Biol. **246**, 583–603 (2007)
29. N. Goda, H.E. Ryan, B. Khadivi, McNulty, W., Rickert, R.C., Johnson, R.S.: Hypoxia-inducible factor 1alpha is essential for cell cycle arrest during hypoxia. Mol. Cell. Biol. **23**, 359–369 (2003)

30. A. Goldbeter, A minimal cascade model for the mitotic oscillator involving cyclin and cdc2 kinase. Proc. Natl. Acad. Sci. U.S.A. **88**, 9107–9111 (1991)
31. M. Guerrero, X.A. Li, Analysis of a large number of clinical studies for breast cancer radiotherapy: estimation of radiobiological parameters for treatment planning. Phys. Med. Biol. **48**(20), 3307–3326 (2003)
32. S. Gupta, T. Koru-Sengul, S.M. Arnold, G.R. Devi, M. Mohiuddin, M.M. Ahmed, Low-dose fractionated radiation potentiates the effects of cisplatin independent of the hyper-radiation sensitivity in human lung cancer cells. Mol. Cancer Ther. **10**(2), 292–302 (2011)
33. C. Hennequin, V. Favaudon, Biological basis for chemo-radiotherapy interactions. European J. Cancer **38**(2), 223–230 (2002)
34. C. Hennequin, N. Giocanti, V. Favaudon, Interaction of ionizing radiation with paclitaxel (Taxol) and docetaxel (Taxotere) in HeLa and SQ20B cells. Cancer Res. **56**(8), 1842–1850 (1996)
35. A.R . Kansal, S. Torquato, G.R. Harsh IV, E.A. Chiocca, T.S. Deisboeck, Cellular automaton of idealized brain tumor growth dynamics. BioSystems **55**, 119–127 (2000)
36. M.A. Konerding, W. Malkusch, B. Klapthor, C. van Ackern, E. Fait, S.A. Hill, C. Parkins, D.J. Chaplin, M. Presta, J. Denekamp, Evidence for characteristic vascular patterns in solid tumours: quantitative studies using corrosion casts. Br. J. Cancer **80**, 724–732 (1999)
37. F. Levi, A. Okyar, Circadian clocks and drug delivery systems: impact and opportunities in chronotherapeutics. Expert Opin. Drug Deliv. **8**(12), 1535–1541 (2011)
38. F. Levi, A. Okyar, S. Dulong, P.F. Innominato, J. Clairambault, Circadian timing in cancer treatments. Annu. Rev. Pharmacol. Toxicol. **50**, 377–421 (2010)
39. W. Liu, T. Hillen, H. Freedman, A mathematical model for m-phase specific chemotherapy including the g0-phase and immunoresponse. Math. Biosci. Eng. **4**(2), 239 (2007)
40. J.S. Lowengrub, H.B. Frieboes, F. Jin, Y.L. Chuang, X. Li, P. Macklin, S.M. Wise, V. Cristini, Nonlinear modelling of cancer: bridging the gap between cells and tumours. Nonlinearity **23**, R1–R9 (2010)
41. P. Macklin, S. McDougall, A.R. Anderson, M.A. Chaplain, V. Cristini, J. Lowengrub, Multiscale modelling and nonlinear simulation of vascular tumour growth. J. Math. Biol. **58**(4–5), 765–798 (2009)
42. A. Maity, McKenna, W.G., Muschel, R.J.: The molecular basis for cell cycle delays following ionizing radiation: a review. Radiother. Oncol. **31**(1), 1–13 (1994)
43. A. Matzavinos, M.A. Chaplain, V.A. Kuznetsov, Mathematical modelling of the spatio-temporal response of cytotoxic T-lymphocytes to a solid tumour. Math. Med. Biol. **21**(1), 1–34 (2004)
44. Matzavinos, A., Kao, C.Y., Green, J.E., Sutradhar, A., Miller, M., Friedman, A.: Modeling oxygen transport in surgical tissue transfer. Proc. Natl. Acad. Sci. U.S.A. **106**, 12,091–12,096 (2009)
45. S.R. McDougall, A.R. Anderson, M.A. Chaplain, Mathematical modelling of dynamic adaptive tumour-induced angiogenesis: clinical implications and therapeutic targeting strategies. J. Theor. Biol. **241**(3), 564–589 (2006)
46. H.B. Mistry, D.E. MacCallum, R.C. Jackson, M.A. Chaplain, F.A. Davidson, Modeling the temporal evolution of the spindle assembly checkpoint and role of Aurora B kinase. Proc. Natl. Acad. Sci. U.S.A. **105**(51), 20,215–20,220 (2008)
47. B. Novak, J.J. Tyson, Modelling the controls of the eukaryotic cell cycle. Biochem. Soc. Trans. **31**, 1526–1529 (2003)
48. B. Novak, J.J. Tyson, A model for restriction point control of the mammalian cell cycle. J. Theor. Biol. **230**, 563–579 (2004)
49. M.R. Owen, T. Alarcon, P.K. Maini, H.M. Byrne, Angiogenesis and vascular remodelling in normal and cancerous tissues. J. Math. Biol. **58**, 689–721 (2009)
50. M.R. Owen, H.M. Byrne, C.E. Lewis, Mathematical modelling of the use of macrophages as vehicles for drug delivery to hypoxic tumour sites. J. Theor. Biol. **226**, 377–391 (2004)
51. J. Panetta, J. Adam, A mathematical model of cycle-specific chemotherapy. Math. Comput. Model.**22**(2), 67–82 (1995)

52. A.A. Patel, E.T. Gawlinski, S.K. Lemieux, R.A. Gatenby, A cellular automaton model of early tumor growth and invasion. J. Theor. Biol. **213**, 315–331 (2001)
53. Pawlik, T.M., Keyomarsi, K.: Role of cell cycle in mediating sensitivity to radiotherapy. Int. J. Radiat. Oncol. Biol. Phys. **59**(4), 928–942 (2004)
54. Perfahl, H., Byrne, H.M., Chen, T., Estrella, V., Alarcon, T., Lapin, A., Gatenby, R.A., Gillies, R.J., Lloyd, M.C., Maini, P.K., Reuss, M., Owen, M.R.: Multiscale modelling of vascular tumour growth in 3D: the roles of domain size and boundary conditions. PLoS ONE **6**, e14,790 (2011)
55. G. Powathil, M. Kohandel, M. Milosevic, S. Sivaloganathan, Modeling the spatial distribution of chronic tumor hypoxia: implications for experimental and clinical studies. Comput. Math. Meth. Med. **2012**, 410,602 (2012)
56. G.G. Powathil, D.J.A. Adamson, M.A.J. Chaplain, Towards predicting the response of a solid tumour to chemotherapy and radiotherapy treatments: Clinical insights from a computational model. PLOS Computational Biology (To appear) (2013). DOI 10.1371/journal.pcbi.1003120
57. G.G. Powathil, K.E. Gordon, L.A., Hill, M.A. Chaplain, Modelling the effects of cell-cycle heterogeneity on the response of a solid tumour to chemotherapy: Biological insights from a hybrid multiscale cellular automaton model. J. Theor. Biol. **308**, 1–9 (2012)
58. I. Ramis-Conde, M.A. Chaplain, A.R. Anderson, D. Drasdo, Multi-scale modelling of cancer cell intravasation: the role of cadherins in metastasis. Phys. Biol. **6**(1), 016,008 (2009)
59. I. Ramis-Conde, D. Drasdo, A.R. Anderson, M.A. Chaplain, Modeling the influence of the E-cadherin-beta-catenin pathway in cancer cell invasion: a multiscale approach. Biophys. J. **95**(1), 155–165 (2008)
60. B. Ribba, T. Alarcon, K. Marron, P. Maini, Z. Agur, The Use of Hybrid Cellular Automaton Models for Improving Cancer Therapy. Lect. Notes Comput. Sci **3305**, 444–453 (2004)
61. B. Ribba, T. Colin, S. Schnell, A multiscale mathematical model of cancer, and its use in analyzing irradiation therapies. Theor. Biol. Med. Model. **3**, 7 (2006)
62. B. Ribba, K. Marron, Z. Agur, T. Alarcon, P.K. Maini, A mathematical model of Doxorubicin treatment efficacy for non-Hodgkin's lymphoma: investigation of the current protocol through theoretical modelling results. Bull. Math. Biol. **67**(1), 79–99 (2005)
63. B. Ribba, O. Saut, T. Colin, D. Bresch, E. Grenier, J.P. Boissel, A multiscale mathematical model of avascular tumor growth to investigate the therapeutic benefit of anti-invasive agents. J. Theor. Biol. **243**(4), 532–541 (2006)
64. M. Richard, K. Kirkby, R. Webb, N. Kirkby, A mathematical model of response of cells to radiation. Nuclear Instruments and Meth. Phy. Res. Section B: Beam Interactions Mater. Atoms **255**(1), 18–22 (2007)
65. R.K. Sachs, P. Hahnfeld, D.J. Brenner, The link between low-LET dose-response relations and the underlying kinetics of damage production/repair/misrepair. Int. J. Radiat. Biol. **72**(4), 351–374 (1997)
66. G.K. Schwartz, M.A. Shah, Targeting the cell cycle: a new approach to cancer therapy. J. Clin. Oncol. **23**, 9408–9421 (2005)
67. G. Serini, D. Ambrosi, E. Giraudo, A. Gamba, L. Preziosi, F. Bussolino, Modeling the early stages of vascular network assembly. EMBO J. **22**, 1771–1779 (2003)
68. E. Shochat, D. Hart, Z. Agur, Using computer simulations for evaluating the efficacy of breast cancer chemotherapy protocols. Math. Models Meth. Appl. Sci. **9**(4), 599–615 (1999)
69. M. Sturrock, A.J. Terry, D.P. Xirodimas, A.M. Thompson, M.A. Chaplain, Spatio-temporal modelling of the Hes1 and p53-Mdm2 intracellular signalling pathways. J. Theor. Biol. **273**(1), 15–31 (2011)
70. M. Sturrock, A.J. Terry, D.P. Xirodimas, A.M. Thompson, M.A. Chaplain, Influence of the nuclear membrane, active transport, and cell shape on the Hes1 and p53-Mdm2 pathways: insights from spatio-temporal modelling. Bull. Math. Biol. **74**(7), 1531–1579 (2012)
71. I.Tannock, R. Hill, R. Bristow, L. Harrington, *Basic Science of Oncology* (MacGraw Hill, Boston 2005)
72. I. Turesson, J. Carlsson, A. Brahme, B. Glimelius, B. Zackrisson, B. Stenerlow, Biological response to radiation therapy. Acta Oncol **42**(2), 92–106 (2003)

73. S. Turner, J.A. Sherratt, Intercellular adhesion and cancer invasion: a discrete simulation using the extended Potts model. J. Theor. Biol. **216**, 85–100 (2002)
74. J.J. Tyson, B. Novak, Regulation of the eukaryotic cell cycle: molecular antagonism, hysteresis, and irreversible transitions. J. Theor. Biol. **210**, 249–263 (2001)
75. B.G. Wouters, J.M. Brown, Cells at intermediate oxygen levels can be more important than the "hypoxic fraction" in determining tumor response to fractionated radiotherapy. Radiat. Res. **147**(5), 541–550 (1997)
76. M. Wu, H.B. Frieboes, S.R. McDougall, M.A. Chaplain, V.Cristini, J. Lowengrub, The effect of interstitial pressure on tumor growth: coupling with the blood and lymphatic vascular systems. J. Theor. Biol. **320**, 131–151 (2013)
77. L. Zhang, Z. Wang, J.A. Sagotsky, T.S. Deisboeck, Multiscale agent-based cancer modeling. J. Math. Biol. **58**(4–5), 545–559 (2009)

Deterministic Mathematical Modelling for Cancer Chronotherapeutics: Cell Population Dynamics and Treatment Optimization

Jean Clairambault

Abstract In this short review paper, I will present the mathematical models that have been designed in the frame of continuous deterministic cell population dynamics that aim at optimization of cancer treatments using chronotherapeutics. Many authors have dealt with chronobiology of cancer, less with continuous mathematical models and even less with the declared aim to optimize chronotherapeutics. The biological and theoretical bases for these models are sketched, started from a historical viewpoint, and the main theoretical results are presented, with biological suggestions to account for them. Chronotherapeutics that leads to therapeutic optimization with the constraint of limiting unwanted toxicity of anticancer drugs towards healthy cell populations is put in a medical perspective together with the other main pitfall of cancer therapeutics, for which optimization procedures should have little to do with circadian biology, i.e., emergence of drug resistance in cancer cell populations, which is amenable to the use of other sorts of models, that are briefly mentioned.

Keywords Deterministic differential equations • Cell population dynamics • Control • Optimization • Cell and tissue biology • Cancer • Therapeutics

1 Introduction

Chronotherapeutics has been designed and used for more than twenty years as an effective treatment against cancer by a few teams around the world, among whom one of the first is Francis Lévi's at Paul-Brousse hospital (Villejuif, France), in application of circadian clock physiology to determine best infusion times within the 24-h span for anticancer drug delivery. Mathematical models have been called

J. Clairambault (✉)
MAMBA (ex-BANG) project-team, INRIA Paris-Rocquencourt, BP 105, F78153 Rocquencourt, France

Lab. Jacques-Louis Lions, BC 187, UPMC, 4 Place Jussieu, F75252 Paris cedex 05, France
e-mail: jean.clairambault@inria.fr

© Springer Science+Business Media New York 2014

A. d'Onofrio, A. Gandolfi (eds.), *Mathematical Oncology 2013*, Modeling and Simulation in Science, Engineering and Technology, DOI 10.1007/978-1-4939-0458-7_9 in the last ten years to give a rational basis to such optimized treatments, for use in the laboratory and ultimately in the clinic. While actual clinical applications of the theoretical optimization principles found have remained elusive so far to improve chronotherapeutic treatments in use, mathematical models provide proofs of concepts and tracks to be explored experimentally, to progress from theory to bedside.

Starting from a simple ordinary differential equation model that allowed setting and numerically solving a drug delivery optimization problem with toxicity constraints, this modelling enterprise has been extended to represent the division cycle in proliferating cell populations with different molecular targets, to allow for the representation of anticancer drug combinations that are used in clinical oncology.

The main point to be made precise in such a therapeutic optimization problem is to establish, here in the frame of circadian chronobiology, physiologically based differences between healthy and cancer cell populations in their responses to drugs. To this aim, clear biological evidence at the molecular level is still lacking, so that, starting from indirect observations at the experimental and clinical levels and from theoretical considerations on the model, speculations have been made, that will be exposed in this review of cancer chronotherapeutics models with the corresponding optimization problems and their numerical solutions, to represent these differences between the two cell populations, with regard to circadian clock control.

2 Circadian Clocks: Biology and Models

2.1 Short Historical Background

The existence of rhythms in natural phenomena, following a period that is grossly superimposable to the day-night alternation, for instance, the folding and unfolding of leaves of plants, has been known since antiquity, and such rhythms had even been noticed by d'Ortous de Mairan in the eighteenth century to occur also in constant darkness, thus being independent of the light of the Sun and hence intrinsically linked to some proper rhythm of the plants [98]. This was the beginning of *chronobiology*, the field of science that deals with biological rhythms. In particular in this case, a *circadian* rhythm, i.e., a rhythm with approximately 24-h period, had been evidenced in that plant (a *Mimosa*), but of course other rhythms of longer period, yearly (seasonal), monthly (menstrual), had been observed throughout the history of mankind. Circadian is a term that was coined in the late 1950s by the chronologist Halberg (1919–2013) on the basis of the Latin *circa diem*, i.e., about a day, to qualify the period of a rhythm [61]; conventionally among chronobiologists, it means a rhythm with period between 20 and 28 h.

Many circadian rhythms have been found in various organisms, fungi, plants, algae, insects, and more recently mammals, and the first gene known to be expressed according to a circadian rhythm was the *Per* gene, found in the fly *Drosophila melanogaster* in 1971 by Konopka and Benzer [70]. Then more and more genes expressed according to a circadian rhythm were found in cells of organisms that were known to present such rhythmic phenomena (in particular the *Clock* gene in mice, by Takahashi in 1994 [103]), so that eventually the concept emerged of a *molecular circadian clock* constituted of genes and their resulting proteins in a network of activation and inhibition loops. Such a molecular circadian clock was found in mammals in every nucleated cell where it was searched for, and it was also shown all these cell clocks, though having their own periods, were under the control of a central circadian pacemaker located in the hypothalamus, the so-called suprachiasmatic nuclei (SCN), constituted of about 20000 neurons coupled together, giving rise to a common rhythm, itself reset by light through a retino-hypothalamic tract that normally ensures some synchrony between mammals of a same population [38, 52, 63].

Recognition of a circadian rhythm (period searched for between 20 and 28 h) in a biological recording may be done by spectral analysis followed by cosinor analysis [84]. These are signal processing and statistical methods that do not explain any mechanism, but contribute to select biological variables as candidates to be regulated by circadian clocks. After identification of a possible period $T = \frac{2\pi}{\omega}$ in the recorded time series by spectral analysis, the mean, the amplitude and the phase at maximum (acrophase) are determined according to the simple model $x(t_i) = M + A\cos\omega t_i \cos\varphi - A\sin\omega t_i \sin\varphi$ by least squares linear regression to determine $\alpha = M, \beta = A\cos\varphi$ and $\gamma = A\sin\varphi$ in $x(t_i) = \alpha + \beta y(t_i) + \gamma z(t_i)$, hence M, A and φ. Then a F-test is used to determine whether or not zero is in the confidence interval for the amplitude. If yes, the null hypothesis is not rejected, and variations in amplitude of the signal are considered as nonsignificant of an actual periodicity, but part of the background noise; conversely, if the null hypothesis is rejected, the time series is likely to present periodicity with period $T = \frac{2\pi}{\omega}$. Note that in order to accurately detect by spectral analysis a period T in a time series, a sample of length at least $2T$ must be available.

2.2 *Modelling Biological Clocks*

A simple way to design a biological clock, i.e., a periodic mechanism with molecular ingredients in nucleated cells, is to use the following negative feedback loop: transcription (expression of a gene in the nucleus in the form of a messenger RNA) to translation (synthesis in the cytoplasm of a protein in a ribosome from its RNA) and then to inhibition of transcription (by a nuclear form of the translated protein that goes from the cytoplasm into the nucleus, inhibiting its own transcription). Such a simple 3-variable ODE model of transcriptional regulation had been proposed as a general biological clock principle by Goodwin in 1965 [89], and later in 1999 by Didier Gonze, Jean-Christophe Leloup and Albert Goldbeter for the 24-h rhythmic protein FRQ in the mould *Neurospora crassa* [73], after another more complex model for the protein PER in *Drosophila* had been proposed by Albert Goldbeter in 1995 [50]. Many other models of molecular circadian clocks have been published, including a very detailed one by Jean-Christophe Leloup and Albert Goldbeter in 2003 [72], all of them relying on activation and inhibition loops that had been evidenced by biological experiments.

These models are of single-cell oscillators, but it is possible to couple these oscillators, introducing some stochastic variability between cellular clocks, and study their synchronization in the central circadian pacemaker in the SCN [14, 53, 54] and also to propose simple models of this central circadian control on independent peripheral cell clocks [23].

2.3 Influence of Circadian Rhythms on Proliferation

At the individual cell level, clocks have been shown to influence both metabolism and proliferation for those cell population that are committed in the cell division cycle. As regards metabolism, the fact that some intracellular enzymes that process drug activation or detoxication show circadian behavior in their gene expression or intracellular protein should be taken into account when representing time-dependent pharmacokinetics-pharmacodynamics (PK-PD) of anticancer drugs. One may recall here that pharmacokinetics describes by their concentrations the fate of drugs in the organism, from their infusion until their molecular target, while pharmacodynamics evaluates the actual effects of drugs on the organism, modifying its behavior; in other words, according to a widely broadcast motto, "pharmacokinetics is what the body does to the drug, pharmacodynamics is what the drug does to the body." Independently of pharmacological actions, an indirect influence of circadian clocks on the cell division cycle has been evidenced, in particular by Georg Bjarnason in 1999 on the 24 h rhythm of the concentration of cell cycle determinants (Cyclins E and B1) in the oral mucosa of men [19], and also a direct one by Matsuo in 2003, showing that the circadian clock protein Bmal1 controls the G_2/M transition in the cell cycle through the kinase Wee1 [79]. Coupling between the circadian clock and the cell cycle has been modelled by Claude Gérard and Albert Goldbeter [47], based on this finding. Such representation of proliferation control by circadian clocks can also be used in cell population models of the cell cycle (i.e., not only at the individual cell level), as proposed in [23] using a FRQ-like model of the circadian clock, or also by using a simple cosine-like wave for the clock. A recent review on molecular mechanisms linking circadian clocks and the cell division cycle has been published in [78].

2.4 Differences Between Healthy and Diseased Clocks?

Peripheral circadian clocks are synchronized by the SCN [38, 52, 63] and such synchronization may be experimentally disrupted, either by surgical ablation (thermocoagulation) or by out-of-phase non 24-hour periodic jet-lag-like repeated entrainment by artificial light, in laboratory mice [63]. Both experimental conditions, resulting in total loss of circadian rhythmicity in body temperature and rest-activity alternation, led to accelerated tumor progression in B6D2F1 mice, by comparison with a control group of the same strain in which normal entrainment by light on a physiological 24-hour basis (12 h of light, 12 h of darkness) was preserved [42, 43]. These experiments were performed in Francis Lévi's laboratory at Paul-Brousse Hospital in Villejuif, France, a hospital in which at the same time treatments of metastatic colorectal cancer by a combination of cytotoxic drugs (5-Fluorouracil, Oxaliplatin, Irinotecan) delivered according to a circadian schedule [83] designed on a computer and implemented in a programmable pump are conducted. With 3-week autonomy, easily portable by the patients, valuing their quality of life, such pumps allow them to live and work normally. Francis Lévi and his clinical team [83]—but also others [94] —have observed that the more ablated physiological circadian rhythms are in patients—as evidenced by low amplitude of cortisol variations in blood, or of central temperature, the poorer is their prognosis; otherwise said, a preserved physiological circadian rhythm is in favor of a good prognosis in patients with metastatic colorectal cancer under treatment. Additionally to external detrimental environmental factors such as shift work, that enhances the risk of developing several cancers [33, 62, 64, 92], a disrupted central SCN clock may be the result of the cancer disease itself, through circulating cytokines emitted in the tumor tissue by an immune reaction against it [87], and also by some anticancer drugs that have been shown to perturb the clock through a mechanism that is not known [74]. A general and recent review on circadian clocks and cancer, including cancer chronotherapeutics and proposed mechanisms to account for its efficiency, may be found in [44].

3 Using circadian Chronobiology for Cancer Therapeutics

3.1 *The Case of Cancer in Therapeutics*

Cancer is a disease of the physiological control on cell and tissue proliferation. In healthy organisms, normal regeneration of a tissue, based on the cell division cycle (at the term of which one cell becomes two) in renewing cell populations such as intestinal mucosa, haematopoietic bone marrow, skin and others, is physiologically controlled to ensure functional persistence of this tissue. For instance, the production of young red blood cells by the bone marrow compensates without excess the elimination of aging red blood cells in the spleen. In cancer, a defect of control results in overproduction of young cells and unlimited tumor tissue growth. In this respect, the present pharmacology of cancer occupies a special place in the treatment of diseases, since the cytotoxic drugs that constitute its core are not directed towards re-establishing normal physiological control (as is the case for instance with drugs used in cardiology, tending to ensure normal tissue perfusion by our cardiac pump), but are directed towards the elimination of tumor cells, as though these were bacteria or parasites, whereas they are somatic cells endowed with the same basic genome as healthy cells, but in which proliferation control is impaired, usually due to a succession of mutations. Complementary treatments aiming at slowing down the growth of the tumor, either by choking its vascular environment (antiangiogenic agents [65]) or by antagonizing growth factor receptors or blocking elements of intracellular signal transduction cascades downstream of them (monoclonal antibodies [93], tyrosine kinase inhibitors (TKIs) [76], the main weapons for the so-called targeted therapies), may be qualified cytostatic, as they are not used to kill cells (which is what cytotoxics are designed for) but only to slow down entrance or progression in the division cycle. Note however that at high doses cytostatic drugs may become cytotoxic. They are seldom used alone, and cannot control cancer proliferation by themselves.

Understanding where normal proliferation control is impaired is not easy and it is difficult to correct, hence the tough choice to try and eliminate all diseased cells by cytotoxic agents. Making use of non cell killing treatments by re-establishing physiological control of proliferation and letting diseased cells die as non adapted to a healthy organism environment would certainly be better, but these are seldom available. Only in a few cases, in particular chronic myelogenous leukemia (CML; see Sect. 3.2) and in acute promyelocytic leukaemia (APL, a form of acute myeloblastic leukaemia, AML), have mutations been identified (in both these cases in fact chromosome translocations, giving rise to fusion genes BCR-Abl for CML and PML-RARa for APL), that yield chimeric proteins specifically responsible for the disease, which chimeric proteins, additionally, can be eliminated by specific drugs, in particular Imatinib for BCR-Abl [41]. In APL, treatment by ATRA, a non-cell-killing agent targeting the abnormal fusion protein responsible for the blockade of differentiation in myelopoiesis at the promyelocyte stage—nevertheless consolidated by cytotoxic drugs, usually anthracyclines—results in the gradual elimination of diseased cells in the spleen and cure in more than 90 % of cases, an exceptional feature in AML [60].

Only when such so-called druggable targets (such as BCR-Abl or PML-RARa proteins) have been clearly identified as the only cause of the disease is it justified to represent the action of drugs at the single cell level; in all other cases, where more complex mechanisms underlying uncontrolled tissue proliferation are at work, the cell population level is the best one to describe and model the effects of drugs. Wherever modelling considerations may be lead to, such "targeted therapies" are actively searched for in pharmaceutical research, with limited therapeutic success so far, either because most often, not just one, but numerous intracellular pathways are disrupted, or because of treatment complications (see Sect. 3.2).

Conversely, in a whole-body therapeutic perspective, others favor the idea of identifying and enhancing physiological controls on tissue proliferation, with the aim to use them in a preventive way, of course (diet, personal life hygiene, etc.), but also to some extent in therapeutics, taking advantage of their possible added action in present cancer treatments. The circadian system, constituted of the central SCN pacemaker and of all peripheral cell clocks, that receive synchronizing messages from it, is one possible such control system on cell and tissue proliferation, acting by the known controls of G_2/M transition by Bmal1 through kinase Wee1 [79], and also through the cyclin-dependent kinase inhibitor (CDKI) p27 [56] acting on G_1/S transition in the cell division cycle.

An essential component of the molecular circadian clock, the gene Per2, has been shown to be a tumor suppressor gene [44, 45], and the p53 protein, the so-called "guardian of the genome," that controls both checkpoints, G_1/S and G_2/M, has also been shown to be controlled through its inhibitor Mdm2 by ATF4, a component of the circadian clock [66]. Circadian clocks thus exert their action on the cell division cycle by controlling gating (by cyclin-dependent kinases, CDKs) at cell cycles checkpoints, and this may be taken into account by scheduling drug infusion profiles that aim at preserving healthy cell populations, but they may also modulate the action of intracellular drug processing enzymes, and this may also be taken into account by time-scheduling of treatments. What is the most important effect of clocks in adapting to them a rational scheduling of drug delivery flows: by fitting them either to gating at checkpoints in the cell division cycle or to enzymatic intracellular processing (in particular detoxication), is hard to decide, and there are experimental arguments towards taking both effects into account in models. The first choice necessarily involves modelling the cell division cycle divided into phases, while the second one implies modelling of molecular cell and tissue PK-PD for the drugs at stake.

3.2 *Pitfalls Encountered in Cancer Therapeutics*

Since tissue proliferation is necessary to the maintenance of a multicellular organism, drugs that will limit cancer growth by inhibiting mechanisms of cell proliferation that are common to all fast renewing tissues (which is the case so far of most anticancer drugs) will also affect healthy tissue, limiting their use, so that finding and exploiting differences in behavior towards these drugs between healthy and cancer cell populations is a major challenge of cancer therapeutics. In some rare cases where one isolated abnormal, disease-specific, molecule has been identified and may be inhibited by a drug, (two conditions that are fulfilled in particular in the case of the chimeric protein BCR-Abl, responsible for CML, and that can be neutralized by specific drugs the first of which was Imatinib [41]), this can be achieved without being deleterious to healthy cells. This is the grail pursued by so-called targeted therapies, but very often, even if the target is reached, other unpredictable and unwanted targets are also reached in healthy cells, and hoped-for specificity is lost, which may result in withdrawal from the market (as was the case, for instance, for gemtuzumab-ozogamicin that had been proposed in the treatment of acute myelogenous leukaemia and was withdrawn by its manufacturer in the USA, due to unfavorable outcome in clinical trials, although the case is still an object of debates [88]). So that, even when a drug seems well targeted, unpredicted toxic side effects may not be excluded, that limit its use.

Even when a drug seems well targeted without major side effects, another phenomenon may occur in a cancer cell population, limiting or even leading to forsake its use, which is the development of resistance to the drug at stake. Cancer cells, endowed with genomic instability, are often able to overexpress genes that make them able to develop mechanisms, such as drug efflux by ABC transporters [55], a phenomenon that may come with or without genetic mutations. In the former case, it involves selection of an established resistant cell clone at the expense of shrinking the other–drug-sensitive–cells in the tumor population by a Darwinian mechanism, and in the latter, it is due to epigenetic modifications (modifications in genes that control expression of genes coding for proteins) that are reversible, i.e., not irreversibly inscribed in the genome. In either case, the right level to describe this drug-limiting phenomenon is the population of cancer cells, in which biological (phenotypic, even though the population may be genetically homogeneous [48]) variability can be taken into account.

These two obstacles, unwanted toxicity to healthy cells and the development of resistance to the treatment in cancer cells, are the two major pitfalls encountered in cancer therapeutics and they define the major constraints to be fulfilled by anticancer treatments: to avoid toxicity to healthy cell populations and to avoid emergence of drug resistance in cancer cell populations.

3.3 An Optimization Problem Under Constraints

From these considerations results a conception of cancer therapeutics, using cytotoxic and cytostatic drugs, as an optimization problem, where the objective is to contain tumor cell populations within limits compatible with the patients' life and a good quality of life (rather than eradication of tumor cells, an objective less easy to reach), under the constraints to preserve healthy cell populations—according to preservation criteria that have to be appreciated by the physician according to his patient's state of health—and to avoid the development of an uncontrollable drug resistant tumor cell clone. The most difficult elements to define in such a perspective of modelling proliferation in cell populations towards therapeutic optimization are clear differences between healthy and cancer cell populations, and between sensitive and resistant cancer cell populations. As regards contrasts between healthy and tumor cell populations, it has been proposed to characterize them according to the behavior of the two cell populations with respect to circadian control, but so far, only assumptions without biological certainty on mechanisms can be made. Nevertheless, such assumptions allow to draw proofs of concepts by modelling, allowing to numerically solve optimization problems much ahead of validation of these assumed biological characterizations.

In the sequel, I will propose modelling frames that have been used for these cell populations and for physiological and pharmacological control on their proliferation. Most of the work that has been done in this direction was performed in conjunction with Francis Lévi's team and is related to chronotherapeutics, directed towards solving the toxicity constraint problem, but I will also ultimately show more recent results, out of the chronotherapeutic framework (that is not necessarily relevant in this case), in which the drug resistance constraint is taken into account.

4 Drug Delivery Optimization and Chronotherapeutics

4.1 Molecular Pharmacokinetics-Pharmacodynamics

To represent the action of a drug, delivered in the general circulation (this includes oral route, through an absorption mechanism that is usually intestinal and hepatic, but most often it is processed by direct intravenous infusion), on its target, wanted (therapeutic efficacy) or unwanted (toxic side effects), one must represent its fate in the organism from its infusion to its effects by PK-PD models. PK-PD depends on the drug at stake and is usually represented by a system of ODEs for concentrations of the different compounds. The parameters of these ODEs depend on the organism under study, and for some of them, e.g., kinetic constants of drug detoxication enzymes, on circadian rhythms within this organism; in the perspective of personalized medicine, they should ideally be identified in each patient to propose actually individualized treatments.

In the frame of chronobiology, the patient population level, that is, the classic one in clinical PK-PD, is not at the forefront (although, in a pioneering study, chronotoxicity of anticancer drugs has been tested in different mice strains [7]); rather, in molecular PK-PD for chronotherapeutics, a single organism with its different organs and cell populations concerned by the fate of the drug at stake is the object of study. Whole-body physiologically based PK-PD (WBPBPKPD), a term coined by Malcolm Rowland [101]) based on ODEs and exemplified (without circadian clocks) for 5-fluorouracil by Tsukamoto *et al.* [102] is an aim to be pursued, with the addition of circadian influences on drug processing mechanisms when relevant.

4.2 A Simple ODE Model Based on a Simplifying Assumption

With the aim to simultaneously represent the dynamics of a cancer cell population, therapeutic target, and of a healthy cell population, unwanted toxicity target of the same anticancer drug delivery to be subsequently optimized, a simple ODE model has been designed and partly identified on tumor growth curves in mice from Francis Lévi's lab, with and without treatment by oxaliplatin [8, 22]. Focus was put on *chronopharmacodynamics* to define differences between healthy and cancer cell population behavior in response to the treatment: it had been experimentally observed at the lab that injection schedules (oxaliplatin was delivered in boli at fixed times of the 24-hour span) that led to more therapeutic efficacy (as measured by decrease in tumor growth curves) were at the same time those leading to least toxicity (as measured by total body weight loss). Hence this simple modelling assumption: the hour of best therapeutic efficacy should coincide with the hour of least toxicity, i.e., pharmacodynamic effects should be phase-opposed between the two cell populations.

This is of course a very simplified assumption, in favor of which no known biological mechanism exists. It relies only on macroscopic observations and these observations are made according to a rather poor sampling frequency of injections, 6 in the 24-hour span, i.e., a 4-hour time resolution. Nevertheless, establishing a clear difference in behavior between the two cell populations with respect to their responses to the drug infusion, it allowed to put in practice an optimization algorithm for a continuous drug delivery schedule under toxicity constraint, yielding at least a proof of concept for this optimization strategy, provided that actual differences with respect to circadian influence between healthy and cancer cell populations exist. The system of ODEs runs as follows:

A damped harmonic oscillator stands for healthy cell population dynamics:

$$\frac{dP}{dt} = -\lambda P + \frac{i(t)}{V_{dist}} \Phi(t) \tag{1}$$

$$\frac{dC}{dt} = -\mu C + \xi_C P \tag{2}$$

$$\frac{dZ}{dt} = \{-\alpha - f(C,t)\} Z - \beta A + \gamma \tag{3}$$

$$\frac{dA}{dt} = Z - Z_{eq}, \tag{4}$$

where

$$f(C,t) = F\left(1 + \cos(2\pi \frac{t - \varphi_A}{24})\right) \frac{C^{\gamma_A}}{C_{50}^{\gamma_A} + C^{\gamma_A}},$$

and $\lambda, \mu, \xi_C, \alpha, \beta, \gamma, Z_{eq}, F, \varphi_A, \gamma_A, C_{50}$ are positive constants, identified on tumor growth curves or from literature data [22], or else estimated. These equations represent drug diffusion and elimination by first-order pharmacokinetics for concentrations in the plasmatic and target cell compartments (P and C), from infusion in the general circulation according to the instantaneous drug delivery flow $i(t)$ (Φ representing a "tap on-tap off" function), and healthy tissue (normal jejunal mucosa, here) homeostasis by a linear system showing a stable focus at ($Z_{eq}, A_{eq} = \beta^{-1}(\gamma - \alpha Z_{eq})$), perturbed by the drug toxicity function which comes to strengthen the natural autoregulation coefficient α.

A Gompertz model stands for tumor cell population dynamics:

$$\frac{dP}{dt} = -\lambda P + \frac{i(t)}{V_{dist}} \Phi(t) \tag{5}$$

$$\frac{dD}{dt} = -\nu D + \xi_D P \tag{6}$$

$$\frac{dB}{dt} = -aB \ln(\frac{B}{B_{max}}) - g(D,t)B \tag{7}$$

Note that equations (1) and (5) are exactly the same, since they both represent the distribution of the drug in the plasma after infusion, and this the only feature these two systems, representing two cell populations have in common, since they are physically apart from each other: experimentally the tumor, a Glasgow osteosarcoma was implemented under the skin, whereas the main toxicity target in this mouse population was identified to be the jejunal mucosa. In this system of equations, function g, which represents antitumor drug efficacy, is assumed, as is function f for toxicity, to present circadian variations; it is given by

$$g(D,t) = H \left(1 + \cos(2\pi \frac{t - \varphi_B}{24})\right) \frac{D^{\gamma_B}}{D_{50}^{\gamma_B} + D^{\gamma_B}},$$

and $\lambda, \nu, \xi_D, a, B_{max}, H, \varphi_B, \gamma_B, D_{50}$ are positive constants, identified on tumor growth curves or from literature data [22], or else estimated. The difference of behaviors between the two populations of cells with respect to drug response is coded as $\varphi_A - \varphi_B = 12$ hours.

4.3 *Numerical Optimization of Drug Delivery*

Using this simple system of ODEs, it was possible to tackle the problem of drug delivery optimization, i.e., minimization of the tumor cell population under the constraint of minimizing unwanted toxicity on the healthy cell population by keeping it under a prescribed level (to be in future clinical applications defined by the clinician in charge), by a nonlinear conjugate gradient method [8]. Note that this method consists of numerical optimization, and it does not yield an optimal solution, but rather a suboptimal solution (the algorithm searches saddle points of a Lagrangian, and since the problem is not convex, it yields only necessary, not sufficient conditions of optimality), so that one could not completely exclude the existence of local minima in the descent algorithm yielding the best infusion profile [8]. Nevertheless, the existence of a global minimum can be proved, assuming for the evolution of the two cell populations $A(t)$ and $B(t)$ reasonable differentiability conditions with respect to time [8], which amounts to numerically solve a problem for which we know a unique solution to exist. Furthermore, the optimization

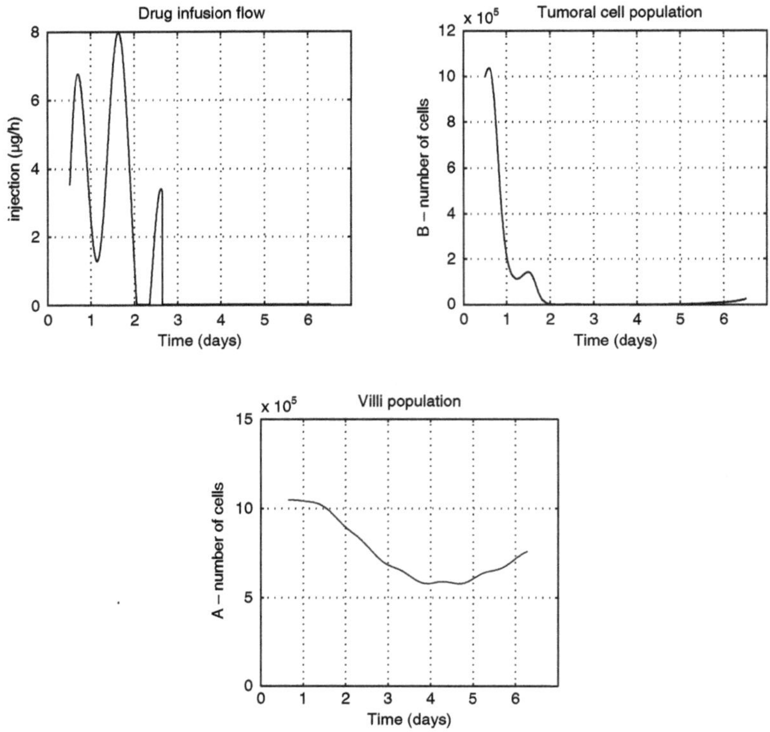

Fig. 1 Quasi-optimal solution of the stabilization (tumor containment by repeated drug delivery courses) problem under absolute constraint of preserving at least 50 % of the equilibrium population of healthy cells, for a duration of 2 days in a 7-day observation window. The optimal drug infusion flow i (left, upper panel) shows peaks at the times of minimal unwanted toxicity that are assumed to be simultaneously those of maximum therapeutic efficacy. The resulting tumor cell population (right, upper panel) is contained—treatment courses being repeated—, but not eradicated, while the healthy cell population (lower panel) is preserved over a prescribed level, here 50 % of the equilibrium population. See Ref. [8] for details

problem may be set in at least two different forms: the eradication problem consists in minimizing the minimum of tumor cells, whereas the stabilization (tumor containment) problem consists in minimizing the maximum of tumor cells in a given observation window. Figure 1 shows the results of such a stabilization procedure.

However, this optimization procedure has two main flaws: it opposes the behaviors of the tumor cell population and of the healthy cell population by an assumption which is far from granted in general (a 12-hour dephasing between their maximal sensitivity to the drug), and it represents the action of a single drug on a single target (a death rate), which excludes the representation of combinations of drugs acting on different biological targets. But in clinical settings, most anticancer treatments combine different drugs, all of them resulting in blocking or slowing down the cell division cycle on which relies all tissue proliferation, healthy or tumor, but acting on different molecular targets to potentiate their combined effects on cancer cell populations: cytotoxic drugs hit the DNA or cell proteins that are essential to cell division, leading cells committed in the division cycle to their inevitable death, while cytostatic drugs only slow down the division cycle, at least at non-massive doses. It is therefore necessary to re-examine the ways by which differences between healthy and tumor cell populations should be represented with respect to their responses to drug treatments, and to design a model of the cell division cycle amenable to represent *at the cell population level* the different molecular targets of the various anticancer drugs that are in use in the clinic.

5 Cell Cycle Modelling Using PDEs in Cell Populations

5.1 An Age-Structured McKendrick Model with Periodic Control

The so-called McKendrick, or Von Foerster-McKendrick, model of growing population dynamics was introduced in an integral form in 1911 by Sharpe and Lotka [95] in demography, and then independently and under its PDE form in 1926 by McKendrick [81], to be rediscovered in 1959 by Von Foerster [104]. It has been studied in detail, e.g., in [6, 67, 82]. Applied to the cell division cycle represented as an age-structured population dynamics model organized in a merry-go-round of subpopulations biologically identified as phases (G_1, S, G_2 and M), it was first proposed in 2003 in [28] under the form

$$\begin{cases} \dfrac{\partial n_i(t,x)}{\partial t} + \dfrac{\partial n_i(t,x)}{\partial x} + d_i(t,x)n_i(t,x) + K_{i\to i+1}(t,x)n_i(t,x) = 0, \\ n_{i+1}(t,0) = \int_0^\infty K_{i\to i+1}(t,x)n_i(t,x)dx, \\ n_1(t,0) = 2\int_0^\infty K_{I\to 1}(t,x)n_I(t,x)dx, \end{cases}$$

together with initial conditions $(n_i(t=0,.))_{1\le i\le I}$. Death rates in phases are noted d_i and transition rates between phases, assumed to be time-periodic, $K_{i\to i+1}$. Phase i $(1 \le i \le I)$ may be one of the classical four G_1, S, G_2 and M, but also an aggregated phase such as $S - G_2$, or even a single proliferating phase $G_1 - S - G_2 - M$, or on the contrary a subdivision inside a phase, e.g., pre- or post-restriction point in G_1; the equation describes the evolution of the densities $n_i(t,x)$ of cells having age x at time t in phase i.

Let me stress here that age x is a 'physiological but abstract' variable, that lumps together complex unidirectional (in time) biological phenomena occurring in the cell machinery, are based on protein synthesis to achieve cell division. Variable x has nothing to do with spatial distribution of cells in their population (space is considered here as irrelevant), since $n_i(t,x)$ is a density of cells that are at universal time t at a stage x of their way in phase i (according to an abstract clock measuring the degree of protein synthesis), starting from 0 to transit to next phase without fixed time limit, but governed by the transition rate function $K_{i\to i+1}$. The protein synthesis-related clock that governs evolution within G_1 and G_2 phases (preparing, respectively, S and M phases) might theoretically be followed by the concentrations of cyclins D and A, respectively, but this is not the way it has been done so far (see below Sect. 5.2 about the FUCCI analysis method). However, phase S may be readily followed in flow cytometry by the synthesis of DNA, from 2n to 4n chromosomes, a method that has been used by Britta Basse and colleagues [9, 10], giving an immediate interpretation of age x in this case.

Such space-independent representation of the distribution of cells in their population is particularly adapted to taking into account proliferation control by drugs assumed to be homogeneously distributed in concentration in the cell population. Taking spatial distribution of drugs into account should lead to more complex models, structured in both space and a physiological variable like age, but this would show useful only when a spatial distribution of cells in the tumor is known, and apart from the case of (very small) avascular tumors organized in spheroids, the topology of tumors is seldom known. *Mutatis mutandis*, the idea of representing tumor growth by an age-structured, rather than spatial, model when the cell division cycle, target of anticancer drugs by different molecular mechanisms, is the most relevant feature to be taken into account, is of the same order as in integration theory choosing the Lebesgue than the Riemann integral: it is just more practical.

The main output of such a linear model is its first eigenvalue λ, the so-called Perron eigenvalue, which, assuming minimal hypotheses on the parameters of the model, is always positive and simple (i.e., the associated eigenspace is generated by a single function that may be normalized to be of unit integral, hence bounded). Moreover, it may be shown, using the Krein-Rutman theorem (an infinite-dimensional version of the Perron-Frobenius theorem) that in each phase the solution converges for large times, in an L^1 sense, to $e^{\lambda t}$ times a fixed multiple of the associated normalized eigenvector [86], i.e., the behavior of the solutions in all phases is governed by an exponential term given by the Perron eigenvalue, which of course is nothing but ln 2 divided by the doubling time of the population. This means that knowledge of the death rates and of transition rates, targets of internal physiological or external pharmacological control, entirely determines the proliferative behavior of the population through its first eigenvalue λ, which is thus the main predictive output of the model.

In the case where transition rates $K_{i\to i+1}$ are time-independent, it is easy to see that the function of age x in phase i

$$f_i : x \mapsto K_{i\to i+1}(x)e^{-\int_0^x K_{i\to i+1}(\xi)d\xi}$$

is the probability density function (p.d.f.) of the duration of this phase, and it has been proved that for a given family of p.d.f.s with varying variance, the first eigenvalue λ increases with increasing variance of their p.d.f.s (one phase is enough) [18]. This result has the simple following interpretation: the more variable the duration of phases (i.e., the overlapping between phases), the faster the cell population proliferation. This may be put in relation with the observations of Sect. 2.4, at least if one admits that a disruption of circadian control on phase transitions should result in enhanced variability of the duration of phases (and hence of overlapping between them). However, this mathematical result has been proved only in the case $K_{i \to i+1} = K_{i \to i+1}(x)$, and not in the case when $K_{i \to i+1}$ is also time-dependent.

In [25–27, 29], the question of the influence of a periodic control–circadian, i.e., physiological, or pharmacological, i.e., external–has been examined. This question arose from the biological observations mentioned in Sect. 2.4, reported in [42, 43], since circadian control is by definition time-periodic. This induced to compare in the McKendrick model the evolution–as measured by its first eigenvalue–of a periodically controlled (on transition or death rates) proliferating cell population model with the version of the same model where periodic functions are replaced by their arithmetic time averages (i.e., no more periodic control). By analogy with the above mentioned biological observations, the authors of Ref. [27] expected the first eigenvalue of the former to be lesser than those of the latter. To their surprise, they proved (Theorem 1 in [27]) that the opposite is true, i.e., periodic control enhances proliferation, at least if the control is exerted only on death rates, and that if one uses an arithmetico-geometric form of the time average (difficult, however, to justify biologically), this result holds true also for transition rates [25, 26], but that without this use of an arithmetico-geometric mean, it is impossible in general to predict how a time-periodic control on the transition rates will affect cell population growth.

These results may be interpreted, in the light of the observations mentioned in Sect. 2.4 in at least two different ways:

Firstly, if one admits that the use of the arithmetico-geometric mean is correct, it might be that only healthy tissues (tumor-surrounding and immune cells) committed in fighting the tumor development are sensitive to the messages of the central circadian clock pacemaker that is disrupted (surgically ablated [43] or perturbed by chronic jet-lag-like light entrainment [42]) and are thus weakened in their proliferation by the disruptions of the clock, whereas tumor tissues, relatively insensitive to circadian commands, proliferate unabashed and with less opposition from tumor-combating healthy tissues.

Secondly, if one leaves aside the use of the arithmetico-geometrical mean and if one admits that tumor cell populations are somehow sensitive to circadian commands, it may mean that circadian messages are not, or little, exerted on death rates, but only on transition rates. It is biologically known, in fact, that clock-controlled genes exert their influences, via Cyclin-CDK complexes that control G_1/S and G_2/M transitions, as mentioned in Sect. 3.1, on transitions between phases (let us recall here that CDK is abridged from cyclin-dependent kinase; the most important CDKs in the cell cycle are CDK1, needs Cyclin B to be activated and let cells process from phase G_2 to phase M, and CDK2, that needs Cyclin E to be activated and let cells process from phase G_1 to phase S).

If the second interpretation is true, it only means that it is useless to look for circadian control on the apoptotic cascade. If the first one is true, it means that taking into account the influence of circadian inputs on healthy cell populations combating tumor development should be represented in a competition model, which remains to be done. Both interpretations may be right, recalling however, as mentioned in Sect. 3.1, that indirect circadian influences on death rates should exist, via circadian control on the activity of some intracellular drug processing enzymes (activation or degradation) and also via circadian control on Mdm2 [66], the main inhibitor of protein p53.

5.2 Another Optimization Problem Under Toxicity Constraints

Now, what is the biological reality of the McKendrick model and how can it be experimentally identified? To answer this question, observation of proliferating cell populations was needed. This was made possible, thanks to a newly released analysis technique coming in 2008 from Miyawaki's lab in Japan, the so-called FUCCI (fluorescence ubiquitination-based cell cycle indicator) analysis method, which made possible recording individual living and proliferating cells in a culture medium [90,91] in two phases of the cell cycle: G_1 and $S-G_2-M$, and also thanks to the European consortium C5Sys (2010–2013), in which such measurements were performed in cultures of proliferating NIH3T3 cells. The FUCCI technology allows to follow individual proliferating cells and to measure the durations of the G_1 phase, and of the complete cycle, using prior hybridization with fluorescent proteins of physiological proteins characteristic of G_1 or of $S-G_2-M$ phase (see Fig. 2).

What simplifies the identification of the McKendrick model in this case is that cells that were followed during their division cycle (most often only one cycle was observed) were by definition living cells from the beginning to the end of the recording, i.e., death terms were nil, and that these NIH3T3 cells were moving freely in a liquid medium, with no communication between them, nor with any external influence applied. This means that transition rates $K_{i \to i+1}$ in the cell population were completely time-independent, representing only the biological variability of the cell population with respect to age x in each of the two phases. Under these conditions, recalling that

$$f_i(x) = K_{i \to i+1}(x) e^{-\int_0^x K_{i \to i+1}(\xi)d\xi}$$

is the p.d.f. of the duration of phase i, which may be experimentally evaluated using FUCCI recordings of individual cells in the population [16, 18], straightforward computation yields the inversion formula

$$K_{i \to i+1}(x) = \frac{f_i(x)}{1 - \int_0^x f_i(\xi)d\xi},$$

so that in the case of this NIH3T3 cell population in culture, the 2-phase McKendrick model is completely identified.

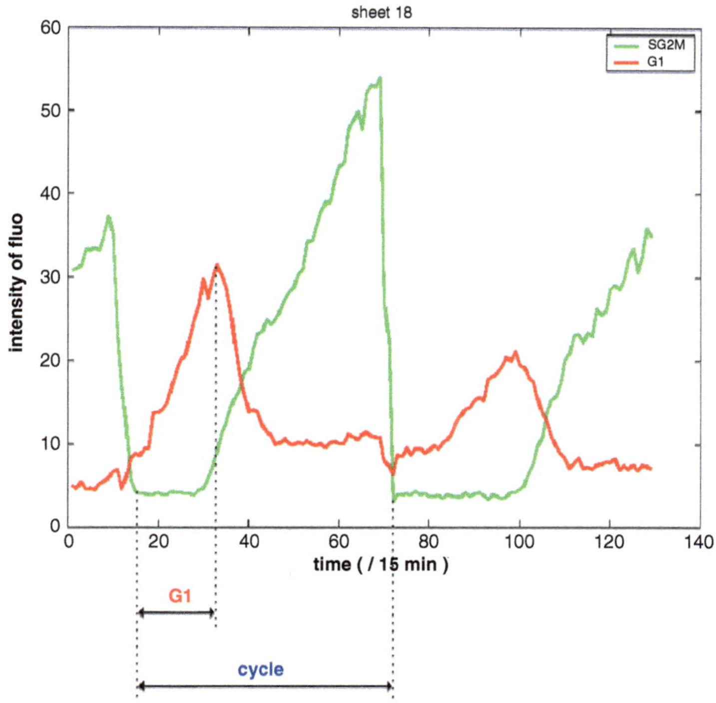

Fig. 2 After [17], method to determine the duration of phases G_1 and $S - G_2 - M$, on FUCCI recordings [90,91], on one cell committed in the division cycle. See References [17,18] for details

Assuming then that the periodic controls on the cell division cycle, both circadian (built-in) and pharmacological (tunable), are exerted on transition rates only, we had to hypothesize (or, better, to experimentally identify, which unfortunately did not prove possible with the FUCCI data provided in the C5Sys consortium) clear differences between healthy and cancer populations, in order to tackle in these new modelling settings the same optimization problem as in Sect. 4.3: maximizing tumor cell kill under the constraint of preserving a healthy cell population. Taking the model identified on NIH3T3 cells as a likely basis for a generic proliferating cell population, we hypothesized that such differences were due only to a difference in the effects of circadian messages on gating by cyclin-CDK complexes at phase transitions: sharp gating in the healthy case, resulting in small overlapping between phases, and loose gating in the tumor case, resulting in much broader overlapping, as sketched on Fig. 3.

These theoretical gating functions (hereafter noted ψ_i) occur in the model as time-dependent multiplicative modulating factors for the $K_{i \to i+1}(x)$, These gating functions thus synchronize cells with respect to cell cycle timing in cell populations

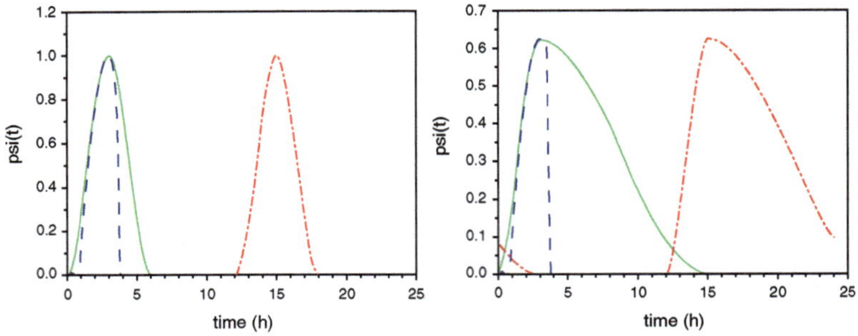

Fig. 3 Gating functions $\psi_i(t)$ at 2 phase transitions representing as functions of time the theoretical activity of the two Cyclin-CDK complexes (for G_1/S and G_2/M) under circadian control, with a dephasing set at 12 h between phase transitions (12 h numerically yielding highest proliferation rates). When gating functions are nil, there is no transition between phases. Here is represented the assumed difference between healthy and cancer cell populations: sharp gating (left panel) for healthy tissues and loose gating (right panel) for tumors. The dashed curve common to both panels represents the corrected (by treatment) gating function $[1-g(t)].\psi_2(t)$, where g is the output of the algorithm, solution to the optimization problem, prescribing to deliver a cancer drug (by the general circulation to both tissues simultaneously) so as to result locally (at the tumor and at the healthy tissue site) in the pharmacodynamic function g. See text and Ref. [17, 18] for details

that are well synchronized when the gating is sharp (gate, open during a brief interval of time), and poorly synchronized when it is loose. This modelling choice relies on the intuitive, not proven, but likely assumption (private conversation with F. Lévi) that healthy cell populations are more synchronized than cancer cell populations with respect to cell cycle timing, and that such synchronization is due to the central circadian clock, i.e., the circadian pacemaker makes healthy cells pass in a "disciplined" and orderly way from one phase to the next, while cancer cells, less "obedient" to messages of the clock, pass in a disorderly way.

In this model setting, the transition functions without control, representing only the biological variability with respect to phase durations in age x within the cell population, are chosen as the functions $\kappa_{i\to i+1}(x)$ identified on the NIH3T3 cell population; the circadian influence on gating is represented by the fixed functions $\psi_i(t)$ sketched on Fig. 3; and if $g(t)$ ($0 \leq g(t) \leq 1$) is the drug infusion flow to be optimized, blocking cell cycle transitions, the complete transition rates are defined in the model as $K_{i\to i+1}(x,t) = [1-g(t)].\psi_i(t).\kappa_{i\to i+1}(x)$ ($1 \leq i \leq 2$), see on Fig. 3 the illustration of the corrected gating function $[1-g(t)].\psi_i(t)$ (dashed curve). As in Sect. 4.3, the optimization algorithm searches for the function g (in fact, a bang-bang one), except that in this case the observed outputs are not cell population numbers, but first eigenvalues representing proliferation rates (see Fig. 4): the first eigenvalue of the cancer cell population is minimized while maintaining the first eigenvalue of the healthy tissue over a prescribed fixed threshold (to be defined in future clinical applications by the clinician in charge), a situation comparable with the optimization problem of Sect. 4.3, where the healthy cell population number

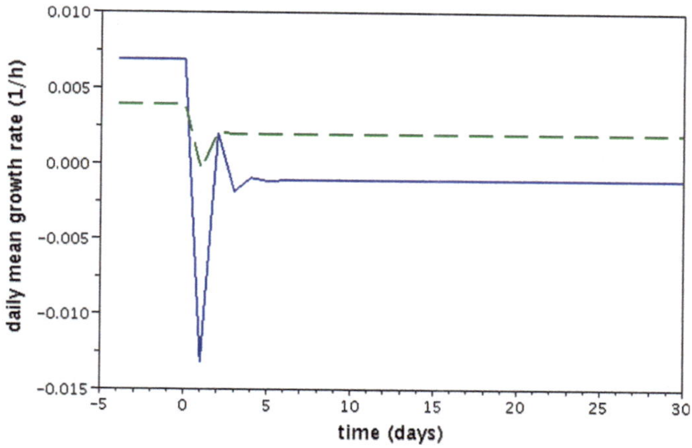

Fig. 4 Results of the optimization method, from [17]. Contrary to the model presented in Sect. 4.3, the outputs are not cell population numbers, but proliferation rates. One can see that after a short transient time interval, the proliferation rate of the healthy cell population, initially disadvantaged, overcomes the tumor proliferation rate. See text and Ref. [17, 18] for details

had to be maintained over a prescribed fixed percentage of its equilibrium level. In a different model setting, this is another proof of concept for the optimization method.

An illustration of this method is presented on Fig. 3. It represents (dashed curve) the optimized delivery of a drug that is active on transition G_2/M only, e.g., 5-Fluorouracil. Note that the PK of such a drug, from infusion in the general circulation until its presence on the target tissue site (tumor or healthy tissue), is not represented here. It should be added to the model equations to allow for accurate optimization of an actual intravenous infusion flow. Note also that whereas the FUCCI recordings should give us access to the M/G_1 transition, we rather assume that the duration of phase M is fixed, all cells in M passing into G_1 in fixed time (about 1 h), and that the variability in age duration of the aggregated phase $S - G_2 - M$ is in fact that of $S - G_2$, so that the p.d.f. $f_2(x)$ gives us access by the inversion formula, under this assumption, to a $K_{2\to 1}(x)$ transition function that is thus in fact related to the G_2/M transition.

5.3 *Possible Extensions of or Alternatives to the McKendrick Model*

One possible nonlinear extension consists in introducing exchanges (in both directions) between the proliferating population, divided or not in phases, and a quiescent population seen as a storage tank of cells that may be recruited in the proliferation cycle when needed, and to which an overflow of proliferating cells may be discharged when they lack energy resources, as introduced by Mats Gyllenberg and Glenn Webb [39, 58, 59] to give a mechanistic explanation to the Gompertz growth curves often encountered in population dynamics, and in a series of articles following [12] ([11, 20, 21, 40]). Note that in a model with several proliferative phases, the exchanges should be located in G_1 before the restriction point and that the input of cytostatic drugs could be represented as slowing down recruitment into proliferation or enhancing way out to quiescence. Such models give rise to biologically realistic situations representing space and nutrient limitations, and a clear difference, relying on the recruitment function from the quiescence phase, may be set between healthy and cancer cell populations [11, 12]. But this difference has no relation with circadian clocks, which could be introduced, as in Sect. 5.1 by their action on phase transitions; this remains a open modelling problem. Note that if the model is no longer linear, one cannot speak of eigenvalues anymore; however, linearizations around particular points of the population numbers may be studied [11, 12].

Another simple way to represent exchanges with a quiescent population in a still linear model is to exclude feedback from quiescence to proliferation, considering quiescence only as a sideway expansion cell tank, as proposed in [46]:

$$\begin{cases} \dfrac{\partial}{\partial t} p(t,x) + \dfrac{\partial}{\partial x} p(t,x) + \{\mu + K(x)\}\, p(t,x) = 0 \ , \\ p(t, x=0) = 2(1-f) \displaystyle\int_{\xi \geq 0} K(\xi)\, p(t,\xi)\, d\xi \ , \\ p(t, x=0) = p_0(x) \ , \\ \dfrac{d}{dt} Q(t) = 2f \displaystyle\int_{\xi \geq 0} K(\xi)\, p(t,\xi)\, d\xi - \nu Q(t) \ , \\ Q(0) = Q_0 \ . \end{cases}$$

In this age-structured McKendrick model designed to theoretically study the action of a cytostatic drug enhancing the way out of proliferating cells with density $p(t, x)$ to quiescent cells with density $Q(t)$, the drug target here is f, rate of escape at mitosis towards the siding phase Q, f to be enhanced by a cytostatic drug. The model [46] (see also [15]) was identified on the human non small cell lung cancer (NSCLC) cell line PC-9 submitted to the cytostatic drug erlotinib. Here again, a division of the proliferating cell population into phases could be added, together with circadian control at phase transitions (another open modelling problem).

Another model [24], also relying on the McKendrick model, but more complex than the previously described ones, has been proposed to take into account both PK-PD models for the two main drugs in use in the clinic of colorectal cancer: 5-Fluorouracil (with added folinic acid, i.e., Leucovorin to potentialize it) and Oxaliplatin and the possibility of repair in cell populations that have been hit by cytotoxic drugs. Involving 3 phases (G_1, $S - G_2$ and M) and additional subpopulations R_1 and R_2, also structured in age and evolving in parallel with the first two, consisting of these cells that are under repair from cytotoxic insult

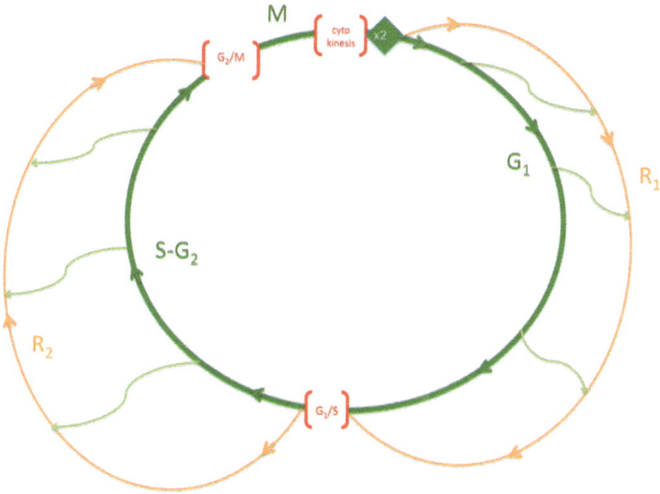

Fig. 5 From [24], illustration of a new model of the cell cycle with repair: cell cycle phases $G_1(i = 1)$, $S - G_2(i = 2)$ and $M(i = 3)$ with age-dependent variables, plus two additional subpopulations, R_1 and R_2 described by age-independent variables, to describe the fate of those cells that have been hit by drug-induced DNA damage and are waiting to be repaired—or sent to apoptosis. See Ref. [24] for details

by the two drugs illustrated on Fig. 5, it has also been used to solve once more the same therapeutic optimization problem (minimizing cancer cell population proliferation while maintaining the proliferation rate of the healthy cell population over a prescribed threshold).

In this model, circadian clocks also control phase transitions as in Sect. 5.2, but cytotoxic drugs are assumed to continuously exert their effects by sending cells to these parallel repair phases (a sort of "delayed death": at any rate, these cells go out of the proliferating phases, but may come back to them at any moment), and not by blocking phase transitions. Then, since it is a linear model, the same optimization principles used in Sect. 5.1 are used, minimizing the proliferation rate of the cancer cell population while maintaining that of the healthy cell population over a prescribed threshold, except that the optimized input functions are the flows of the two drugs in the general circulation, since the model includes a PK-PD representation of their fate in the organism from infusion until arrival on the proliferating cell population sites. The results, although good (see Ref. [24]), are less spectacular than in the previous model case, which again induces to speculate that the strongest drug effects on proliferation should occur on phase transitions, rather than on death rates, as has been assumed in this last model. A is shown a theoretically optimal combination of the drugs 5Fluorouracil, Leucovorin and Oxaliplatin, another proof of concept of the method is shown in Fig. 6. Improving the PK-PD model and the representations of the modes of action of the drugs

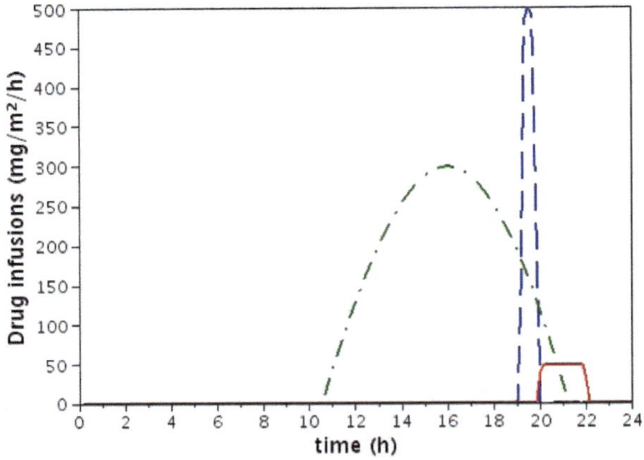

Fig. 6 From [24], illustration of an optimal infusion strategy proposing drug infusion flows on a 24-hour basis by a combination of Leucovorin (*dash-dotted line*), 5-fluorouracil (*dotted line*) and Oxaliplatin (*solid line*), by infusions repeated every day in order to minimize proliferation of the cancer cell population while maintaining the growth rate of the healthy cell population above a prescribed toxicity threshold. See Ref. [24] for details

(possibly adding the representation of a cytostatic drug like Cetuximab) remains to be done to put it in realistic clinical settings.

Other models of the cell cycle in proliferating cell populations with control by circadian clocks have been published, firstly to establish likely mechanisms for such control, and then to propose optimal chronotherapeutic strategies with cytotoxic drugs, all in collaboration with Francis Lévi. In particular, Samuel Bernard and Hanspeter Herzel [13] used deterministic models with delays, while Attila Altinok and Albert Goldbeter [1–4] used a clock-controlled cellular automaton model of the cell cycle to justify the chronotherapeutic strategies used by Francis Lévi in the clinic. Although these models do not propose optimization algorithms for the drug delivery time schedules, they present interesting ideas from different points of view to guide the determination of such optimized regimens.

6 Future prospects

6.1 *Need for More knowledge on Cell Cycle Control Mechanisms*

Now, to what extent are these theoretical models applicable in the clinic of cancers? Firstly, they are all based on the prediction of best drug delivery time schedules, and demand adapted technological appliances to put them in practice. While such devices are used in clinical chronotherapy, they should be adapted to preclinical trials on animals, which is not a simple problem–drug infusion in rodents, for instance, is usually performed only in boli, by lack of adapted devices–, not to mention the cost of programmable pumps, mundane limitations that so far have not given the possibility to experimentally test theoretically optimal drug delivery schedules.

Furthermore, much remains to be elicited in different fields of research related to chronotherapy: identification of PK-PD models and involvement of circadian control in them (a different model for each drug and each disease, to be further personalized according to patients' specificities, in particular gender [49]); identification of actual mechanisms of synchronization between cells with respect to cell cycle timing, both in healthy and in cancer cell populations (is it true that cancer cell populations are poorly synchronized? what is the role of circadian clocks in cell synchronization? does there exist synchronization mechanisms that rely on intercellular communication, e.g., using gap junctions?), identification of gating mechanisms at cell cycle phase transitions by cyclin-CDK complexes (shape of the gating function, with and without control by circadian clocks? this might be performed by FUCCI analysis, if one can find a fluorescent protein to be hybridized with activated CDKs), by p53 and other cell cycle fate determinant proteins, etc.

However, it is clear that before therapeutics with drugs should be called, with the complexity of their mechanisms of action needed to be represented, when the disease has reached a life-threatening level, preventive medicine must be more broadcast and popularized among healthy people (or supposed to be so), since cancers usually take a long time to develop in living organisms and can be more easily combated when taken at an early stage of their development. In this respect, disruptions of the circadian clock have been shown to enhance the development of tumors, experimentally in laboratory rodents [42, 43], but also by large epidemiological studies in humans [33, 62, 64, 92]. Conversely, re-establishing regular periodicity in daily rhythms by adapted (restricted) food intake regimens seems to go in the opposite and favorable direction in rodents [42, 105].

Both from the aforementioned common sense and well established rules for a better quality of life (not always satisfactorily explained with respect to their mechanisms, but that nevertheless should be more widely broadcast), and from the aforementioned pending questions on molecular mechanisms involving circadian clocks, cell cycle determinants and anticancer drugs, one may see that the field of research in modelling for cancer chronotherapeutics is vast and has only begun to be explored.

6.2 *Cancer Chronotherapeutics and the Immune System*

In particular, modelling the immune response in cancer is still in its infancy. Pioneering models of the immune response, using immunotherapy associated with chemotherapy, have been published in recent years, in particular by Lisette de Pillis and Amy Radunskaya [34–36], and also by Peter Kim and colleagues [68, 80] and by Marcello Delitala and Tommaso Lorenzi [37], among others. However, none of them relates to an influence of circadian clocks, although it might exist, since it has been shown by Rune Smaaland and colleagues, using observations on samples from their own bone marrows, that DNA synthesis in bone marrow follows a circadian rhythm [96]. Furthermore, circadian rhythms of circulating lymphocytes have been evidenced in Man [75, 97].

Conversely, an influence of the immune response on the central circadian pacemaker is likely, since it has been observed that patients with high levels of circulating cytokines–which are emitted by immune cells surrounding the tumor, mainly T-lymphocytes–also show high degrees of fatigue and other even clearer signs of a disrupted central clock (ablated rhythms of blood cortisol and of rest-activity alternation) [87]. If such detrimental influence of the immune response on the clock is established, taking into account the tumor growth-enhancing effect of clock disruption mentioned in Sect. 2.4, this may mean that tumors use part of the immune response to their own advantage, a phenomenon to be potentially taken into account in immunotherapy models involving circadian biology.

If both the immune system and circadian clocks are to be taken into account in modelling for cancer chemotherapeutics, it should also be mentioned that some anticancer drugs exert a detrimental influence on circadian clocks [74, 100], and as regards the immune system, some of them may be detrimental to the immune system, and some may be beneficial to it, even so that their anti-tumor efficacy may be due to stimulation of the immune response [106–108].

6.3 Taking Drug Resistance Into Account

Is the other main pitfall of anticancer therapeutics, emergence of drug resistance, related to circadian clocks? ABC transporter activity [55] is one of the main mechanisms on which drug resistance relies, and it has been shown that some ABC transporters (P-gp and Abcc2) show circadian rhythms in their gene expression and protein concentration [5, 85]. Other mechanisms (e.g., enhanced activity of drug degradation enzymes), show circadian rhythmicity, may also be responsible for the emergence of drug resistance in individual cells.

However, it does not seem that overcoming drug-induced drug resistance, which is a phenomenon occurring on a time scale that is not related with the 24-hour span—much longer in fact, due to mutations or epigenetic modifications, as mentioned in Sect. 3.2—may benefit from chronotherapeutics. Mathematical models aiming at representing drug resistance and optimization methods of drug delivery have been proposed for some time already, distinguishing between a sensitive cell subpopulation and a resistant one [30–32, 51, 69, 99].

Quite recently, continuous models structured according to a phenotypic trait and based on integrodifferential equations, are common in ecology and are based on Darwinian selection principles, applied to the problem of emergence—and its overcoming by combinations of cytotoxic and cytostatic drugs—of drug resistant subpopulations in cancer cell populations, have been proposed [57, 71, 77]. They offer the advantage of allowing the possibility to represent slow evolution, according to the expression of a phenotype rather than by jumps due to pointwise mutations, of a cancer cell population, a feature which makes them amenable to represent epigenetic modifications on which drug-induced drug resistance is likely to rely when it is reversible. These models do not take into account so far the cell division cycle, since the drug targets are simply a proliferation rate and a death rate, without molecular support, but more complex models taking the cell cycle into account might be relevant to describe different molecular drug effects on proliferation and death.

How should one balance and put in perspective the two pitfalls of unwanted toxicity and of emergence of drug resistance for therapeutic applications? Most likely, the clinic will put a hierarchy between them, and this will depend on each cancer and each drug delivery problem. However, it is possible to take both problems simultaneously into account, as shown in [77]. In any case, it is not obvious how chronotherapeutics may be relevant in this perspective.

Conclusion

I have presented some mathematical models of cell population dynamics designed in the last ten years, aiming at optimizing cancer chronotherapeutics. As long as the constraint chosen for the optimization problem is the limitation of unwanted toxic side effects, proofs of concepts have been achieved, showing the interest of chronotherapeutics, even though many unknowns still remain to identify before such theoretical models may be applicable in the clinic, reinforcing those that are already in use. As regards the other main pitfall of cancer therapeutics, drug resistance, different cell population dynamics models, transposed from mathematical ecology and set at a different time scale and based on Darwinian selection principles, have begun to emerge, and so far chronotherapeutics has not proved relevant for them.

Acknowledgements The author is gratefully indebted to his young colleagues Frédérique Billy and Olivier Fercoq for the work achieved together in modelling, model identification and theoretical therapeutic optimization, and to Francis Lévi for his long-lasting collaboration and frequent and fruitful discussions. This work has been supported by a grant from the European Research Area in Systems Biology (ERASysBio+) to the French National Research Agency (ANR) #ANR-09-SYSB-002 for the research network Circadian and Cell Cycle Clock Systems in Cancer (C5Sys) coordinated by Francis Lévi (INSERM U776, Villejuif, France).

References

1. A. Altinok, D. Gonze, F. Lévi, A. Goldbeter, An automaton model for the cell cycle. Interface focus **1**, 36–47 (2011)
2. A. Altinok, F. Lévi, A. Goldbeter. A cell cycle automaton model for probing circadian patterns of anticancer drug delivery. Adv. Drug Deliv. Rev. **59**, 1036–1010 (2007)
3. A. Altinok, F. Lévi, A. Goldbete, Optimizing temporal patterns of anticancer drug delivery by simulations of a cell cycle automaton. In M. Bertau, E. Mosekilde, and H. Westerhoff, editors, Biosimulation in Drug Development, pp. 275–297. (Wiley, 2008)
4. A. Altinok, F. Lévi, A. Goldbeter, Identifying mechanisms of chronotolerance and chronoefficacy for the anticancer drugs 5-fluorouracil and oxaliplatin by computational modeling. Eur. J. Pharm. Sci. **36**, 20–38 (2009)
5. H. Ando, H. Yanagihara, K.-i. Sugimoto, Y. Hayashi, S. Tsuruoka, T. Takamura, S. Kaneko, A. Fujimura, Daily rhythms of p-glycoprotein expression in mice. Chronobiol. Int. **22**(4), 655–665 (2005)
6. O. Arino. A survey of structured cell population dynamics. Acta Biotheor. **43**, 3–25 (1995)
7. A. Ballesta, J. Clairambault, S. Dulong, F. Lévi, in *A Systems Biomedicine Approach for Chronotherapeutics Optimization: Focus on the Anticancer Drug Irinotecan*, ed. by A. d'Onofrio, P. Cerrai, A. Gandolfi. New Challenges for Cancer Systems Biomedicine, part V, SIMAI Lecture Notes (Springer, New York, 2012), pp. 301–327
8. C. Basdevant, J. Clairambault, F. Lévi, Optimisation of time-scheduled regimen for anti-cancer drug infusion. Math. Model. Numerical Anal. **39**, 1069–1086 (2006)
9. B. Basse, B. Baguley, E. Marshall, G. Wake, D. Wall, Modelling the flow cytometric data obtained from unperturbed human tumour cell lines: Parameter fitting and comparison. Bull. Math. Biol. **67**, 815–830 (2005)
10. B. Basse, B.C. Baguley, E.S. Marshall, W.R. Joseph, B. van Brunt, G. Wake, and D.J.N. Wall, A mathematical model for analysis of the cell cycle in cell lines derived from human tumors. J. Math. Biol. **47**, 295–312 (2003)
11. F. Bekkal Brikci, J. Clairambault, B. Perthame, Analysis of a molecular structured population model with possible polynomial growth for the cell division cycle. Math. Comput. Model. **47**(7), 699–713 (2008)
12. F. Bekkal Brikci, J. Clairambault, B. Ribba, B. Perthame, An age-and-cyclin-structured cell population model for healthy and tumoral tissues. J. Math. Biol. **57**(1), 91–110 (2008)
13. S. Bernard, B.C. Bernard, F. Lévi, H. Herzel, Tumor growth rate determines the timing of optimal chronomodulated treatment schedules. PLoS Comput. Biol. **6**(3), e1000712 (2010)
14. S. Bernard, D. Gonze, B. Čajavec, H. Herzel, A. Kramer, Synchronization-induced rhythmicity of circadian oscillators in the suprachiasmatic nucleus. PLoS Comput. Biol. **3**(4), e68 (2007)
15. F. Billy J. Clairambault, Designing proliferating cell population models with functional targets for control by anti-cancer drugs. Discrete and Continuous Dynamical Systems - Series B, **18**(4), 865–889 (2013)
16. F. Billy, J. Clairambault, F. Delaunay, C. Feillet, N. Robert, Age-structured cell population model to study the influence of growth factors on cell cycle dynamics. Math. Biosci. Eng. **10**, 1–17 (2012)
17. F. Billy, J. Clairambault, O. Fercoq, in *Optimisation of Cancer Drug Treatments Using Cell Population Dynamics*, ed. by A. Friedman, E. Kashdan, U. Ledzewicz, H. Schättler. Mathematical Methods and Models in Biomedicine, part 4 (Springer, New-York, 2013), pp. 265–309
18. F. Billy, J. Clairambault, O. Fercoq, S. Gaubert, T. Lepoutre, T. Ouillon, S.Saito, Synchronisation and control of proliferation in cycling cell population models with age structure. *Mathematics and Computers in Simulation*, **96**, 66–94 (2014)
19. G. Bjarnason, R. Jordan, R. Sothern. Circadian variation in the expression of cell-cycle proteins in the human oral epithelium. Am. J. Pathol **154**, 613–622 (1999)

20. R. Borges, À. Calsina, S. Cuadrado, Equilibria of a cyclin structured cell population model. Discrete Continuous Dynam, Syst. Series B **11**, 613–627 (2009)
21. R. Borges, À. Calsina, S. Cuadrado, Oscillations in a molecular structured cell population model. Nonlinear Anal.: Real World Appl. **12**(4), 1911–1922 (2011)
22. J. Clairambault, Modelling oxaliplatin drug delivery to circadian rhythm in drug metabolism and host tolerance. Adv. Drug Deliv. Rev. **59**, 1054–1068 (2007)
23. J. Clairambault, A step toward optimization of cancer therapeutics. physiologically based modelling of circadian control on cell proliferation. IEEE-EMB Magazine **27**, 20–24 (2008)
24. J. Clairambault, O.Fercoq. in *Mathematical Modelling of Cancer Growth and Treatment*, eds by M. Bachar, J. Batze, and M. Chaplain, LNMBIOS SubseriesPhysiologically structured cell population dynamic models with applications to combined drug delivery optimisation in oncology. (Springer, New York, 2013) To appear, 2013. Available as preprint at http://hal.archives-ouvertes.fr/hal-00750633.
25. J. Clairambault, S. Gaubert, T. Lepoutre, Comparison of Perron and Floquet eigenvalues in age structured cell division models. Math. Model. Nat. Phenomena **4**, 183–209 (2009)
26. J. Clairambault, S. Gaubert, T. Lepoutre, Circadian rhythm and cell population growth. Math. Comput. Model. **53**, 1558–1567 (2011)
27. J. Clairambault, S. Gaubert, B. Perthame, An inequality for the Perron and Floquet eigenvalues of monotone differential systems and age-structured equations. C. R. Acad. Sci. (Paris) Ser. I Mathématique **345**, 549–554 (2007)
28. J. Clairambault, B. Laroche, S. Mischler, B. Perthame, A mathematical model of the cell cycle and its control. Technical report, Number 4892, INRIA, Domaine de Voluceau, BP 105, 78153 Rocquencourt, France (2003)
29. J. Clairambault, P. Michel, B. Perthame, Circadian rhythm and tumour growth. C. R. Acad. Sci. (Paris) Ser. I Mathématique (Équations aux dérivées partielles) **342**, 17–22 (2006)
30. A. Coldman, J. Goldie, A model for the resistance of tumor cells to cancer chemotherapeutic agents. Math. Biosci. **65**(2), 291–307 (1983)
31. M. Costa, J. Boldrini, R. Bassanezi, Chemotherapeutic treatments involving drug resistance and level of normal cells as a criterion of toxicity. Math. Biosci. **125**(2), 211–228 (1995)
32. M. Costa, J. Boldrini, R. Bassanezi, Drug kinetics and drug resistance in optimal chemotherapy. Math. Biosci. **125**(2), 191–209 (1995)
33. S. Davis, D.K. Mirick, Circadian disruption, shift work and the risk of cancer: a summary of the evidence and studies in seattle. Cancer Causes Cont. **17**(4), 539–545 (2006)
34. L.G. de Pillis, W. Gu, A.E. Radunskaya, Mixed immunotherapy and chemotherapy of tumors: modeling, applications and biological interpretations. J. Theor. Biol. **238**(4), 841–862 (2006)
35. L.G. de Pillis, A. Radunskaya, A mathematical tumor model with immune resistance and drug therapy: an optimal control approach. Computat. Math. Meth. Medicine **3**(2), 79–100, (2001)
36. L.G. de Pillis, A.E. Radunskaya, C.L. Wiseman, A validated mathematical model of cell-mediated immune response to tumor growth. Cancer Res. **65**(17), 7950–7958 (2005)
37. M. Delitala, T. Lorenzi, Recognition and learning in a mathematical model for immune response against cancer. Discrete and Cont. Dynam. Syst.-Series B **18**(4), 891–914 (2013)
38. C. Dibner, U. Schibler, U. Albrecht, The mammalian circadian timing system: organization and coordination of central and peripheral clocks. Annual Rev. Physiol **72**, 517–549 (2010)
39. A. d'Onofrio, A. Fasano, B. Monechi, A generalization of Gompertz law compatible with the gyllenberg-webb theory for tumour growth. Math. Biosci. **230**(1), 45–54 (2011)
40. M. Doumic, Analysis of a population model structured by the cells molecular content. Math. Model. Nat. Phenom **2**(3), 121–152 (2007)
41. B. Druker, M. Talpaz, D. Resta, B. Peng, E. Buchdunger, J. Ford, N. Lydon, H. Kantarjian, R. Capdeville, S. Ohno-Jones, C.Sawyers, Efficacy and safety of a specific inhibitor of the bcr-abl tyrosine kinase in chronic myeloid leukemia. N. Engl. J. Med. **344**, 1031–1037 (2001)
42. E. Filipski, P.F. Innominato, M.Wu, X.-M. Li, S.Iacobelli, L.-J. Xian, F.Lvi, Effects of light and food schedules on liver and tumor molecular clocks in mice. J. Natl. Cancer Inst. **97**(7), 507–517 (2005)

43. E. Filipski, V.M. King, X.Li, T.G. Granda, M.-C. Mormont, X. Liu, B. Claustrat, M.H. Hastings, F. Lvi, Host circadian clock as a control point in tumor progression. J. Natl. Cancer Inst. **94**(9), 690–697 (2002)
44. L. Fu, N.M. Kettner; The circadian clock in cancer development and therapy. Prog. Mol. Biol. Transl. Sci. **119**, 221–282 (2013)
45. L. Fu, H. Pelicano, J. Liu, P. Huang, C. Lee, The circadian gene *per2* plays an important role in tumor suppression and DNA damage response in vivo. Cell **111**, 41–50 (2002)
46. P. Gabriel, S.P. Garbett, V. Quaranta, D.R. Tyson, G.F. Webb, The contribution of age structure to cell population responses to targeted therapeutics. J. Theor. Biol. **311**, 19–27 (2012)
47. C. Gérard, A. Goldbeter, Entrainment of the mammalian cell cycle by the circadian clock: modeling two coupled cellular rhythms. PLoS Comput. Biol. **8**(5), e1002516 (2012)
48. M. Gerlinger, A.J. Rowan, S.Horswell, J. Larkin, D.Endesfelder, E.Gronroos, P. Martinez, N. Matthews, A. Stewart, P. Tarpey, I. Varela, B. Phillimore, S. Begum, N.Q. McDonald, A. Butler, D. Jones, K. Raine, C. Latimer, C.R. Santos, M. Nohadani, A.C. Eklund, B. Spencer-Dene, G.Clark, L. Pickering, G. Stamp, M. Gore, Z. Szallasi, J. Downward, P.A. Futreal, C. Swanton, Intratumor heterogeneity and branched evolution revealed by multiregion sequencing. N. Engl. J. Med. **366**(10), 883–892 (2012)
49. S. Giacchetti, P.A. Dugue, P.F. Innominato, G.A. Bjarnason, C. Focan, C. Garufi, S. Tumolo, B. Coudert, S. Iacobelli, R. Smaaland, et al., Sex moderates circadian chemotherapy effects on survival of patients with metastatic colorectal cancer: a meta-analysis. Annal. Oncol. **23**(12), 3110–3116 (2012)
50. A. Goldbeter, A model for circadian oscillations in the drosophila period protein (per). Proc. Royal Soc. London. Series B: Biol. Sci. **261**(1362), 319–324 (1995)
51. J. Goldie, A. Coldman, Quantitative model for multiple levels of drug resistance in clinical tumors. Cancer Treat. Reports **67**(10), 923–931 (1983)
52. D.A. Golombek, R.E. Rosenstein, Physiology of circadian entrainment. Physiol. Rev. **90**(3), 1063–1102 (2010)
53. D. Gonze, S. Bernard, C. Waltermann, A. Kramer, H. Herzel, Spontaneous synchronization of coupled circadian oscillators. Biophys. J. **89**(1), 120–129 (2005)
54. D. Gonze, J. Halloy, A. Goldbeter, Robustness of circadian rhythms with respect to molecular noise. Proc. Natl. Acad. Sci. U S A **99**(2), 673–678 (2002)
55. M.M. Gottesman, T.Fojo, S.E. Bates, Multidrug resistance in cancer: role of atp-dependent transporters. Nat. Rev. Cancer **2**(1), 48–58 (2002)
56. A. Gréchez-Cassiau, B. Rayet, F. Guillaumond, M. Teboul, and F. Delaunay, The circadian clock component Bmal1 is a critical regulator of p21 (**WAF1/CIP1**) expression and hepatocyte proliferation. J. Biol. Chem. **283**, 4535–42 (2008)
57. J. Greene, O. Lavi, M.M. Gottesman, D. Levy, The impact of cell density and mutations in a model of multidrug resistance in solid tumors. Bulletin of Mathematical Biology, **76**, 627–653 (2014)
58. M. Gyllenberg, G.F. Webb, Quiescence as an explanation of gompertzian tumor growth. Growth Dev. Aging **53**, 25–33 (1989)
59. M. Gyllenberg, G.F. Webb, A nonlinear structured population model of tumor growth witsh quiescence. J. Math. Biol. **28**, 671–694 (1990)
60. T. Haferlach, Molecular genetic pathways as therapeutic targets in acute myeloid leukemia. Hematology **2008**, 400–411 (2008) Am. Soc. Hematol. Educ. Program.
61. F. Halberg, Chronobiology. Annual Rev. Physiol. **31**(1), 675–726 (1969)
62. J. Hansen, Risk of breast cancer after night-and shift work: current evidence and ongoing studies in denmark. Cancer Causes Cont. **17**(4), 531–537 (2006)
63. M.H. Hastings, A.B. Reddy, E.S. Maywood, A clockwork web: circadian timing in brain and periphery, in health and disease. Nat. Rev. Neurosci. **4**(8), 649–661 (2003)
64. E. Haus M. Smolensky, Biological clocks and shift work: Circadian dysregulation and potential long-term effects. Cancer Causes Cont. **17**(4), 489–500 (2006)
65. E.C. Hayden, Cutting off cancer's supply lines. Nature **458**(7239), 686 (2009)

66. M. Horiguchi, S. Koyanagi, A.M. Hamdan, K. Kakimoto, N. Matsunaga, C. Yamashita, S. Ohdo, Rhythmic control of the *ARF-MDM2 pathway by ATF4* underlies circadian accumulation of p53 in malignant cells. Cancer Res. **73**(8), 2639–2649 (2013)
67. B.L. Keyfitz, N. Keyfitz, The McKendrick partial differential equation and its uses in epidemiology and population study. Math. Comput. Model. **26**(6), 1–9 (1997)
68. P.S. Kim P.P. Lee, Modeling protective anti-tumor immunity via preventative cancer vaccines using a hybrid agent-based and delay differential equation approach. PLoS Computat. Biol. **8**(10), e1002742 (2012)
69. M. Kimmel, A. Swierniak, Control theory approach to cancer chemotherapy: Benefiting from phase dependence and overcoming drug resistance, in *Tutorials in Mathematical Biosciences III*, (Springer, Newyork, 2006) pp. 185–221
70. R. Konopka, S.Benzer, Clock mutants of drosophila melanogaster. Proc. Natl. Acad. Sci. U S A **68**, 2112–16 (1971)
71. O. Lavi, J. Greene, D. Levy, M.M. Gottesman, The role of cell density and intratumoral heterogeneity in multidrug resistance. Cancer Research, **73**, 7168–71775 (2013)
72. J.-C. Leloup, A. Goldbeter, Toward a detailed computational model for the mammalian circadian clock. Proc. Natl. Acad. Sci. U S A **100**(12), 7051–7056 (2003)
73. J.C. Leloup, D.Gonze, A.Goldbeter. Limit cycle models for circadian rhythms based on transcriptional regulation in drosophila and neurospora. J. Biol. Rhythms **14**(6), 433–448 (1999)
74. F. Lévi, A. Okyar, S. Dulong, P. Innominato, J. Clairambault, Circadian timing in cancer treatments. Annual Rev. Pharmacol. Toxicol. **50** 377–421 (2010)
75. F.A. Lévi, C. Canon, Y. Touitou, J. Sulon, M. Mechkouri, E. Demey-Ponsart, J.P. Touboul, J.M. Vannetzel, I. Mowzowicz, A. Reinberg, Circadian rhythms in circulating T lymphocyte subtypes and plasma testosterone, total and free cortisol in five healthy men. Clin. Exp. Immunol. **71**(2), 329–335 (1988)
76. A. Levitzki, Tyrosine kinase inhibitors: views of selectivity, sensitivity, and clinical performance. Annual Rev. Pharmacol. Toxicol. **53**, 161–185 (2013)
77. A. Lorz, B. Perthame, T. Lorenzi, M.E. Hochberg, J. Clairambault. Populational adaptive evolution, chemotherapeutic resistance and multiple anti-cancer therapy. ESAIM: Math. Model. Numer. Anal. **47**(1), 377–399 (2013)
78. S. Masri, M. Cervantes, P. Sassone-Corsi, The circadian clock and cell cycle: interconnected biological circuits. Curr. Opin. Cell. Biol. (2013)
79. T. Matsuo, S. Yamaguchi, S. Mitsuia, A. Emi, F. Shimoda, H. Okamura, Control mechanism of the circadian clock for timing of cell division in vivo. Science **302**, 255–259 (2003)
80. F. Mazenc, P.S. Kim, S.-I. Niculescu, Stability of an imatinib and immune model with delays. IMA J. Math. Cont. Inform. **28**(4), 447–462 (2011)
81. A. McKendrick, Applications of mathematics to medical problems. Proc. Edinburgh Math. Soc. **54**, 98–130 (1926)
82. J. Metz, O. Diekmann, The dynamics of physiologically structured populations, vol. 68 *Lecture Notes in Biomathematics* (Springer, New York, 1986)
83. M.C. Mormont, F. Lévi, Cancer chronotherapy: principles, applications, and perspectives. Cancer **97**(1), 155–169 (2003)
84. W. Nelson, Y. L. Tong, J. K. Lee, F. Halberg, Methods for cosinor-rhythmometry. Chronobiologia **6**(4), 305–323 (1979)
85. A. Okyar, E. Piccolo, C. Ahowesso, E. Filipski, V. Hossard, C. Guettier, R. La Sorda, N. Tinari, S. Iacobelli, F. Lévi, Strain- and sex-dependent circadian changes in ABCC2 transporter expression: implications for irinotecan chronotolerance in mouse ileum. PLoS One **6**(6), e20393 (2011)
86. B. Perthame, Transport Equations in Biology. *Frontiers in Mathematics series* (Birkhäuser, Boston, 2007)
87. T. Rich, P. Innominato, J. Boerner, M.-C. Mormont, S. Iacobelli, B. Baron, C. Jasmin, F. Lévi, Elevated serum cytokines correlated with altered behavior, serum cortisol rhythm, and dampened 24-hour rest-activity pattern in patients with metastatic colorectal cancer. Clin. Cancer Res. **11**, 1757–64 (2005)

88. J.M. Rowe B. Löwenberg, Gemtuzumab ozogamicin in acute myeloid leukemia: a remarkable saga about an active drug. Blood **121**(24), 4838–4841 (2013)
89. P. Ruoff, M. Vinsjevik, C. Monnerjahn, L. Rensing. The goodwin model: simulating the effect of light pulses on the circadian sporulation rhythm of neurospora crassa. J. Theor. Biol. **209**(1), 29–42 (2001)
90. A. Sakaue-Sawano, H. Kurokawa, T. Morimura, A. Hanyu, H. Hama, H. Osawa, S. Kashiwagi, K. Fukami, T. Miyata, H. Miyoshi, T. Imamura, M. Ogawa, H. Masai, A. Miyawaki, Visualizing spatiotemporal dynamics of multicellular cell-cycle progression. Cell **32**, 487–498 (2008)
91. A. Sakaue-Sawano, K. Ohtawa, H. Hama, M. Kawano, M. Ogawa, A.Miyawaki, Tracing the silhouette of individual cells in S/G2/M phases with fluorescence. Chem Biol. **15**, 1243–48 (2008)
92. E.S. Schernhammer, F. Laden, F.E. Speizer, W.C. Willett, D.J. Hunter, I. Kawachi, C.S. Fuchs, G.A. Colditz, Night-shift work and risk of colorectal cancer in the nurses health study. J. National Cancer Instit. **95**(11), 825–828 (2003)
93. A.M. Scott, J.D. Wolchok, L.J. Old, Antibody therapy of cancer. Nat. Rev. Cancer **12**(4), 278–287 (2012)
94. S.E. Sephton, R.M. Sapolsky, H.C. Kraemer, D.Spiegel, Diurnal cortisol rhythm as a predictor of breast cancer survival. J. Natl. Cancer Instit. **92**(12), 994–1000, (2000)
95. F.R. Sharpe, A.J. Lotka, L. A problem in age-distribution. Philosophical Magazine Series 6 **21**, 435–438 (1911)
96. R. Smaaland, O. Laerum, K. Lote, O. Sletvold, R. Sothern, R. Bjerknes. DNA synthesis in human bone marrow is circadian stage dependent. Blood **77**, 2603–2611 (1991)
97. S. Suzuki, S. Toyabe, T. Moroda, T. Tada, A. Tsukahara, T. Iiai, M. Minagawa, S Maruyama, K. Hatakeyama, K.Endoh, T.Abo, Circadian rhythm of leucocytes and lymphocytes subsets and its possible correlation with the function of the autonomic nervous system. Clin Exp. Immunol. **110**(3), 500–508 (1997)
98. B. Sweeney, *Rhythmic phenomena in plants* (Academic Press, New York 1969)
99. A. Świerniak, M. Kimmel, J. Smieja, Mathematical modeling as a tool for planning anticancer therapy. Eur. J. Pharmacol. **625**(1), 108–121 (2009)
100. H. Terazono, A. Hamdan, N. Matsunaga, N. Hayasaka, H. Kaji, T. Egawa, K. Makino, Y. Shigeyoshi, S. Koyanagi, S. Ohdo, Modulatory effects of 5-fluorouracil on the rhythmic expression of circadian clock genes: a possible mechanism of chemotherapy-induced circadian rhythm disturbances. Biochem. pharmacol. **75**(8), 1616–1622 (2008)
101. T. Tozer, M. Rowland, *Introduction to Pharmacokinetics and Pharmacodynamics: the Quantitative Basis of Drug Therapy* (Lippincott 2006)
102. Y. Tsukamoto, Y. Kato, M. Ura, I. Horii, H. Ishitsuka, H. Kusuhara, Y. Sugiyama, A physiologically based pharmacokinetic analysis of capecitabine, a triple prodrug of 5-FU, in humans: the mechanism for tumor-selective accumulation of 5-FU. Pharm. Res. **18**(8), 1066, 1190–1202 (2001)
103. M.H. Vitaterna, D.P. King, A.-M. Chang, J.M. Kornhauser, P.L. Lowrey, J.D. McDonald, W.F. Dove, L.H. Pinto, F.W. Turek, J.S. Takahashi, Mutagenesis and mapping of a mouse gene, Clock, essential for circadian behavior. Science **264**(5159), 719–725 (1994)
104. H. Von Foerster, in *Some Remarks on Changing Populations,* ed. by F. Stohlman Jr. The kinetics of Cellular Proliferation (Grune and Stratton, New York, 1959), pp. 382–407
105. M. Wu, X.Li, L.Xian, F.Lévi, Effects of meal timing on tumor progression in mice. Life Sci. **75**(10), 1181–1193 (2004)
106. L. Zitvogel, L. Apetoh, F. Ghiringhelli, G. Kroemer, Immunological aspects of cancer chemotherapy. Nat. Rev. Immunol. **8**, 59–73 (2008)
107. L. Zitvogel, L. Galluzzi, M.J. Smyth, G. Kroemer, Mechanism of action of conventional and targeted anticancer therapies: reinstating immunosurveillance. Immunity **39**(1), 74–88 (2013)
108. L. Zitvogel, O. Kepp, G. Kroemer, Immune parameters affecting the efficacy of chemotherapeutic regimens. Nat. Rev. Clin. Oncol. **8**(3), 151–160 (2011)

Tumor Microenvironment and Anticancer Therapies: An Optimal Control Approach

Urszula Ledzewicz and Heinz Schättler

Abstract In this paper, results about the structure of cancer treatment protocols that can be inferred from an analysis of mathematical models with the methods and tools of optimal control are reviewed. For homogeneous tumor populations of chemotherapeutically sensitive cells, optimal controls are bang-bang corresponding to the medical paradigm of maximum tolerated doses (MTD). But as more aspects of the tumor microenvironment are taken into account, such as heterogeneity of the tumor cell population, tumor angiogenesis and tumor-immune system interactions, singular controls which administer agents at specific time-varying reduced dose rates become optimal and give an indication of what might be the biologically optimal dose (BOD).

Keywords Optimal control • Cancer chemotherapy • Tumor microenvironment • Antiangiogenic treatments • Tumor-immune system interactions

1 Introduction

In any cancer treatment, the question arises how therapeutic agents (various drugs, radiation dosages, antiangiogenic biological agents, cancer vaccines, ...) should be given in order to be at the same time reasonably safe and effective. Mathematically, the scheduling of therapeutic agents over time in order to minimize some objective related to tumor burden (e.g., tumor volume) and quality of life of the patient (e.g., some measure of the toxic side effects of treatment) while the underlying system follows some dynamics (in this case determined by the processes of tumor

U. Ledzewicz (✉)
Department of Mathematics and Statistics, Southern Illinois University,
Edwardsville, Illinois 62026, USA
e-mail: uledzew@siue.edu

H. Schättler
Department of Electrical and Systems Engineering, Washington University,
St. Louis, Missouri 63130, USA
e-mail: hms@wustl.edu

© Springer Science+Business Media New York 2014
A. d'Onofrio, A. Gandolfi (eds.), *Mathematical Oncology 2013*,
Modeling and Simulation in Science, Engineering and Technology,
DOI 10.1007/978-1-4939-0458-7_10

development and treatment interactions) is an optimal control problem. In this paper, we review some results about the structure of treatment protocols that can be inferred from mathematical models with the methods and tools of optimal control.

Our emphasis will be on models for cancer chemotherapy. For most tumors, it is a standard medical practice to give chemotherapy at *maximum tolerated doses* (MTD) with rest periods in between. The underlying rationale simply is that when the disease has progressed into an advanced stage, it is imperative to kill as many of the cancer cells as possible and this has to be done right now. Since drugs are rarely selective in their activation mechanisms, chemotherapy also severely damages other proliferating cells that are essential for survival like bone marrow. This necessitates the introduction of rest periods for the patient to recover from the strong toxic attack. We are interested in questions of the following type: Under what kind of conditions is an MTD approach the optimal treatment strategy? When should different protocols be favored? If resistance to chemotherapeutic agents is present, is a metronomic scheduling of chemotherapeutic agents (essentially, a continuous-type treatment at low doses) which avoids the high toxicities associated with MTD doses equally effective? Naturally, answers to such questions depend on the type of tumor. Simple dividing characteristics that are important in the scheduling of treatment are given by the tumor doubling time, the growth fractions of tumor cells, and much more.

A tumor consists not just of cancerous cells but of a full array of other structures that in various ways aid and abet the tumor, but also fight it. The most important structure that sustains the tumor is its vasculature which provides the tumor with the oxygen and nutrients needed for further growth; an example of an endogenous structure that fights the tumor is the body's immune system. The *tumor microenvironment* consists of these components and much more (e.g., macrophages and fibroblast cells that form the intracellular matrix), all still residing in healthy tissue. In modern oncology thus the point of view of the *tumor as a system of interacting components* has become the more common one and modern treatments are multi-targeted therapies that not only aim to kill cancer cells but often include antiangiogenic therapy, immunotherapy, and other options. Yet, the complex interactions between these and other treatment modalities still are not fully understood and are the topic of active current medical research (e.g., see [1]).

In clinical trials, the scheduling of therapeutic agents is pursued in medically guided, exhaustive trial-and-error approaches of simple strategies. Hardly ever are nonstandard protocols pursued in this research since complex protocols are relatively difficult, if not impossible to test in a laboratory setting, or at a minimum at great cost. The analysis of mathematical models can be of benefit here by giving some theoretical suggestions for treatment protocols through an alternative noninvasive tool or by establishing benchmarks for medically realizable protocols. As of today, the question how chemotherapeutic agents should best be administered if a more wholistic approach to treatment is taken that takes the structures of the tumor microenvironment into account still has not been answered (e.g., see [21]).

This paper is organized as follows: In Sect. 2 we give a brief introduction to the main tools and results from optimal control theory that are needed in the analysis of mathematical models for cancer treatment. Especially, the distinction between *bang-bang controls (which correspond to maximum dose treatment periods interlaced with rest periods*) and *singular controls (which correspond to time-varying administration schedules at lower dose rates*) will be emphasized. Bang-bang controls directly relate to the MTD strategies of medical practice while singular controls are of special interest in the search for the *biologically optimal dose* (BOD). This is an effective dose which has minimal or at least low side effects [26]. In Sect. 3, we start with a discussion of optimal treatment protocols for compartmental models of cancer chemotherapy. It is easily seen that optimal controls indeed support the traditional MTD paradigm if it is assumed that the tumor consists of a homogeneous population of chemotherapeutically sensitive cells. However, as compartments of varying sensitivities or even full resistance are introduced into the model, this no longer is valid and singular controls along with the associated lower dose rates become candidates for optimality. Optimal administration of antiangiogenic agents also is done by means of singular controls and will be discussed in Sect. 4, both as stand-alone approach and in combination therapy with chemotherapy. Once tumor-immune system interactions are taken into account, optimal administration of cytotoxic agents no longer follows an MTD approach, but a so-called "chemo-switch" regimen: after an initial interval of maximum dose treatment, in optimal solutions dose rates are reduced and given by singular controls. These results are given in Sect. 5.

Overall, an optimal control analysis of mathematical models for cancer chemotherapy as it is presented here leads to results that provide information about the qualitative structure of treatment protocols that can be of use in the design of practical treatment protocols.

2 Optimal Control–A Brief Introduction

We briefly review the main results of optimal control theory. However, rather than considering the general case, we restrict the mathematical structure to a model of the form that most examples in biomedical applications have: a multi-input control-affine system. This simply reflects the fact that "controls" represent structures imposed on an existing dynamical system from the outside to influence its behavior and that these are naturally set up in a way so that these effects are most easily analyzed. This generally leads to linear terms in the controls. For such systems, the so-called bang-bang and singular controls become the prime candidates for optimality. We describe the principal tools for analyzing singular controls which include Lie brackets for computing derivatives of the switching function and the Legendre-Clebsch condition as the main necessary condition for optimality.

2.1 Control Affine Systems as Mathematical Models for Biomedical Models

We say a control system is *control-affine* with drift vector field f and control vector fields g_i, $i = 1, \dots, m$, if the dynamics takes the following form:

$$\dot{x} = f(x) + \sum_{i=1}^{m} g_i(x)u_i, \qquad x \in M, \qquad u \in U. \tag{1}$$

The vector x is the *state* of the system and takes values in an open and connected subset M of $\mathbb{R}^n$; the vector u represents the *controls* and takes values in a control set $U \subset \mathbb{R}^m$. In the biomedical models we shall be considering, the controls represent dose rates or concentrations of some therapeutic agents and all take nonnegative values that lie in prescribed ranges. We therefore take the control set U as an m-dimensional interval of the form

$$U = [0, u_1^{\max}] \times \cdots \times [0, u_m^{\max}]. \tag{2}$$

The class $\mathcal{U}$ of *admissible controls* is given by Lebesgue-measurable functions u defined on some interval I with values in the control set (almost everywhere), $u : I \to U, t \mapsto u(t)$. The differential equation (1) represents the *dynamics* which connects the controls with the state of the system. Given an admissible control $u \in \mathcal{U}$, it follows from classical results about solutions to ordinary differential equations that for any initial condition x_0, there exists a unique solution x to (1) with initial condition $x(0) = x_0$. We call this solution x the trajectory corresponding to the control u and call the pair (x, u) an admissible controlled trajectory.

An optimal control problem then consists in finding, among all admissible controlled trajectories, one that minimizes an objective, possibly subject to additional constraints. Here we only consider constraints of a fixed terminal time T or on the final state $x(T)$ of the system. The former correspond to therapy over an a priori specified horizon (Sect. 3) and the latter arise if therapy with an a priori given amount of therapeutic agents is considered (Sect. 4). We assume that such constraints have a regular geometric structure and are given in the form $N = \{x \in M : \psi(x) = 0\}$ with $\psi : M \to \mathbb{R}^{n-k}$ a continuously differentiable mapping and the matrix $D\psi$ of the partial derivatives of ψ with respect to x of full rank everywhere on N. We choose the functional form of the objective to be consistent with the control-affine structure of the dynamics, i.e., we take the functional to be minimized in the form

$$\mathcal{J}(u) = \int_0^T \left(L(x(s)) + \sum_{i=1}^{m} \theta_i u_i(s) \right) ds + \varphi(x(T)) \tag{3}$$

with $L : M \to \mathbb{R}$, $x \mapsto L(x)$ the Lagrangian and $\varphi : N \to \mathbb{R}$, $x \mapsto \varphi(x)$ a penalty term on the final state. Both L and φ are continuously differentiable functions. The terminal time T can be fixed or free. We choose the functional dependence of the objective on the controls to be linear since the integrals $\int_0^T u_i(t)dt$ have an immediate interpretation in terms of the total dose of agents given and thus are biomedically meaningful. It would be mathematically simpler to choose quadratic terms for the controls in the objective, but such terms are imposed arbitrarily. We thus consider the following optimal control problem:

[OC] minimize the objective $\mathcal{J}(u)$ over all admissible controlled trajectories (x, u) subject to the terminal constraint $x(T) \in N$.

2.2 *Necessary Conditions for Optimality: The Pontryagin Maximum Principle*

The fundamental necessary conditions for a controlled trajectory (x, u) to be optimal are given by the Pontryagin maximum principle [59]. (We refer the reader to [3, 4, 61] for some modern treatments of the subject.) We consistently write tangent vectors as column vectors and multipliers as row vectors denoting the space of row vectors by $(\mathbb{R}^n)^*$. The *Hamiltonian* function H of the optimal control problem [OC] is defined as

$$H = \lambda_0 \left(L(x) + \sum_{i=1}^{m} \theta_i u_i \right) + \left\langle \lambda,\ f(x) + \sum_{i=1}^{m} g_i(x) u_i \right\rangle \tag{4}$$

Theorem 2.1 (Pontryagin Maximum Principle [59]). *Let* (x_*, u_*) *be an optimal controlled trajectory for the problem [OC] defined over the interval* $[0, T]$*. Then there exist a constant* $\lambda_0 \geq 0$*, a multiplier* $\nu \in (\mathbb{R}^{n-k})^*$*, and a co-vector* $\lambda : [0, T] \to (\mathbb{R}^n)^*$*, the so-called* adjoint variable*, such that the following conditions are satisfied:*

1. Nontriviality *of the multipliers:* $(\lambda_0, \lambda(t)) \neq 0$ *for all* $t \in [0, T]$.
2. Adjoint equation: *the adjoint variable* λ *is a solution to the time-varying linear differential equation*

$$\dot{\lambda}(t) = -\lambda_0 \nabla L(x_*(t)) - \lambda(t) \left(Df(x_*(t) + \sum_{i=1}^{m} u_i^*(t) g_i(x_*(t)) \right) \tag{5}$$

with terminal condition

$$\lambda(T) = \lambda_0 \frac{\partial \varphi}{\partial x}(x_*(T)) + \nu \frac{\partial \psi}{\partial x}(x_*(T)). \tag{6}$$

3. Minimum condition: *almost everywhere in* $[0, T]$ *we have that*

$$H(\lambda_0, \lambda(t), x_*(t), u_*(t)) = \min_{v \in U} H(\lambda_0, \lambda(t), x_*(t), v) \tag{7}$$

and the Hamiltonian is constant along λ *and* (x_*, u_*). *If the terminal time* T *is free, the value of this constant is* 0.

Controlled trajectories (x, u) for which there exist multipliers λ_0 and λ such that the conditions of the maximum principle are satisfied are called *extremals* and the triples $(x, u, (\lambda_0, \lambda))$ including the multipliers are called *extremal lifts*. The constant multiplier λ_0 can be zero and in this case the extremal is called *abnormal* while it is called *normal* if $\lambda_0 > 0$. In this case, since the conditions are linear in the multipliers, it is always possible to normalize $\lambda_0 = 1$.

In the original formulation of the theorem by Pontryagin et al. [59], the minimum condition (7) was formulated as a maximum condition and gave the result its name. In fact, depending on the choice of the signs associated with the multipliers λ_0 and λ, the maximum principle can be stated in four equivalent versions. Since the problems we will be considering are all cast as minimization problems, we prefer this formulation, but retain the classical name. The minimum condition contains the essence of the result and states that in order to solve the minimization problem on the function space of admissible controls, the control u_* needs to be chosen so that for some extremal lift it minimizes the Hamiltonian H pointwise over the control set U, i.e., for every $t \in [0, T]$ the control $u_*(t)$ is a minimizer of the function $v \mapsto H(\lambda_0, \lambda(t), x_*(t), v)$ over the control set U.

2.3 Bang-Bang and Singular Controls

In our case, since U is an m-dimensional interval, the minimum condition splits into m scalar minimization problems that are easily solved. Defining the functions

$$\Phi_i(t) = \lambda_0 \theta_i + \langle \lambda(t), g_i(x_*(t)) \rangle, \tag{8}$$

it follows that the optimal controls satisfy

$$u_i^*(t) = \begin{cases} 0 & \text{if } \Phi_i(t) > 0, \\ u_i^{\max} & \text{if } \Phi_i(t) < 0. \end{cases} \tag{9}$$

A priori, the control is not determined by the minimum condition at times when $\Phi(\tau) = 0$. In such a case, all controls trivially satisfy the minimum condition and, in principle, are candidates for optimality. Naturally, if the derivative $\dot{\Phi}(\tau)$ exists and does not vanish, then the control switches between $u_i = 0$ and $u_i = u_i^{\max}$ with the order depending on the sign of $\dot{\Phi}(\tau)$. Such a time τ is called a bang-bang switch. On the other hand, if $\Phi(t)$ were to vanish identically on an open interval I, then, although the minimization property by itself gives no information about the control, in this case also all the derivatives of $\Phi(t)$ must vanish and this, except for some degenerate situations, generally does determine the control. Controls of this kind are called *singular* while the constant controls $u_i = 0$ and $u_i = u_i^{\max}$ are called *bang* controls and controls that only switch between 0 and the maximum control values are *bang-bang controls*. Strictly speaking, to be singular is not a property of the control, but of the extremal lift since it also depends on the multiplier λ defining the function Φ_i. This function is called the *switching function* for the control u_i.

The terminology "singular" has its historical origin in the fact that the switching functions can be expressed as

$$\Phi_i(t) = \frac{\partial H}{\partial u_i}(\lambda_0, \lambda(t), x_*(t), u_*(t)) \tag{10}$$

and thus the condition $\Phi(t) = 0$ formally is the first-order necessary condition for the Hamiltonian to have a minimum in the interior of the control set. For singular controls, the Hessian matrix $\frac{\partial^2 H}{\partial u^2}$ corresponding to second order necessary conditions for optimality is singular. In fact, for a control-affine system this matrix is identically zero.

If the control corresponds to the application of some therapeutic agent, then bang-bang controls represent treatment strategies that switch between maximum dose therapy sessions and rest periods, the typical MTD-type applications on chemotherapy. Singular controls on the other hand represent time-varying administrations of the agent at intermediate and often significantly lower doses. Although administration of such time-varying schedules may be difficult in practice, there is growing interest in such structures in the medical community because of mounting evidence that "more is not necessarily better" [18,55] and that a biologically optimal dose (BOD) with the best overall response should be sought. In this direction, the concept of metronomic chemotherapy as well as other approaches like chemo-switch protocols [57] or adaptive therapy [12] have been introduced. We shall say more about these medical connections later on. But the question whether optimal controls are bang-bang or singular has an immediate interpretation and relevance for the structure of optimal treatment protocols. While the terminology is somewhat misleading, these singular structures indeed are the more natural candidates for optimality.

2.4 The Legendre-Clebsch Condition for Optimality of Singular Controls

In the solution of any optimal control problem, it becomes necessary to determine singular controls and then synthesize optimal controls from the primary candidates—bang and singular controls. In order to do so, we need to analyze the derivatives of the switching functions. In these formulas, the notion of the Lie bracket of vector fields arises naturally: given two differentiable vector fields f and g defined on some open set $M \subset \mathbb{R}^n$, $f, g : M \to \mathbb{R}^n$, their Lie bracket $[f, g]$ is another vector field defined on G by

$$[f, g](x) = Dg(x)f(x) - Df(x)g(x). \tag{11}$$

Its importance in optimal control is because of the following simple formula that is verified by a direct computation:

Proposition 2.1. *Let $x(\cdot)$ be a solution of the dynamics (1) for the controls u_i and let λ be a solution of the corresponding adjoint equation (5). For a continuously differentiable vector field h, the derivative of the function*

$$\Psi(t) = \langle \lambda(t), h(x(t)) \rangle = \lambda(t)h(x(t))$$

is given by

$$\dot{\Psi}(t) = \left\langle \lambda(t), \left[f + \sum_{i=1}^{m} u_i^*(t) g_i, h \right] (x(t)) \right\rangle - \lambda_0 \nabla L(x(t))h(x(t)).$$

Singular controls are computed by differentiating the switching functions until the controls explicitly appear and then solving the resulting equations for the controls. We demonstrate the procedure for the simpler case of a single-input control system of the form $\dot{x} = f(x) + g(x)u$. Since $[g, g] \equiv 0$, the derivative of the switching function Φ is given by

$$\dot{\Phi}(t) = \langle \lambda(t), [f, g](x(t)) \rangle - \lambda_0 \nabla L(x(t))g(x(t)), \tag{12}$$

does not depend on the control, and thus is once more differentiable. In the second derivative $\ddot{\Phi}(t)$, the control appears linearly and, expressing the switching function as $\Phi = \frac{\partial H}{\partial u}$, the term multiplying the control is given by $\frac{\partial}{\partial u} \frac{d^2}{dt^2} \frac{\partial H}{\partial u}(\lambda_0, \lambda(t), x_*(t), u_*(t))$. A singular control (more precisely, the singular lift) is said to be of order 1 over an open interval I if this expression does not vanish on I and in this case, we can solve the equation $\ddot{\Phi}(t) = 0$ for the control as a function of the state and multiplier. Essentially, the sign of this expression distinguishes between locally minimizing and maximizing controls. This is the interpretation of the Legendre-Clebsch condition, the fundamental necessary condition for optimality of singular controls which states that for minimizing controls we must have that

$$\frac{\partial}{\partial u} \frac{d^2}{dt^2} \frac{\partial H}{\partial u}(\lambda_0, \lambda(t), x_*(t), u_*(t)) \leq 0 \quad \text{for all} \quad t \in I. \tag{13}$$

If this expression vanishes over an interval, then it becomes necessary to differentiate the switching function further. This leads to the concept of singular controls of higher order and the generalized Legendre-Clebsch condition. In some special, but common circumstances, it follows from Lie algebraic identities that the control can only appear for the first time in an even order derivative. Then the singular control is said to be of intrinsic order k if this is the $2k$th derivative and one then has the following necessary condition for optimality for singular controls of finite order:

Theorem 2.2 (Generalized Legendre-Clebsch Condition). *Suppose the controlled trajectory (x_*, u_*) defined over the interval $[0, T]$ is optimal for the optimal control problem [OC] and the control u_* is singular of intrinsic order k over an open interval $I \subset [0, T]$. Then there exists an extremal lift $\Gamma = ((x_*, u_*), \lambda)$ with the property that*

$$(-1)^k \frac{\partial}{\partial u} \frac{d^{2k}}{dt^{2k}} \frac{\partial H}{\partial u} (\lambda_0, \lambda(t), x_*(t), u_*(t)) \geq 0 \quad \text{for all} \quad t \in I. \tag{14}$$

2.5 *Sufficient Conditions for Optimality*

The optimality conditions discussed so far are all necessary and do not guarantee that a controlled trajectory that satisfies them is optimal. The theory of sufficient conditions for optimality is more intricate. Essentially, to guarantee local optimality properties, it becomes necessary to embed a reference extremal (i.e., controlled trajectory and associated multiplier) into a family of extremals in such a way that the controlled trajectories cover a neighborhood of the reference controlled trajectory. If this can be done globally in the form of what is called a *regular synthesis*, then the associated controls all are globally optimal. These concepts are related to classical ideas from the calculus of variations about fields of extremals or, in a more modern language, to dynamic programming and solutions of the Hamilton-Jacobi-Bellman equations. However, the details are too involved to even be outlined here and we refer the interested reader to the literature on the subject, such as, for example, our text [61].

3 Compartmental Models for Cancer Chemotherapy

In this section we formulate a general bilinear version of the optimal control problem [OC] that serves as the mathematical framework for compartmental models for cancer chemotherapy. Applications of optimal control to mathematical models for cancer chemotherapy have a long history (e.g., [6, 43, 64, 66]), but generally early models were noncompartmental. The models we consider here were formulated and first analyzed in the work of Swierniak and coworkers (e.g., [24, 67, 71]) and then reconsidered in our work [33, 34, 69]. Compartments may be comprised of various phases of the cell cycle (Sect. 3.2) or may correspond to different subpopulations of cancer cells of varying chemotherapeutic sensitivities (Sect. 3.3). While optimal controls are bang-bang with upfront dosing for homogeneous cell populations of chemotherapeutically sensitive cells and thus agree with the medical MTD paradigm of scheduling chemotherapy, as resistance effects come into play, this is no longer the case and singular controls with associated lower dose rates become candidates for optimality.

3.1 A General Bilinear Model

We consider a mathematical model with a finite number n of compartments and use the first orthant $M = \mathbb{P}$ in $\mathbb{R}^n$ as state space; $N = (N_1, \dots, N_n)^T$ denotes the state with N_i the average number of cancer cells in the ith compartment, $i = 1, \dots, n$. The control is a vector $u = (u_1, \dots, u_m)^T$ with u_i denoting various drug concentrations in the blood stream. For simplicity, in our language we identify the drug dose rates with their concentrations. Indeed, standard linear pharmacokinetic equations are easily incorporated within the general structure below at the expense of increasing the dimension of the state space, but they do not alter the results we obtain [36] and thus we use this simplified approach here. As before, the control set U is the m-dimensional interval $U = [0, u_1^{\max}] \times \cdots \times [0, u_m^{\max}]$ and admissible controls are Lebesgue-measurable functions u that take values in the control set. The dynamics consists of balance equations that describe the inflows and outflows between the various compartments and is assumed to be of the form

$$\dot{N}(t) = \left(A + \sum_{j=1}^{m} u_j B_j \right) N(t), \qquad N(0) = N_0, \tag{15}$$

where the A and B_j, $j = 1, \dots, m$, are constant $n \times n$ matrices, $A, B_j \in \mathbb{R}^{n \times n}$. The matrix A describes the transitions between the various compartments in the absence of treatment and the matrices B_j represent the effects of the jth drug on the system. An equation of the form (15) is called a *bilinear control system* since it is linear both in the state N and the control u. Note, however, that there exist quadratic terms since the controls u_i are multiplied with the states N_j and thus overall this equation is not linear in all the variables (N, u).

The dynamics represents in- and outflows of the various compartments, and for this reason, no matter what the control is, all diagonal entries of the matrix $A + \sum_{j=1}^{m} u_j B_j$ are negative (there always is a positive outflow from each compartment) and all the off-diagonal entries (which model the inflows) are nonnegative. Zero values may occur when there are no connections between some of the compartments, but every row will have at least one positive entry. In mathematics, matrices with these properties are called M-matrices (named so in honor of Minkowski) and their structure implies the positive invariance properties for the state space $\mathbb{P}$ required for the model to be consistent [69].

(M) For all $u \in U$ the matrices $A + \sum_{j=1}^{m} u_j B_j$ have negative diagonal entries and nonnegative off-diagonal entries, $A + \sum_{j=1}^{m} u_j B_j \in \mathcal{M}$.

Let $r = (r_1, \dots, r_n)$ and $q = (q_1, \dots, q_n)$ be n-dimensional row vectors of positive numbers and let $s = (s_1, \dots, s_m)$ be a nonzero m-dimensional row vector of nonnegative numbers. These coefficients represent subjective weights which define the objective as

$$J = rN(T) + \int_0^T (qN(t) + su(t))\, dt \to \min \tag{16}$$

The term $su = \sum_{i=1}^{m} s_i u_i$ in the integral is a weighted average of the amounts of the various drugs given and the coefficients s_i represent the degrees of toxicity of the drugs. Side effects generally depend on the specific cytotoxic agent used and may be more severe than those of a cytostatic or recruiting agent. This would be reflected in the choice of these weights. Similarly, the second integral term $qN = \sum_{i=1}^{m} q_i N_i$ represents a weighted average of the number of cancer cells in the respective compartments during treatment and the penalty term $rN(T) = \sum_{i=1}^{m} r_i N_i(T)$ represents a weighted average of the number of cancer cells in the respective compartments at the end of treatment. The inclusion of the term qN in the Lagrangian is important since otherwise optimization will lead to protocols that put all the emphasis on the end of the therapy interval ignoring the behavior in between. While relevant biological information should be taken into account when selecting the parameters, it generally is also useful to modulate these parameters within specified ranges to obtain otherwise desired features of the optimal solutions. We then consider the following optimal control problem:

[CC] for a fixed therapy horizon $[0, T]$, minimize the objective (16) over all Lebesgue-measurable functions $u : [0, T] \to U$ subject to the dynamics (15).

There are no constraints on the terminal state $N(T)$ in this formulation and by the nontriviality of the multipliers this implies that all extremals are normal. We thus normalize $\lambda_0 = 1$ and drop it in the notation. The adjoint equation and terminal condition then take the form

$$\dot{\lambda} = -q - \lambda \left(A + \sum_{j=1}^{m} u_j B_j \right), \qquad \lambda(T) = r. \tag{17}$$

Under assumption (M), the positive orthant $\mathbb{P}^*$ in the dual space $(\mathbb{R}^n)^*$ also is negatively invariant for the adjoint equation (17), i.e., if $\lambda(t_0) \in \mathbb{P}^*$, then all components of λ are positive for all times $t < t_0$. We thus have the following fact:

Proposition 3.1 ([69]). *Under assumption (M), all states N_i and multipliers λ_i are positive over* $[0, T]$.

This is useful information in evaluating the signs of various expressions that arise in an analysis of optimal controls. We recall that the switching functions are given by $\Phi_j(t) = s_j + \lambda(t)B_jN(t)$, $j = 1, \ldots, m$, with singular controls possible if one of them vanishes over an open interval. Proposition 2.1 simplifies to the following statement:

Proposition 3.2. *Suppose M is a constant matrix and let $\Psi(t) = \lambda(t)MN(t)$, where N is a solution to the system equation (15) corresponding to the control u and λ is a solution to the corresponding adjoint equation (17). Then*

$$\dot{\Psi}(t) = \lambda(t)\left[A + \sum_{j=1}^{m} u_j B_j, M\right] N(t) - qMN(t), \tag{18}$$

with $[X, Y] = YX - XY$ the commutator of the matrices X and Y.

Whether or not optimal controls can be singular depends on the properties of the matrices A and B_i and needs to be evaluated on a case-by-case basis. Here we briefly describe two models: one for homogeneous and the other for heterogeneous tumor populations and point out the differences in the structures of optimal controls that result from these assumptions.

3.2 Cell-Cycle-Specific Models for Homogeneous Tumor Populations

We consider the problem of administering a single cytotoxic agent that is active in the G_2/M phase of the cell cycle such as, for example, paclitaxel. This model was originally considered by Swierniak in [67] and has been analyzed further by us in [33]. Taking into account the phase sensitivity of the drug, the cell cycle is broken up into two compartments with one combining the second growth phase G_2 and mitosis M and the other compartment simply made up of the remaining phases of the cell cycle. The state N of the system can then be described by a 2-dimensional vector with $N_1(t)$ denoting the average number of cancer cells in the first compartment at time t (comprised of the phases G_0, G_1 and S) and $N_2(t)$ the average number of cancer cells in the second compartment at time t (comprised of G_2 and M).

Cell division is a stochastic process with the individual cells determining the sample paths and the transit times following some empirical distribution. Various probabilistic models such as χ^2- or Weibull distributions can be used to describe these transit times. In the approach by Swierniak, an exponential distribution (a special case of the Weibull distribution) is used. This leads to balance equation for the compartments that are linear in the states: the outflow of the first compartment equals the inflow into the second compartment and thus we have that

$$\dot{N}_2(t) = -a_2N_2(t) + a_1N_1(t)$$

with a_i the inverse mean transit time through the ith compartment. In the second compartment cell division occurs and thus, while the outflow from the second compartment is still given by $a_2 N_2(t)$, the inflow into the first compartment doubles to $2a_2 N_2(t)$ giving

$$\dot{N}_1(t) = -a_1 N_1(t) + 2a_2 N_2(t).$$

Since the differential equations are linear, quotients of the variables obey Riccati differential equations and it follows that in steady state, i.e., in the "long" run, fixed proportions of the cells will lie in the respective compartments: if

$$x = \frac{N_1}{N_1 + N_2} \qquad \text{and} \qquad y = \frac{N_2}{N_1 + N_2}$$

denote the average proportions of cells in the two compartments, $x, y > 0$, $x + y = 1$, then y satisfies the Riccati equation

$$\dot{y} = a_1 - (a_1 + a_2)y - a_2 y^2 \tag{19}$$

and has a well-defined steady state (i.e., a unique, globally asymptotically stable equilibrium point) y_* in the open interval $(0, 1)$ given by

$$y_* = \frac{1}{2}\left(\sqrt{\left(1 + \frac{a_1}{a_2}\right)^2 + 4\frac{a_1}{a_2}} - \left(1 + \frac{a_1}{a_2}\right)\right). \tag{20}$$

All solutions approach this value as $t \to \infty$. We only remark that these proportions x_* and y_* can be measured using cell cycle flow cytometry. If we write $C(t) = N_1(t) + N_2(t)$ for the average total number of cancer cells, then the differential equations imply that

$$\dot{C}(t) = a_2 N_2(t) = a_2 y(t) C(t) \approx a_2 y_* C(t)$$

and thus, in steady state, the total tumor population grows exponentially at about rate $a_2 y_*$. This allows us to relate the coefficients a_i to the tumor doubling time and these steady states as follows:

Proposition 3.3. *With T denoting the tumor doubling time and x_* and y_*, the steady-state proportions of cells in the $G_0/G_1 + S$ and G_2/M phases of the cell cycle, respectively, we have that*

$$a_1 = (1 + y_*)\frac{\ln 2}{T x_*} \qquad \text{and} \qquad a_2 = \frac{\ln 2}{T y_*}.$$

Drug treatment influences the cell cycle in many ways and in the model considered here only the most fundamental aspect is considered, cell killing of a cytotoxic agent in the G_2/M phase. It is implicitly assumed that all cancer cells are drug sensitive. Recall that the control variable u represents the drug concentration in the blood stream and in accordance with the *log-kill hypothesis*, we assume that the drug concentration $u(t)$ kills a fraction of the outflow $a_2N_2(t)$ of cells from the G_2/M compartment. Thus the number of cells killed is given by $\varphi u(t)a_2N_2(t)$ with φ a constant chemotherapeutic killing parameter. The control set is a compact interval $[0, u_{\max}]$ with $u_{\max}$ denoting the maximum dose rate/concentration. In the model, the control u always appears in conjunction with the constant φ and thus, in order to keep the number of free parameters to a minimum, we combine it with the maximum dose rate into one quantity that we still denote with $u_{\max}$ under the assumption that $u_{\max} \leq 1$. If the concentration is high enough, then indeed $u_{\max} = 1$ is realistic: almost all the cancer cells in that compartment can be killed. Cells which are killed in G_2/M leave this compartment, i.e., are counted as outflows from the second compartment, but they no longer enter the first compartment. Only the remaining fraction $(1 - u)a_2N_2$ undergoes cell division. Thus the controlled mathematical model becomes

$$\dot{N}_1 = -a_1N_1 + 2(1-u)a_2N_2,$$
$$\dot{N}_2 = a_1N_1 - a_2N_2,$$

or, in matrix form, $\dot{N}(t) = (A + uB)N(t)$, with A and B given by

$$A = \begin{pmatrix} -a_1 & 2a_2 \\ a_1 & -a_2 \end{pmatrix} \quad \text{and} \quad B = \begin{pmatrix} 0 & -2a_2 \\ 0 & 0 \end{pmatrix}. \tag{21}$$

For this model, singular controls are not optimal. Denoting the coefficient at the control u in the objective by s, the switching function is given by $\Phi(t) = s + \lambda(t)BN_*(t)$. If the control is singular on an open interval I, then, using Proposition 3.2 it follows that

$$\dot{\Phi}(t) = \{\lambda(t)[A, B] - qB\}\, N_*(t) \equiv 0 \tag{22}$$

and

$$\begin{aligned}\ddot{\Phi}(t) = {} & \{\lambda(t)[A,[A,B]] - q[A,B] - qBA\}\, N_*(t) \\ & + u(t)\left\{\lambda(t)[B,[A,B]]N_*(t) - qB^2\right\} N_*(t).\end{aligned} \tag{23}$$

Hence the Legendre-Clebsch condition is determined by the expression

$$\frac{\partial}{\partial u}\frac{d^2}{dt^2}\frac{\partial H}{\partial u}(\lambda(t), N_*(t), u_*(t)) = \left\{\lambda(t)[B,[A,B]]N_*(t) - qB^2\right\} N_*(t). \tag{24}$$

It is clear that $B^2 \equiv 0$ and a direct computation verifies that

$$[B,[A,B]] = 8a_1a_2^2\begin{pmatrix} 0 & 1 \\ 0 & 0 \end{pmatrix} = -4a_1a_2B. \tag{25}$$

Furthermore, $\Phi(t) \equiv 0$ implies that $\lambda(t)BN_*(t) \equiv -s$ and thus

$$\frac{\partial}{\partial u}\frac{d^2}{dt^2}\frac{\partial H}{\partial u}(\lambda(t), N_*(t), u_*(t)) = 4a_1a_2s > 0 \tag{26}$$

violating the Legendre-Clebsch condition for optimality of a singular control.

Theorem 3.1. *If (N_*, u_*) is an optimal controlled trajectory for the optimal control problem [CC] with matrices (21), then there does not exist an interval on which the control u_* is singular.*

We give a couple of examples of locally optimal bang-bang controls. For the cell cycle parameters we have chosen the values $a_1 = 0.197$ and $a_2 = 0.356$ used in [67] and in all computations the initial condition is taken as the steady-state proportions defined by Eq. (20), normalizing the total number of initial cancer cells to 1 (times 10^{10}), i.e., $N_1(0) = 0.7012$ and $N_2(0) = 0.2988$. This would be representative of conditions where the cancer has been growing exponentially for some time without treatment; even if chemotherapy has been given earlier, in the rest periods the cells redistributed over the compartments and their proportions are given by these values. The control limit is taken as $u_{\max} = 0.90$, but is just meant for illustrative purposes. Figure 1 shows two examples of controls and corresponding trajectories when the coefficients in the objective have been chosen as $r = (3,3)$, $q = (0.1, 0.1)$ and $s = \frac{1}{2}$. The examples shown are for time horizons of $T = 21$ and $T = 60$ [days]. In all cases, extremals are bang-bang trajectories with exactly one switching from $u = u_{\max}$ to $u = 0$. The total reductions in cancer cells at the end of the therapy horizon are given by $N_1(T) + N_2(T) = 0.5297$ and 0.4799, respectively.

All extremals shown here are *strong local minima*; that is, there exists a neighborhood W of the graph of the corresponding trajectory in $[0,T] \times \mathbb{P}$ such that the controls are optimal with respect to any other control u for which the graph of its corresponding trajectory N lies in W [33]. In fact, for this 2-compartment model we have consistently seen that extremal bang-bang trajectories that have more than one switching are not optimal and the examples shown are expected to be globally optimal. This simply means that we can take the neighborhood W as the full space $[0,T] \times \mathbb{P}$, but we have not verified this.

Analogous results have been obtained for multidrug 3-compartment models when the actions of a G_2/M-specific cytotoxic agent were combined either with a cytostatic agent that was slowing down the progression of cells during the synthesis phase [34, 62] or with a recruiting agent that was applied to entice dormant cells to reenter the active cell cycle from the compartment G_0 [35]. In each case, singular controls can be excluded from optimality using the Legendre-Clebsch condition and optimal controls are bang-bang with one switching for the cytotoxic agent giving the dose upfront. In the model formulation, however, it is implicitly assumed that the tumor population is homogeneous and consists of chemotherapeutically sensitive cells. Also, the problem considered here corresponds to one particular chemotherapy session only. The steady-state proportions of the uncontrolled system reestablish very quickly during the rest periods and thus multiple chemotherapy

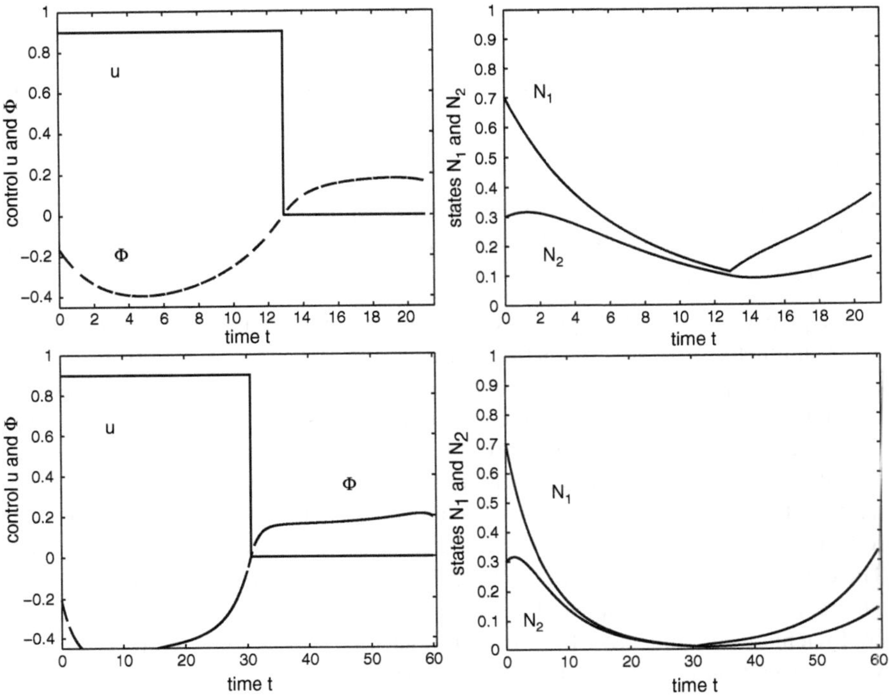

Fig. 1 Examples of locally optimal controls (*left*) and their corresponding trajectories (*right*) for $T = 21$ (*top*) and $T = 60$ (*bottom*) from the steady-state solution

sessions reduce to repetitions of the structure obtained above. Overall, for tumor populations that are homogeneous and consist of chemotherapeutically sensitive cells, these mathematical models therefore confirm the prevailing paradigm that *chemotherapy should be given in an MTD scheme upfront*. However, this no longer is so clear cut once tumor heterogeneity is taken into account.

3.3 Compartmental Models for Heterogeneous Tumor Populations

Malignant cancer cell populations are genetically unstable and coupled with fast proliferation rates, this leads to a great variety in the structure of the cells within one tumor—the number of genetic errors present within one cancer cell can lie in the thousands [42]. Consequently, tumors often consist of a heterogeneous mixtures of various subpopulations that show widely varying sensitivities towards the actions of a particular chemotherapeutic agent [13, 14]. In medicine, the *Norton-Simon hypothesis* [44] postulates that tumors consist of faster growing cells that are sensitive to chemotherapy and slower growing populations of cells that exhibit lower sensitivities or, with time, become resistant to the chemotherapeutic agent (*acquired drug resistance*). There may even exist small subpopulations of cells for which the specific activation mechanism of a chemotherapeutic agent does not work at all and which thus are not sensitive to the treatment from the beginning (ab initio, *intrinsic resistance*). Given such a scenario, over time, as the drugs kill sensitive tumor cells, resistant subpopulation of cancer cells may emerge that will make an MTD-style therapy less and less effective [37, 41, 71]. Even if the fraction of intrinsically resistant tumor cells is tiny (undetectable) after the sensitive cells have been killed by the treatment, it may then grow in time to become a fully developed tumor of chemotherapeutically resistant cells leading to the failure of therapy, possibly only after many years of seeing remission of the cancer.

Compartmental models of the type (15) can also be used to investigate the structure of optimal controls if the tumor population is heterogeneous. In [17], Hahnfeldt, Folkman and Hlatky compare the effects of MTD and metronomic chemotherapy (when given by bolus-type injections) on sensitive and resistant tumor populations. Optimizing the maximum asymptotic factor reduction in tumor size between periods in an infinite cycle of periodic therapy periods, the authors come to the conclusion that a metronomic, regular scheduling of the drugs has better long-term effects. We here consider the same underlying dynamics in a continuous-time formulation and explore the structure of optimal protocols that minimize the tumor burden as measured by the average over one, but possibly very large therapy interval. Since we want to explore the effects that heterogeneity has, we distinguish three subpopulations which, for simplicity of terminology, are labeled "sensitive," S, "partially sensitive," P, and "resistant," R. The terminology is only meant to indicate that these populations have different sensitivities towards a chemotherapeutic agent with S the highest and R the lowest. We assume that these subpopulations grow at growth rates α_1, α_2, and α_3, respectively. Generally we do not make assumptions on the order of the growth rates, but an ordering $\alpha_1 > \alpha_2 > \alpha_3$ would be consistent with the "Norton-Simon hypothesis." We allow for transitions between the compartments, i.e., we include the typical effects that sensitive cells can become more resistant, but we also allow for resensitizations which make cells less resistant to the chemotherapeutic agent [15]. We denote the transition rates from the sensitive to the partially sensitive and resistant compartments by σ_P and σ_R, respectively, and use analogous notations for the other transition rates. Thus, for example, ρ_P denotes the transition rates from resistant to partially sensitive cells. These rates are assumed to be constant and *we assume* they all are *positive*. This corresponds to an ergodic structure in which all compartments are repeatedly visited by cells. Cell kill by a chemotherapeutic agent is expressed by the standard linear log-kill hypothesis: if we denote the concentration of the drug in the bloodstream by u, then the rate of cells eliminated is given by $\varphi_i u$, $i = 1, 2, 3$, with the coefficients φ_1, φ_2, and φ_3 representing the effectiveness of the drug on the sensitive, partially sensitive and resistant subpopulations, respectively. Thus $\varphi_1 > \varphi_2 > \varphi_3 \geq 0$. The case $\varphi_3 = 0$ corresponds to the situation of a fully resistant subpopulation R. We again do not include the standard pharmacokinetic model on the agent here and treat u as the control of the system with maximum concentration given by $u_{\max}$. The controlled dynamics is then simply determined by the inflows and outflows from the various compartments and is given by the following 3-dimensional linear system of equations:

$$\dot{S} = (\alpha_1 - \sigma_P - \sigma_R - \varphi_1 u)\, S + \pi_S\, P + \rho_S\, R, \tag{27}$$

$$\dot{P} = \sigma_P\, S + (\alpha_2 - \pi_S - \pi_R - \varphi_2 u)\, P + \rho_P\, R, \tag{28}$$

$$\dot{R} = \sigma_R S + \pi_R P + (\alpha_3 - \rho_S - \rho_P - \varphi_3 u)\, R. \tag{29}$$

Even if initially no partially sensitive or resistant cells are present, they will immediately appear because of the ergodic nature of the underlying Markov chain and resulting transitions between the compartments. Without loss of generality, we thus assume that all initial conditions S_0, P_0, and R_0 are positive. Admissible controls are Lebesgue-measurable functions with values in a compact interval $[0, u_{\max}]$, $u : [0, T] \to [0, u_{\max}]$, $t \mapsto u(t)$.

We denote the proportions of the respective populations by

$$x = \frac{S}{S+P+R}, \qquad y = \frac{P}{S+P+R} \qquad \text{and} \qquad z = \frac{R}{S+P+R};$$

it then follows that x, y, and z obey Riccati equations and direct computations verify that

$$\dot{x} = \nu_S x + \pi_S y + \rho_S z - x(\alpha_1 x + \alpha_2 y + \alpha_3 z), \tag{30}$$

$$\dot{y} = \sigma_P x + \nu_P y + \rho_P z - y(\alpha_1 x + \alpha_2 y + \alpha_3 z), \tag{31}$$

$$\dot{z} = \sigma_R x + \pi_R y + \nu_R z - z(\alpha_1 x + \alpha_2 y + \alpha_3 z). \tag{32}$$

with the system evolving on the unit simplex

$$\Sigma = \{(x, y, z) : x \geq 0,\ y \geq 0,\ z \geq 0,\ x + y + z = 1\}.$$

Proposition 3.4 ([28]). *The dynamics (30)–(32) has exactly one equilibrium point $(x_*, y_*, z_*) \in \Sigma$ which is globally asymptotically stable in Σ and defines the steady-state proportions.*

Thus, given an estimate C_0 on the tumor size, there once more exists a well-defined initial condition $S_0 = x_* C_0$, $P_0 = y_* C_0$ and $R_0 = z_* C_0$ for the optimal control problem [OC]. Setting $N = (S, P, R)^T$, we have a 3-dimensional single-input control system of the form $\dot{N} = (A + uB)\, N$ with the matrices determined by equations (30)–(32); the objective is the same as defined in (16) before. The necessary conditions for optimality thus take the same form as for the 2-compartment model considered above. It is easily seen that also for this system, although the dynamics is not described by an $\mathcal{M}$-matrix, all states and multipliers

λ_i, $i = 1, 2, 3$ are positive over the interval $[0, T]$ and we have the same formulas (22) and (23) for the derivatives of the switching function $\Phi(t) = s + \lambda(t)BN_*(t)$ with the Legendre-Clebsch condition again given by (24). Here

$$[B,[A,B]] = -\begin{pmatrix} 0 & (\varphi_2-\varphi_1)^2\pi_S & (\varphi_3-\varphi_1)^2\rho_S \\ (\varphi_1-\varphi_2)^2\sigma_P & 0 & (\varphi_3-\varphi_2)^2\rho_P \\ (\varphi_1-\varphi_3)^2\sigma_R & (\varphi_2-\varphi_3)^2\pi_R & 0 \end{pmatrix} \tag{33}$$

so that $\lambda(t)[B,[A,B]]N_*(t) \le 0$ while

$$qB^2N_*(t) = q_1\varphi_1^2 S_*(t) + q_2\varphi_2^2 P_*(t) + q_3\varphi_3^2 R_*(t) > 0.$$

Hence $\left\{\lambda(t)[B,[A,B]]N_*(t) - qB^2\right\} N_*(t) < 0$ and the strengthened Legendre-Clebsch condition is always satisfied. Essentially, this is just a consequence of having different sensitivities.

Proposition 3.5. *For the compartmental model defined by equations (27)–(29), singular controls are of order* 1 *and the strengthened Legendre-Clebsch condition for minimality is satisfied.*

Thus, in this case it is expected that singular controls are locally optimal. Solving equation (23) for u gives the following formula for the singular control:

$$u_{\text{sing}}(t) = \frac{\{\lambda(t)\,[A,[A,B]] - q\,[A,B] - qBA\}\,N_*(t)}{\{-\lambda(t)\,[B,[A,B]] + qB^2\}\,N_*(t)}. \tag{34}$$

This singular control actually does not depend on the values S, P, and R of the state, but only on the values of the proportions x, y, and z. In order to be admissible, the control values need to lie in the control set $[0, u_{\max}]$. It follows from the strengthened Legendre-Clebsch condition that the denominator is positive. In the numerator, all terms in the vector $-qBA$ are positive, but there exist coefficients in the matrices $[A,[A,B]]$ and in the vector $-q\,[A,B]$ that are negative, but just a few. Thus generally, and this is what we have seen consistently in numerical computations, the values of the expression (34) are positive and thus admissible for suitable upper bounds $u_{\max}$.

Analyzing optimal concatenations between bang and singular controls is difficult and this analysis has not been carried out yet. However, it is not difficult to give some numerical samples of singular controls and corresponding trajectories. Along a singular arc, the multiplier λ satisfies $\Phi(t) = s + \lambda(t)BN_*(t) \equiv 0$ and $\dot{\Phi}(t) = \{\lambda(t)[A,B] - qB\}\,N_*(t) \equiv 0$ and is determined by these conditions up to a positive scalar multiple. In principle, here singular controls are possible everywhere in the state space and in Fig. 2 we give an example of an extremal controlled trajectory for which the control is given by the maximum dose rate for an initial interval $[0, \tau_b]$ and then is singular over the remaining period $[\tau_b, T]$. In this simulation the parameter values defining the dynamics are $\alpha_1 = 1$, $\alpha_2 = 0.5$, and $\alpha_3 = 0.1$ with transition rates $\sigma_P = 0.05$, $\sigma_R = 0.01$, $\pi_S = 0.03$, $\pi_R = 0.01$, $\rho_S = 0.01$, and $\rho_P = 0.03$.

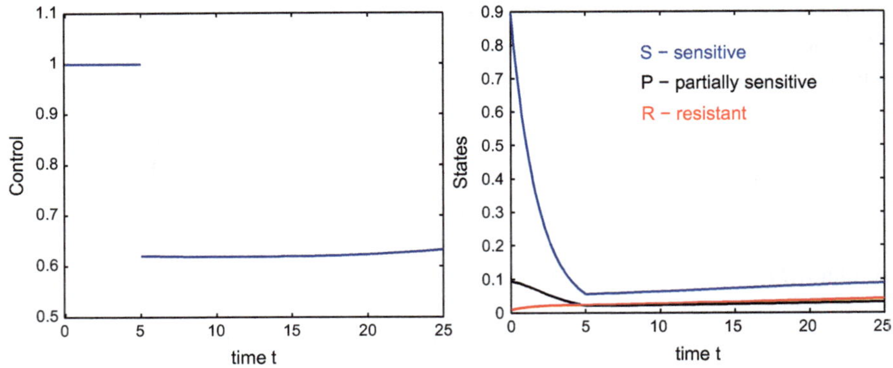

Fig. 2 Example of an extremal control and associated states for a bang-singular controlled trajectory

Normalizing the initial cancer burden to $C(0) = 1$, the corresponding steady-state proportions are given by $S_0 = x_* = 0.8954$, $P_0 = y_* = 0.0933$, and $R_0 = z_* = 0.0112$ and we used these as initial condition. The maximum dose rate is normalized to $u_{\max} = 1$ and the pharmacodynamic coefficients are $\varphi_1 = 1.5$, $\varphi_2 = 1$, and $\varphi_3 = 0.1$. All these values are for illustrative purpose only. In the objective we chose all weights q_i equal to 0.01 and we used $\tau_b = 5$ and $T = 25$, so that a full dose is given for 20% of the time. Over this time horizon the lower dose rates of the singular controls are able to maintain a lower cancer burden, but eventually the resistant population will become dominant. However, this will happen regardless of the specific administration protocol of the drug.

In the medical literature protocols like these are referred to as "chemo-switch" protocols and our computations show that, *as differing chemotherapeutic sensitivities and even drug resistance come into play, lower dose rates become a valid alternative to MTD protocols.*

4 Mathematical Models for Antiangiogenic Treatments

The most important structure of a tumor's microenvironment is its vasculature. In order to grow beyond a small size, a tumor needs to develop its own network of blood vessels and capillaries that will provide it with nutrients and oxygen. This process is called angiogenesis and was already pointed out as a therapeutic target by J. Folkman in the early 1970s [8, 9]. Antiangiogenic treatments aim at depriving the tumor of this needed vasculature by either disrupting the signaling process that the tumor uses to recruit surrounding, mature, host blood vessels or by directly inhibiting the growth of endothelial cells that form the lining of the newly developing blood vessels and capillaries. Ideally, without an adequate support network, the tumor's further development is halted and it even shrinks. Rather than fighting the fast duplicating, genetically unstable, and continuously mutating tumor cells, this indirect treatment approach targets the genetically stable endothelial cells. As a consequence, no clonal resistance to angiogenic inhibitors has been observed in experimental cancer [2] and for this reason, after the discovery of antiangiogenic mechanisms that the tumor uses to control its vasculature in the 1990s [5, 10, 25], antiangiogenic treatments were a new hope in the war on cancer. Unfortunately, these high hopes have not been realized, mostly due to the maintenance only character of the treatment [22]. However, antiangiogenic approaches have become a valuable component in the treatment of many cancer types in connection with other traditional approaches like chemo- or radiotherapy that directly attack tumor cells.

A widely influential population-based mathematical model for tumor development under angiogenic signaling was developed and biologically validated in 1999 by Hahnfeldt, Panigrahy, Folkman, and Hlatky [16]. This model has become an object of strong interest also in the mathematical literature and to this date is still undergoing vigorous development. It has been analyzed from a dynamical systems perspective (e.g., by d'Onofrio and Gandolfi [47, 48], Forys et al. [11]) as well as from an optimal control point of view (by the authors and coworkers [38,39,52] and by Swierniak [68,70]) with numerous generalizations and variations of the underlying model that have been proposed (e.g., [7, 46, 49, 51, 58, 60]). In Sect. 4.1, for the original mathematical model, we describe a complete solution of how to administer an a priori given amount of antiangiogenic agents in order to achieve the best possible tumor reduction. In this solution, an optimal singular arc and its associated singular control determine the structure of optimal controls which are largely defined by a singular segment. These feedback controls, however, are difficult to implement. Yet, the solution is fully robust and excellent simple suboptimal controls that come within 1% of the optimal value exist and will be discussed in Sect. 4.2. Since antiangiogenic therapy only targets cancer cells indirectly, in order to be effective, it needs to be combined with therapies that also kill the cancer cells. In Sect. 4.3 we show how the solution for the antiangiogenic monotreatment therapy presented in Sect. 4.1 provides the basis for the solutions for such combination therapy problem.

4.1 Synthesis of Optimal Controlled Trajectories for the Monotherapy Problem

In the model by Hahnfeldt et al. [16], the spatial aspects of angiogenesis are incorporated into a nonspatial 2-compartment model with the *primary tumor volume*, p, and the *carrying capacity* of the vasculature, q, as its principal variables. Intuitively, the latter can be thought of as the ideal tumor volume sustainable by the vascular network and is closely related to the volume of endothelial cells that form the lining of the existing and newly forming capillaries. The dynamics consists of two ODEs that describe the evolution of the tumor volume and its carrying capacity, which, with u denoting the action of an antiangiogenic agent, is given by the following equations:

$$\dot{p} = -\xi p \ln\left(\frac{p}{q}\right), \qquad p(0) = p_0, \tag{35}$$

$$\dot{q} = bp - \left(dp^{\frac{2}{3}} + \mu\right) q - \gamma u q, \qquad q(0) = q_0, \tag{36}$$

In equation (35) a Gompertzian model with ξ a constant parameter is chosen to model tumor growth (other choices are equally possible). Note that the carrying capacity and tumor volume are balanced for $p = q$ and thus $\dot{p} = 0$ in this case while the tumor volume shrinks for inadequate endothelial support ($p > q$) and increases if this support is plentiful ($p < q$). Different from conventional approaches, in this model the carrying capacity is not a constant, but itself becomes a state variable whose evolution is governed by a balance of stimulatory and inhibitory effects given in equation (36). Based on an asymptotic analysis of the underlying consumption-diffusion process and the facts that angiogenic inhibitors have a more systemic effect while stimulators, on the other hand, act locally, the functional forms $S(p,q) = bp$ and $I(p,q) = dp^{\frac{2}{3}}q$ for stimulators and inhibitors are proposed in [16]. The term μq, $\mu \geq 0$, that has been separated describes the loss to the endothelial cells through natural causes (death etc.) and $\gamma q u$ models the loss to the vasculature due to outside administration of antiangiogenic agents using a standard log-kill term. The control u represents the concentration in the plasma of such an agent with $u_{\max}$ denoting an a priori set maximum dose rate/concentration.

Different from the previous model formulations, we here assume that a fixed amount A of angiogenic inhibitors is given. Mathematically this represents an isoperimetric constraint and is modeled as

$$\dot{y} = u, \qquad y(0) = 0, \qquad y(T) \leq A. \tag{37}$$

The question then becomes how to use the given amount of agents in the best possible way. Here we choose to minimize the tumor volume. In this formulation, there is no fixed therapy horizon $[0, T]$, but rather the terminal time T is free and it merely represents the time when the minimum tumor volume is being realized. Such models are of practical interest and give an important alternative to the formulations considered earlier. We thus consider the following optimal control problem:

[A] for a free terminal time T, minimize the terminal value $p(T)$ of the tumor volume subject to the dynamics (35)–(37) over all Lebesgue-measurable functions $u : [0, T] \to [0, u_{\max}]$ for which the corresponding trajectory (p, q, y) satisfies the terminal constraint $y(T) = \int_0^T u(t)dt \leq A$.

We denote the 3-dimensional state by $z = (p, q, y)^T$ and write the dynamics in the form

$$\dot{z} = f(z) + ug(z) \tag{38}$$

with

$$f(z) = \begin{pmatrix} -\xi p \ln\left(\frac{p}{q}\right) \\ bp - \left(dp^{\frac{2}{3}} + \mu\right) q \\ 0 \end{pmatrix} \quad \text{and} \quad g(z) = \begin{pmatrix} 0 \\ -\gamma q \\ 1 \end{pmatrix}.$$

All coefficients are positive parameters and we also assume that $\gamma u_{\max} > b - \mu > 0$. The first inequality implies that a constant does rate u_{max} eradicates the tumor [47], but is only made in order not to have to distinguish cases the second inequality is always satisfied for the underlying medical problem. Under these assumptions in [38] we gave a complete global solution to this optimal control problem in the form of a regular synthesis for all initial data (p_0, q_0, A) that are *well posed*. This simply means that there are enough antiangiogenic agents available to realize a terminal value $p(T) < p_0$ since otherwise the optimal terminal time T is given by $T = 0$.

Necessary conditions for optimality of a control u are again given by the Pontryagin maximum principle. It is not difficult to see that all available inhibitors will be exhausted along an optimal trajectory (p_*, q_*, y_*), $y_*(T) = A$ and that $p_*(T) = q_*(T)$ holds at the final time. For this problem, the *switching function* Φ is given by

$$\Phi(t) = \langle \lambda(t), g(z(t)) \rangle = \lambda_3 - \lambda_2(t)\gamma q_*(t) \tag{39}$$

and, compared with the models considered in Sect. 3, here the computation of singular controls simplifies since the Lagrangian L is identically zero. It follows from Proposition 2.1 that the derivative of a function of the form $\Psi(t) = \langle \lambda(t), h(z(t)) \rangle$ is given by $\dot{\Psi}(t) = \langle \lambda(t), [f + ug, h](z(t)) \rangle$ and for the switching function $\Phi(t)$ we thus obtain that $\dot{\Phi}(t) = \langle \lambda(t), [f, g](z(t)) \rangle$ and

$$\ddot{\Phi}(t) = \lambda(t)[f + ug, [f, g]](z(t)), \tag{40}$$

with the control u once more only appearing in the second derivative. If u is singular on some open interval I, then these derivatives all vanish on I and if $\langle \lambda(t), [g, [f, g]](z(t)) \rangle \neq 0$, then (40) can formally be solved for u as

$$u_{\text{sing}}(t) = -\frac{\langle \lambda(t), [f, [f, g]](z(t)) \rangle}{\langle \lambda(t), [g, [f, g]](z(t)) \rangle}. \tag{41}$$

The strengthened Legendre-Clebsch condition for optimality of the singular control here takes the form

$$\langle \lambda(t), [g, [f, g]](z(t)) \rangle < 0 \qquad \text{for all } t \in I. \tag{42}$$

The determination of singular controls and the analysis of their local optimality properties thus reduces to the computation of the Lie brackets $[f,[f,g]]$ and $[g,[f,g]]$ and their inner products with the multiplier λ. For the model [A], the control vector field g and the Lie brackets $[f,g]$ and $[g,[f,g]]$ are linearly independent and thus the Lie bracket $[f,[f,g]]$ can be written as a linear combination of this basis with coefficients that are smooth functions of the state z, say

$$[f,[f,g]](z) = \rho(z)g(z) + \varphi(z)[f,g](z) + \psi(z)[g,[f,g]](z).$$

Along a singular extremal (z,u,λ), the inner products $\langle\lambda(t), g(z(t))\rangle$ and $\langle\lambda(t), [f,g](z(t))\rangle$ vanish identically and thus

$$\langle\lambda(t), [f,[f,g]](z(t))\rangle = \psi(z(t))\,\langle\lambda(t), [g,[f,g]](z(t))\rangle\,.$$

If the singular control is of order 1, we therefore simply have that

$$u_{\text{sing}}(t) = -\psi(z(t)) \tag{43}$$

and the singular control is given in feedback form, i.e., as a function only of the state z alone which does not depend on the multiplier. Naturally, whether this feedback is admissible still needs to be determined separately.

However, this feedback does not define a singular control everywhere, but only on a thin subset. The reason for this lies in the fact that along extremals also the Hamiltonian H needs to vanish identically and thus, along a singular arc, we also have that $\langle\lambda(t), f(z(t))\rangle \equiv 0$ for all $t \in I$. Consequently the multiplier $\lambda(t)$ vanishes against the vector fields f, g and $[f,g]$ along a singular trajectory. Since $\lambda(t) \neq 0$, it follows that these vector fields must be linearly dependent along the singular arc. Thus (43) only defines a singular control on the surface

$$\mathcal{S} = \{z \in \mathbb{R}^3 : \det(f(z), g(z), [f,g](z)) = 0\}$$

where $\det(f(z), g(z), [f,g](z))$ denotes the determinant of the matrix whose columns are formed by the ordered vectors $f(z)$, $g(z)$, and $[f,g](z)$. Evaluating this formula gives that

$$\det(f(z), g(z), [f,g](z)) = \xi\gamma p\left[bp\left(1-\ln\left(\frac{p}{q}\right)\right) - \left(dp^{\frac{2}{3}} + \mu\right)q\right].$$

In particular, $\mathcal{S}$ is a vertical surface independent of y over a base curve $\mathcal{S}_0$ in (p,q)-space given by $\mu + dp^{\frac{2}{3}} = bx(1-\ln x)$ with $x = \frac{p}{q}$. For the singular control, we have the following explicit formulas:

Proposition 4.1. *If the control u is singular on an open interval (α,β) with corresponding trajectory (p,q), then the singular control is determined in feedback form by*

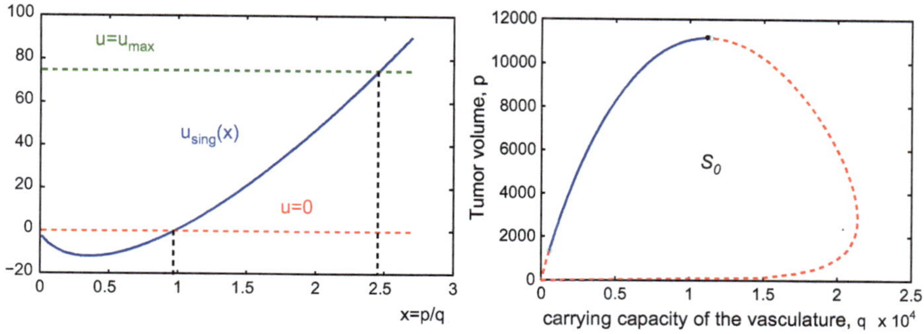

Fig. 3 The singular control u_{sing} plotted as a feedback function of the quotient $x = \frac{p}{q}$ (*left*) and the singular base curve $\mathcal{S}_0$ plotted in (p,q)-space (*right*) with the admissible part marked by the solid portion of the curve. Away from this solid segment the singular control is either negative or exceeds the limit $u_{\max}$

$$\gamma u_{\text{sing}}(t) = \Psi(p(t),q(t)) = \xi \ln\left(\frac{p(t)}{q(t)}\right) + b\frac{p(t)}{q(t)} + \frac{2}{3}\xi\frac{d}{b}\frac{q(t)}{p^{\frac{1}{3}}(t)} - \left(\mu + dp^{\frac{2}{3}}(t)\right) \tag{44}$$

There exists exactly one connected arc on the singular base curve $\mathcal{S}_0$ along which the control is admissible, i.e., satisfies the bounds $0 \leq u_{\text{sing}} \leq u_{\max}$.

Figure 3 illustrates the petallike singular curve $\mathcal{S}_0$ for $u_{\max} = 75$ with the admissible portion marked as a solid curve for the parameter values $\xi = 0.2$, $b = 5$, $d = 0.01$, and $\mu = 0$. The qualitative structure shown in this figure is generally valid, but the admissible portion shrinks with smaller values $u_{\max}$.

The structure of optimal controls and trajectories is summarized in the following theorem:

Theorem 4.1 ([38]). *Given well-posed initial data (p_0, q_0, A), optimal controls are at most concatenations of 4 pieces in the form* $\mathbf{bsu}_{\max}\mathbf{0}$ *with* $\mathbf{0}$ *denoting an arc along the constant control $u = 0$,* $\mathbf{u}_{\max}$ *denoting an arc along the constant control $u = u_{\max}$,* $\mathbf{b}$ *standing for either* $\mathbf{u}_{\max}$ *or* $\mathbf{0}$*, and* $\mathbf{s}$ *denoting an arc in the singular surface $\mathcal{S}$.*

This result provides an upper bound on the number of segments for optimal controls and it significantly limits the structure of possible concatenations. For the medically most relevant case of initial conditions (p_0, q_0) that represent a growing tumor with high carrying capacity, $p_0 < q_0$, and ample supply A of inhibitors, typically optimal controls have the following structure: initially they are given by a segment of full dose therapy, $u \equiv u_{\max}$, until the corresponding trajectory meets the singular surface $\mathcal{S}$. At this point, the optimal control changes to the singular control and antiangiogenic agents are administered at these singular dose rates until all angiogenic inhibitors have been exhausted. During that phase, the corresponding trajectory evolves on the singular surface $\mathcal{S}$. Since the singular surface lies in the region $p > q$, after termination of therapy, the tumor volume will still be decreasing (due to after effects) even if no more agents are administered as long as the trajectory remains in the region $p > q$. The minimum tumor volume will then be realized as the trajectory reaches the diagonal, $p = q$. Thus, for these cases optimal controls follow the shorter concatenation sequence $\mathbf{u}_{\max}\mathbf{s0}$. This is the typical structure of optimal controlled trajectories for medically relevant initial conditions, but it depends on two facts: (i) the overall amount of inhibitors is large enough to reach the singular arc in its admissible range, but (ii) it is not so large that the singular control would saturate along the singular arc, i.e., would reach the limit $u_{\max}$. If (i) is violated and trajectories either do not reach $\mathcal{S}$ at all or reach $\mathcal{S}$ in its inadmissible part, then the singular control never becomes an option and in this case optimal controlled trajectories will simply be given by up-front administration of all antiangiogenic agents at full dose rates. In such a case, optimal controls are bang-bang with exactly one switching from $u = u_{\max}$ to $u = 0$, i.e., of the type $\mathbf{u}_{\max}\mathbf{0}$. If condition (ii) is violated, then optimal concatenation sequences of the forms $\mathbf{0su}_{\max}\mathbf{0}$ and $\mathbf{u}_{\max}\mathbf{su}_{\max}\mathbf{0}$ arise.

The synthesis of optimal trajectories then consists of a unique covering of the full state space by controlled trajectories with the optimal control $u_{\mathrm{opt}} = u_{\mathrm{opt}}(p, q, y)$ identifying the optimal dose rates as a function of an arbitrary point $(p, q; y)$ of the state. Intuitively, a *synthesis* acts like *a "GPS system" showing for every possible state of the system how optimal protocols are administered*, both qualitatively and quantitatively. The variable y merely accounts for the amount of inhibitors that already has been used and it is more convenient, and more illustrative, to show the projections of trajectories into the (p, q)-plane. With only a slight abuse of terminology, we do not distinguish in our language between the trajectories in $(p, q; y)$-space and their projections onto the (p, q)-coordinates. Figure 4 shows this projection and also identifies a typical optimal control of the form $\mathbf{u}_{\max}\mathbf{s0}$ described above.

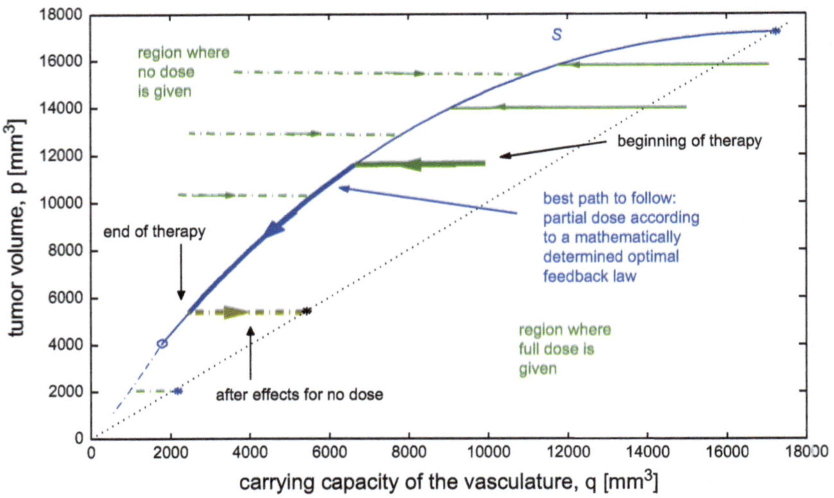

Fig. 4 Synthesis of optimal controlled trajectories for the problem [A]

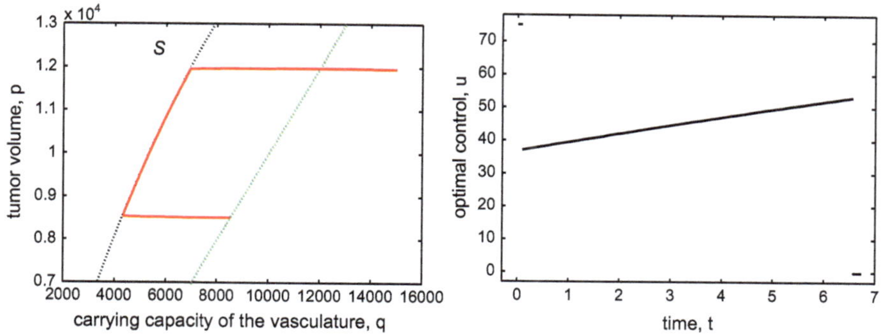

Fig. 5 Example of an optimal $\mathbf{u}_{\max}\mathbf{s0}$ controlled trajectory (*left*) and associated control as function of time (*right*) for initial data $(p_0, q_0, A) = (12000, 15000, 300)$

In Fig. 5, as an example, we show the optimal controlled trajectory (on the left) and its corresponding control (on the right) for the initial condition $(p_0, q_0) = (12000\ [mm^3], 15000\ [mm^3])$ and the values $\xi = 0.084$, $b = 5.85$, $d = 0.00873$ taken from [16] and $\mu = 0.2$. The optimal control is of the type $\mathbf{u}_{\max}\mathbf{s0}$: it takes the maximal value $u = u_{\max}$ for a short interval from 0 to $t_1 = 0.0905$ [days] when the trajectory reaches the singular arc. At this point, the control switches to the time-varying singular control defined by the singular feedback (44) until all inhibitors are exhausted at time $t_2 = 6.5579$ [days]. Then, due to after effects, the minimum value of the tumor volume is realized a short period later at the final time $T = 6.7221$ [days] when the trajectory for $u = 0$ reaches the diagonal. Note the extremely fast q-dynamics away from the singular arc. Partly this is caused by the numerical values that were used which are based on Lewis lung carcinoma in mice, a fast growing cancer. Although the almost horizontal trajectory segments along the controls $u = 0$ and $u = u_{\max}$ are sizable, the time spent along these pieces is small. Most of the time the control is singular and the trajectory follows the associated singular arc (whose projection in the (p, q)-space is a subset of the base curve S_0), but this dynamics is much slower. The optimal final value is given by $p_*(T) = 8533.4\ [mm^3]$. The optimal trajectory is shown as a solid curve in Fig. 5 and the singular curve $\mathcal{S}$ and the diagonal $\mathcal{D}_0$ are shown as dotted curves.

4.2 *Robustness Properties and Realizable Suboptimal Protocols*

Singular controls play an essential role in determining the overall structure of optimal controlled trajectories for this problem. While Lie algebraic computations provide an elegant framework in which the singular controls and corresponding arcs can be determined analytically, these formulas are given as feedback controls that administer time-varying partial doses that are determined by the current state of the system, that is, the tumor volume p and its carrying capacity q. Even at the initial time, while a reasonably reliable estimate for the tumor volume p_0 may be available, the carrying capacity of the vasculature, q_0, is a highly idealized quantity and there exist no methods to measure it. The value of the theoretical optimal solution that was derived, apart from giving interesting qualitative insights into the underlying system, does not primarily lie in providing a feasible strategy, but in clarifying what in principle is possible—in fact, for many practical problem, this precisely is the contribution that optimal control methodologies provide. Then, based on the benchmarks that the theoretically optimal solutions provide, it becomes of importance to formulate *simple*, *easily implementable*, but also *robust strategies* that could be employed even in the face of great uncertainty in the parameters and the state of the system [31, 39]. The solution described above indeed exhibits strong robustness properties with respect to parameter values and this in particular is valid with respect to the initial values q_0 of the carrying capacity. Excellent approximations to the theoretically optimal solution are obtained by simply taking a constant control whose dose rate is given by the *averaged optimal dose rate protocol*, i.e.,

$$\bar{u} \equiv \frac{1}{T_{\mathrm{opt}}} \int_0^{T_{\mathrm{opt}}} u_{\mathrm{opt}}(t)dt = \frac{A}{T_{\mathrm{opt}}},$$

where u_{opt} denotes the optimal control as a function of time, T_{opt} is the time when all antiangiogenic agents have been used up and, as before, A denotes the a priori specified overall amount of agents to be given. Since all antiangiogenic agents are used along the optimal control, the integral is simply given by this total amount A. The final interval when the tumor volume still decreases because of after effects is not included in this computation. Figure 6, on the left, shows a comparison of the graphs of the minimum tumor volumes realized as a function of the initial carrying capacity q_0 by the optimal control (solid red curve), a full dose rate protocol where

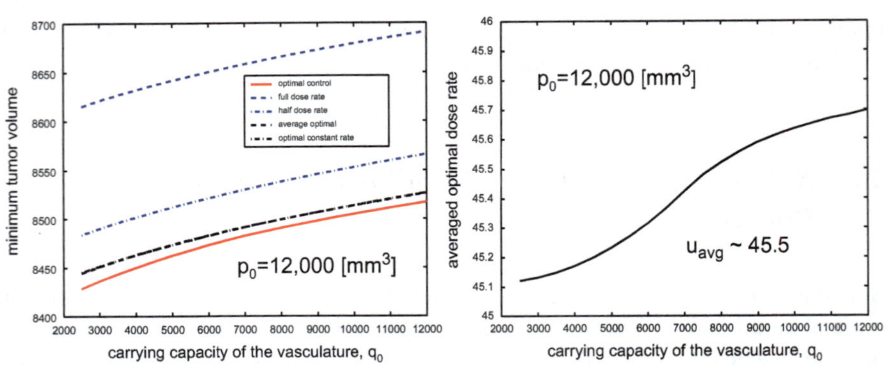

Fig. 6 The minimum tumor volumes and controls for $p_0 = 12,000$ [mm^3]

antiangiogenic agents are given at maximum dose rate $u_{\max}$ (dashed blue curve), the half-dose rate protocol (dash-dotted blue curve), the averaged optimal control protocol (dashed black curve), and the best constant dose rate protocol (dash-dotted black curve) for the fixed initial tumor volume $p_0 = 12,000\ [mm^3]$. The curves for the averaged optimal control protocols and the best constant dose protocol are very close and basically lie on top of each other in the figure with only minute differences for very low and very high tumor volumes. On the right of the same figure we show a graph of the averaged optimal control as a function of q_0. These constant dose rates only vary between 45.1 and 45.7 showing the strong robustness of the solutions with respect to q_0.

For the initial condition (p_0, q_0)=(12,000 [mm^3];15,000 [mm^3]), all antiangiogenic agents are used up at time 6.558 [days] along the optimal solution. It is not difficult to compute the best protocol that would give the same total amount in 6 constant daily doses and these dose rates are given by

$$u_1 = 46.61,\ u_2 = 45.31,\ u_3 = 48.15,\ u_4 = 50.71,\ u_5 = 53.20, \text{ and } u_6 = 56.02.$$

The values closely mimic the structure of the theoretically optimal control shown in Fig. 5. There is a small dip in the dosage from the first to the second day which is caused by the fact that the piece along which the optimal dose rate is $u_{\max}$ is small and thus the first daily value is significantly lower than $u_{\max} = 75$, but still higher than the second daily dose. Then the dosages gradually increase over the remaining days. This reflects the dose intensification along the optimal singular arc. Yet, specifying the time structure by restricting to daily doses reduces the quality of the approximation somewhat.

4.3 Combination of Antiangiogenic and Chemotherapy

Antiangiogenic therapy only attacks tumor growth indirectly through the vasculature and it is natural to combine it with a second therapy that directly attacks the tumor cells such as radio- or chemotherapy. We still consider a model that adds the action of a cytotoxic agent v, but again, rather than including the drug dosage as a penalty term in the objective, limits the overall amount of drugs given.

[AC] For a free terminal time T, minimize the tumor volume $p(T)$ subject to the dynamics

$$\dot{p} = \xi p \ln\left(\frac{q}{p}\right) - \varphi p v, \qquad p(0) = p_0, \tag{45}$$

$$\dot{q} = bp - \left(dp^{\frac{2}{3}} + \mu\right) q - \gamma q u - \eta q v, \qquad q(0) = q_0, \tag{46}$$

over all Lebesgue-measurable functions $u : [0, T] \to [0, u_{\max}]$ and $v : [0, T] \to [0, v_{\max}]$ for which the corresponding trajectory satisfies the terminal constraints

$$\int_0^T u(t)dt \leq A \qquad \text{and} \qquad \int_0^T v(t)dt \leq B. \tag{47}$$

An important feature of the optimal solution—and one that is not at all obvious—is that it builds in a modular way on the solution of the antiangiogenic monotherapy problem [A] already given [52]. Indeed, for a typical initial condition with $p_0 < q_0$, optimal controls for the combination therapy problem [AC] have the following structure: *optimal controls for the antiangiogenic agent follow the optimal solution for the monotherapy problem and then, at a specific time, chemotherapy becomes active and is given in one full dose session.* Both controls cannot be singular simultaneously and the formulas given above for the singular control and singular arc need to be adjusted to the presence of chemotherapy, but this is readily done and we have the following result:

Proposition 4.2 ([52]). *If the optimal antiangiogenic dose rate u follows the singular control u_{sing} on an open interval I, then the chemotherapeutic agent v is bang-bang on I with at most one switching from $v = 0$ to $v = v_{\max}$, and the following relation holds between the controls u and v:*

$$\gamma u_{\text{sing}}(t) + (\eta - \varphi)\, v(t) = \Psi(p(t), q(t)) \tag{48}$$

with Ψ defined by equation (44). Given v, this determines the anti-angiogenic dose rate with a jump discontinuity when chemotherapy becomes active.

This structure allows to set up a simple minimization problem over a 1-dimensional parameter τ that denotes the time when chemotherapy becomes active. We illustrate this for an initial condition (p_0, q_0) with $p_0 < q_0$ where the antiangiogenic agent is immediately applied with full dose. In principle, this time τ when chemotherapy is activated can lie anywhere in $[0, T]$. For example, if the amount $z_{\max}$ of chemotherapeutic agents is high, then it is possible that chemotherapy already becomes active along the interval when the antiangiogenic dose rate is at maximum. On the other hand, if this amount is very low, this activation may only occur after all antiangiogenic agents have been exhausted. The typical case, however, is that this time τ lies somewhere in the interval where the control u follows the singular monotherapy structure. Figure 7 shows an example of numerically computed optimal controls for the combination therapy problem [30].

This structure of optimal controls for the combination therapy has interesting medical interpretation: optimization leads to the conclusion that it is best to follow specific "paths" along which maximum tumor reductions are achieved. This holds for both the monotherapy problem [A] and the combination therapy problem [AC] and these paths, as expressed in the formula (48), are closely linked with the optimal singular arc from the monotherapy problem. Note that the singular curve $\mathcal{S}_0$ lies in the region where the tumor volume p is higher than its carrying capacity q, but

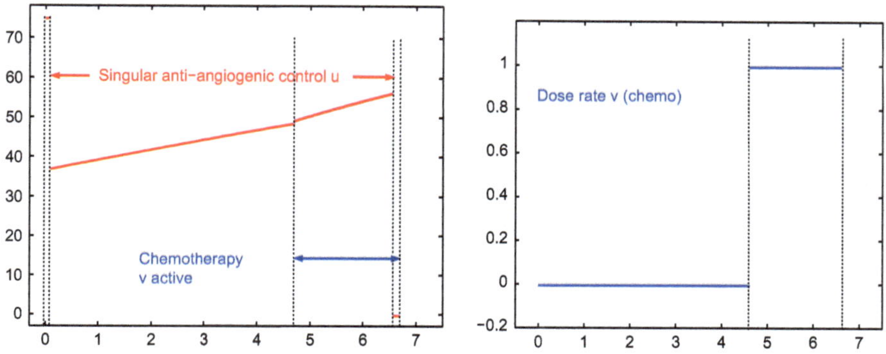

Fig. 7 An optimal solution for the combination therapy problem [AC]

there exists a specific relation between these variables. Clearly q is not pushed to zero too fast, but a definite balance between these two variables is maintained along the optimal solution. Since the vascular network of the tumor is needed to deliver the chemotherapeutic agents, this perfectly makes sense. In the medical literature, similar features have been observed and are know as "pruning" [19, 20]. It has been argued by Jain in [19] that *the preliminary delivery of antiangiogenic agents may regularize a tumor's vascular network with beneficial consequences for the successive delivery of cytotoxic chemotherapeutic agents.* Although no " pruning" aspects have been taken into account in the model (e.g., see [50] for such a model), it is interesting to note that an optimization approach for a rather small and minimally parameterized high-level mathematical model leads to very much the same conclusion: *give antiangiogenic agents until an optimal relation between tumor volume and carrying capacity has been established and then apply full dose chemotherapy while still maintaining the optimal relation between p and q through the administration of antiangiogenic agents.* Even when antiangiogenic treatment is combined with radiotherapy, this feature seems to persist with the optimal monotherapy solution once more playing a major role in the structure of optimal controls for the combination [40].

5 Tumor-Immune System Interactions

A second major component of a tumor's microenvironment is the immune system. The immune system's first response to its environment is on the basis of a discrimination between "own" and "foreign" objects and some tumor cells will simply be classified as "own" and thus tolerated [54]. However, tumor cells also exhibit a large number of abnormalities (such as mutated proteins, under- or over-expressed normal proteins and many more) that lead to the appearance of specific antigens, some of which will be classified as "foreign" and thus do trigger reactions by both the innate and adaptive immune system [23, 65]. In fact, the empirical hypothesis of *immune surveillance*, i.e., that the immune system may act to eliminate or control tumors, is well established in the medical community. The competitive interaction between tumor cells and the immune system is extremely complex and strongly nonlinear. The possible outcome of this interplay is not only constituted by tumor suppression or tumor outbreak but by a multitude of dynamic properties that include the persistence of both benign and malignant scenarios (e.g., see [45, 53]). Here we still consider a classical mathematical model by Stepanova [63] that captures these features of tumor-immune interactions in a low-dimensional, minimally parameterized model. In Sect. 5.1 we describe the model and consider the uncontrolled multi-stable dynamics which has both a benign and malignant region [32, 53]. We then in Sect. 5.2 set up an optimal control problem that induces the system to move from the malignant into the benign region under chemotherapy. After a brief administration of maximum dose chemotherapy, optimal protocols switch to singular controls and significantly lower dose rates [32]. In the medical literature such protocols are sometimes referred to as *"chemo-switch" protocols* [57].

5.1 Multi-stability and Regions of Attractions

We briefly recall Stepanova's model. Let x denote the tumor volume with a fixed carrying capacity $x_\infty < \infty$ and let y be a non-dimensional order of magnitude variable related to the activities of various types of T-cells activated during the immune reaction. We shall refer to y as the immunocompetent cell density. While Stepanova uses an exponential model for the growth of the tumor, here, as in [72], we consider a Gompertzian tumor growth models. The dynamical equations of the model are given by

$$\dot{x} = -\mu_C x \ln\left(\frac{x}{x_\infty}\right) - \gamma x y, \tag{49}$$

$$\dot{y} = \mu_I \left(x - \beta x^2\right) y - \delta y + \alpha, \tag{50}$$

with all Greek letters denoting constant coefficients. The second equation summarizes the main features of the immune system's reaction to cancer. Several organs contribute to the development of immune cells in the body and the parameter α models a combined rate of influx of T-cells generated through these primary organs; δ is simply the rate of natural death of the T-cells. The first term in this equation models the proliferation of lymphocytes. For small tumors, it is stimulated by the tumor antigen which can be assumed to be proportional to the tumor volume x. It is argued in [63] that large tumors suppress the activity of the immune system. The reasons lie in an inadequate stimulation of the immune forces as well as a general suppression of immune lymphocytes by the tumor (see [63] and the references therein). This feature is expressed in the model through the inclusion of the term

$-\beta x^2$. Thus $1/\beta$ corresponds to a threshold beyond which the immunological system becomes depressed by the growing tumor. The coefficients μ_I and β are used to calibrate these interactions and in the product with y collectively describe a state-dependent influence of the cancer cells on the stimulation of the immune system. The first equation models tumor growth. The coefficient γ denotes the rate at which cancer cells are eliminated through the activity of T-cells and the term γxy thus models the beneficial effect of the immune reaction on the cancer volume. Lastly, μ_C simply is a tumor growth coefficient.

For our numerical computations we use the following parameter values that are based on the paper [27] by Kuznetsov, Makalkin, Taylor, and Perelson who estimate these parameters based on in vivo experimental data for B-lymphoma BCL_1 in the spleen of mice: $\alpha = 0.1181$, $\beta = 0.00264$, $\gamma = 1$, $\delta = 0.37451$, $\mu_C = 0.5618$ and $\mu_I = 0.00484$. In that paper, a classical logistic growth term is used for cancer growth and we therefore adjusted the growth rates to account for Gompertzian growth using linear data fitting. Also, the functional form $\left(x - \beta x^2\right) y$ used in Stepanova's model in equation (50) is a quadratic expansion of the term used in [27]. Following [27], x is given in multiples of 10^6 cells and y is a dimensionless quantity that describes the immunocompetent cell density on an order of magnitude basis relative to base value 1. The time scale is taken relative to the tumor cell cycle and is in terms of 0.11 days [27]. As always, we simply use this particular values to illustrate our analytical results.

There always exists a *disease-free equilibrium point* at $(x_f, y_f) = (0, \frac{\alpha}{\delta})$ which is unstable. For the parameter values given above, there exist three equilibria with positive tumor volumes and Fig. 8 shows the phase portrait of the system. There is an asymptotically stable focus at $(x_b, y_b) = (72.961, 1.327)$ (marked by a green star), a saddle point at $(x_s, y_s) = (356.174, 0.439)$ (marked by a black star), and an asymptotically stable node at $(x_m, y_m) = (737.278, 0.032)$ (marked by a red star). In the diagram we have also marked the stable manifold of the saddle as a dashed red curve. The regions of attraction of the stable equilibria are the open regions that are separated by this stable manifold of the saddle.

We call a locally asymptotically stable equilibrium point (x_*, y_*) of the equations (49) and (50) malignant if the corresponding tumor volume x_* is close to the carrying capacity of the system, benign if it is by an order of magnitude smaller. The corresponding regions of attraction are the malignant and benign regions, respectively. In case of a microscopic benign equilibrium, this region can be interpreted as the set of all states of the system where the immune system is able to control the cancer and this is one possible way of describing what medically has been called *immune surveillance*. The region of attraction of the macroscopic equilibrium point, on the other hand, corresponds to conditions when the system has escaped from this immune surveillance and the disease will be lethal. Obviously, an interesting structure is the boundary between these two behaviors that is formed by the stable manifold of the saddle point. The natural therapeutic question then becomes how to move the state back into the benign region if it has been displaced into the malignant region.

5.2 Optimal Control for Tumor-Immune Interactions with Strongly Targeted Drugs

We now consider equations (49) and (50) with a cytotoxic agent u and a rudimentary immune boost v. As a simpler scenario, we assume that the chemotherapeutic agent is strongly targeted towards the tumor cells and therefore neglect its effects on the immunocompetent cell densities. Once more employing the standard log-kill assumption, this leads to the following equations:

$$\dot{x} = -\mu_C x \ln\left(\frac{x}{x_\infty}\right) - \gamma xy - \kappa xu, \qquad x(0) = x_0, \tag{51}$$

$$\dot{y} = \mu_I\left(x - \beta x^2\right) y - \delta y + \alpha + \rho yv, \qquad y(0) = y_0. \tag{52}$$

Given the multi-stable scenario, the practical aim of therapy is to move an initial state (x_0, y_0) that lies in the malignant region into the region of attraction of the benign equilibrium point while keeping side effects tolerable. For this, we consider the following optimal control problem:

[CI] for a free terminal time T, minimize the objective

$$J = Ax(T) - By(T) + \int_0^T (Cu(t) + Dv(t) + S)\, dt, \tag{53}$$

over all Lebesgue-measurable functions $u : [0, T] \to [0, 1]$ and $v : [0, T] \to [0, 1]$ subject to the dynamics (51) and (52).

The choice of the weights aims at striking a balance between the benefit at the terminal time T, $Ax(T) - By(T)$, and the overall side effects measured by the total

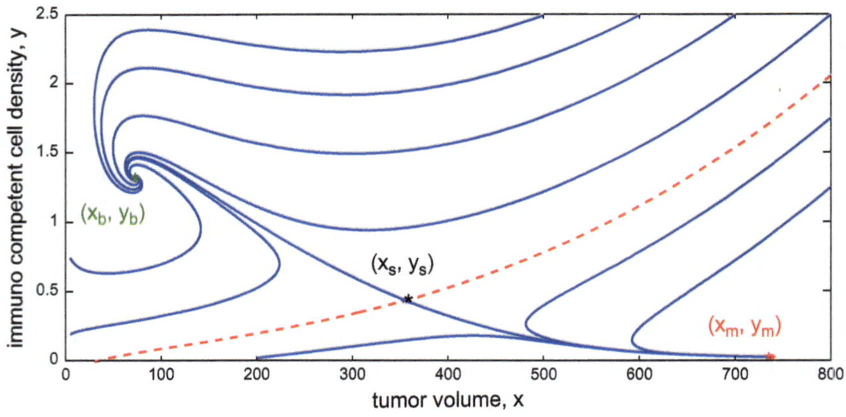

Fig. 8 Phase portrait of the uncontrolled system (49) and (50)

amounts of drugs given, while it at the same time guarantees the existence of an optimal solution by also penalizing the free terminal time T. The most important piece is the penalty term $Ax(T) - By(T)$ at the final time that is designed to induce the state of the system to move from the malignant into the benign region. In order to accomplish this, it may no longer be adequate to simply minimize the tumor volume since, as can be seen in Fig. 8, small tumor volumes are possible that lie in the malignant region if the immune system is depressed. Rather, the geometric shape of the separatrix matters. While it is generally not possible to give an analytic description for this surface, the tangent space to the saddle is easily computed and its normal vector can serve as a reasonable direction in which we want the system to move. This is what we have done here giving the numerical values $A = 0.00192$ and $B = 1$ for the data used earlier.

Once more, optimal controls for the cytotoxic agent consist of concatenations of bang and singular pieces. It can be shown that optimal administration of the immune boost v is bang-bang [29] and analytical formulas for a singular control u and arc can be derived, albeit with slightly different reasoning than above [29, 32]. The typical optimal control u_* is a concatenation of four pieces of the type **1s01**: therapy starts with a short maximum dose therapy session followed by a segment where the control is singular. Along this segment, the system moves along the singular arc from the malignant into the benign region. It is this transfer that matters and the tumor volume may actually increase along this segment. Once safely into the benign region, at one point therapy stops, i.e., the optimal control switches to $u = 0$. This portion of the trajectory closely follows the unstable manifold of the saddle for the uncontrolled system and leads to a "free pass," a trajectory along which no cost is incurred if $S = 0$. (The existence of such structures leads to issues about the existence of optimal controls and for this reason, we generally impose a small penalty S on the terminal time.) Along this portion of the controlled trajectory, the actions of the immune system take over. Quite frequently, after a prolonged rest period, optimal controls still give a short maximum dose chemotherapy and immune boost towards the end.

Figure 9 shows the optimal controlled trajectory for $C = 0.01$, $D = 0.025$, and $S = 0.001$ [29]. In the figure of the controlled trajectory switching points for the cytotoxic agent are indicated by a red asterisk and those for the immune boost with a green asterisk. Initially chemotherapy is given at full dose without immune boost. Already after a brief time interval, as the state of the system nears the separatrix, chemotherapy is reduced drastically and is only administered at lower dose rates according to the singular control u_{sing} and we clearly see the "chemo-switch"-type behavior of administration of a chemotherapeutic agent as optimal. In these solutions, the tumor microenvironment plays a major role: the initial chemotherapy is only designed to bring the state of the system into a region where the immune system is potent enough to control (not necessarily eliminate or eradicate) the cancer. If possible, this aim is achieved with low doses of chemotherapy. In fact—but such a structure is not included in the model—higher doses may be harmful in that they might adversely effect the immune system which otherwise would have come to the assistance in combating the tumor.

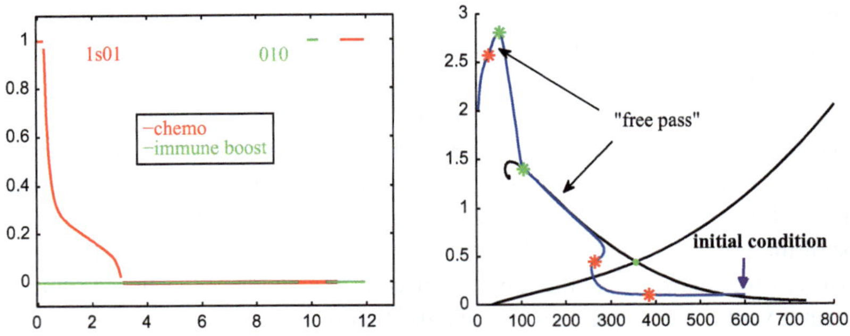

Fig. 9 Optimal control (*left*) and corresponding controlled trajectory (*right*) for $C = 0.01$, $D = 0.025$, and $S = 0.001$. (Reproduced with permission from [29], ©2013, AIMS)

6 Conclusion

We have outlined the qualitative type of results that can be obtained about cancer treatment protocols from an optimal control analysis of high-level mathematical models. Initially, the focus was on the cancerous cells progressing from mathematical models for homogeneous tumor populations of chemotherapeutically sensitive cells to heterogeneous structures of cell populations with varying sensitivities and resistance. From an optimal control point of view, optimal treatment schedules change from bang-bang solutions with upfront dosing (the classical MTD approaches in medicine) to administrations that favor singular controls (time-varying dosing schedules at less than maximum rates) as heterogeneity of the tumor population becomes more prevalent. Once the main components of the tumor microenvironment, its vasculature, and the immune system, are taken into account, in optimal solutions, more is not necessarily better. In this context, and in view of the fact that a properly calibrated dose (which does not waste agents nor have excessive side effects) can deliver the best outcomes, in medical research the search for a "biologically optimal dose" (BOD) is being pursued. In the model for antiangiogenic treatments it becomes clear that full dose therapies do waste agents that can be used more effectively when spread out at lower doses over prolonged time periods. The mathematical solution supports the idea of a normalization of the vasculature prior to the administration of chemotherapy, but then cytotoxic agents are given at the appropriate time in an MTD fashion. In a certain sense, an ideal tumor size-vasculature pattern is sought first which leads to an optimal tumor kill potential that then is exploited by maximum dose chemotherapy. However, as also the immune system is taken into account, *chemo-switch protocols* become optimal. The reason simply is that when the system is in a state where the actions of the immune system are able to control cancer growth, it is overall preferable (in view of the toxic side effects of chemotherapy) to administer lower doses.

Clearly, the models considered here are simplified, and this is natural at the high level of agglomeration that underlies their construction. While biological and medical research prefers to be as detailed as possible in their models, this also makes them amenable to the pitfall of Borges's "exactitude in science." The question simply is to what extent a model needs to be accurate to make significant and realistic predictions. In our view, the smaller the model is to give the relevant conclusion, the better it is. The conclusions that we obtain from these minimally parameterized models would suggest that these models lead to realistic statements about the structure of optimal treatment protocols that should be of interest in medical practice. In fact, the question how to schedule chemotherapeutic drugs in order to optimize their antitumor, antivasculature, and proimmune effects is far from being answered and there are concerted efforts in medical research to explore the benefits of metronomic scheduling in this respect [1, 56]. Qualitative mathematical results about optimal protocols that take into account a tumor's microenvironment can be of assistance in these efforts.

Acknowledgements This material is based upon work supported by the National Science Foundation under collaborative research Grants Nos. DMS 1311729/1311733. Any opinions, findings, and conclusions or recommendations expressed in this material are those of the author(s) and do not necessarily reflect the views of the National Science Foundation.

References

1. N. André, L. Padovani, E. Pasquier, Metronomic scheduling of anticancer treatment: the next generation of multitarget therapy?. Fut. Oncol. **7**(3), 385–394 (2011)
2. T. Boehm, J. Folkman, T. Browder, M.S. O'Reilly, Antiangiogenic therapy of experimental cancer does not induce acquired drug resistance. Nature **390**, 404–407 (1997)
3. B. Bonnard, , M. Chyba, Singular trajectories and their role in control theory. *Mathématiques & Applications*, vol. 40 (Springer, Paris 2003)
4. A. Bressan, A. Piccoli, Introduction to the Mathematical Theory of Control, American Institute of Mathematical Sciences (2007)
5. S. Davis, G.D. Yancopoulos, The angiopoietins: Yin and Yang in angiogenesis. Cur. Top. Microbio. Immun. **237**, 173–185 (1999)
6. M. Eisen, Mathematical Models in Cell Biology and Cancer Chemotherapy, *Lecture Notes in Biomathematics*, vol. 30 (Springer, NewYork 1979)
7. A. Ergun, K. Camphausen, L.M. Wein, Optimal scheduling of radiotherapy and angiogenic inhibitors, Bull. Math. Biol. **65**, 407–424 (2003)
8. J. Folkman, Tumor angiogenesis: therapeutic implications. New Engl. J. Med. **295**, 1182–1196 (1971)
9. J. Folkman, Antiangiogenesis: new concept for therapy of solid tumors, Ann. Surg. **175**, 409–416 (1972)
10. J. Folkman, M. Klagsburn, Angiogenic factors. Science **235**, 442–447 (1987)
11. U. Forys, Y. Keifetz, Y. Kogan, Critical-point analysis for three-variable cancer angiogenesis models. Math. Biosci. Eng. **2**, 511–525 (2005)
12. R.A. Gatenby, A.S. Silva, R.J. Gillies, B.R. Frieden, Adaptive therapy. Canc. Res. **69**, 4894–4903 (2009)
13. J.H. Goldie, Drug resistance in cancer: a perspective. Canc. Meta. Rev. **20**, 63–68 (2001)
14. J.H. Goldie, A. Coldman, *Drug Resistance in Cancer* (Cambridge University Press, Cambridge 1998)

15. P. Hahnfeldt, L.Hlatky, Cell resensitization during protracted dosing of heterogeneous cell populations. Radiat. Res. **150**, 681–687 (1998)
16. P. Hahnfeldt, D. Panigrahy, J. Folkman, L. Hlatky, Tumor development under angiogenic signaling: a dynamical theory of tumor growth, treatment response, and postvascular dormancy. Can. Res. **59**, 4770–4775 (1999)
17. P. Hahnfeldt, J. Folkman, L. Hlatky, Minimizing long-term burden: the logic for metronomic chemotherapy dosing and its angiogenic basis. J. Theo. Biol. **220**, 545–554 (2003)
18. D. Hanahan, G. Bergers, E. Bergsland, Less is more, regularly: metronomic dosing of cytotoxic drugs can target tumor angiogenesis in mice. J. Clin. Invest. **105**, 1045–1047 (2000)
19. R.K. Jain, Normalizing tumor vasculature with anti-angiogenic therapy: a new paradigm for combination therapy. Nat. Med., **7**, 987–989 (2001)
20. R.K. Jain, L.L. Munn, Vascular normalization as a rationale for combining chemotherapy with antiangiogenic agents, Princ. Pract. Oncol. **21**, 1–7 (2007)
21. B. Kamen, E. Rubin, J. Aisner, E. Glatstein, High-time chemotherapy or high time for low dose? J. Clin. Oncol. **18**, Editorial, 2935–2937 (2000)
22. R.S. Kerbel, Tumor angiogenesis: past, present and near future, Carcinogensis, **21**, 505–515 (2000)
23. T.J. Kindt, B.A. Osborne, R.A. Goldsby, *Kuby Immunology* (W.H. Freeman, New York 2006)
24. M. Kimmel, A. Swierniak, Control theory approach to cancer chemotherapy: benefiting from phase dependence and overcoming drug resistance, in *Tutorials in Mathematical Biosciences III: Cell Cycle, Proliferation, and Cancer*. Lecture Notes in Mathematics, vol. 1872 (Springer, Newyork, 2006), pp. 185–221
25. M. Klagsburn, S. Soker, VEGF/VPF: the angiogenesis factor found?. Curr. Biol. **3**, 699–702, (1993)
26. G. Klement, S. Baruchel, , Rak, J., Man, S., Clark, K., Hicklin, D.J., Bohlen, P., Kerbel, R.S.: Continuous low-dose therapy with vinblastine and VEGF receptor-2 antibody induces sustained tumor regression without overt toxicity, J. Clin. Invest. **105**, R15–R24 (2000)
27. V.A. Kuznetsov, I.A. Makalkin, M.A. Taylor, A.S Perelson, Nonlinear dynamics of immunogenic tumors: parameter estimation and global bifurcation analysis. Bul. Math. Biol. **56**, 295–321 (1994)
28. U. Ledzewicz, K. Bratton, H. Schättler, A 3-compartment model for chemotherapy of heterogeneous tumor populations. Acta Appl. Matem. (2014) doi: 10.1007/s10440-014-9952-6
29. U. Ledzewicz, M.S. Faraji Mosalman, H. Schättler, Optimal controls for a mathematical model of tumor-immune interactions under targeted chemotherapy with immune boost, Discr. Cont. Dyn. Syst. Ser. B **18**, 1031–1051 (2013)
30. U. Ledzewicz, A. d'Onofrio, H. Schättler, Tumor development under combination treatments with anti-angiogenic therapies. in *Mathematical Methods and Models in Biomedicine* (Springer, NewYork, 2012), pp. 311–337
31. U.Ledzewicz, J. Marriott, H. Maurer, H. Schättler, Realizable protocols for optimal administration of drugs in mathematical models for novel cancer treatments, Math. Med. Biol. **27**, 157–179, (2010).
32. U. Ledzewicz, M. Naghnaeian, H. Schättler, Optimal response to chemotherapy for a mathematical model of tumor-immune dynamics. J. Math. Biol. **64**, 557–577 (2012)
33. U. Ledzewicz, H. Schättler, Optimal bang-bang controls for a 2-compartment model in cancer chemotherapy. J. Optim. Th. Appl. **114**, 609–637 (2002)
34. U. Ledzewicz, H. Schättler, Analysis of a cell-cycle specific model for cancer chemotherapy. J. Biol. Syst. **10**, 183–206 (2002)
35. U. Ledzewicz, H. Schättler, Optimal control for a bilinear model with recruiting agent in cancer chemotherapy, Proc. of the 42nd IEEE Conference on Decision and Control (CDC), Maui, Hawaii, 2762–2767 (2003)
36. U. Ledzewicz, H. Schättler, The influence of PK/PD on the structure of optimal control in cancer chemotherapy models, Math. Biosci. Engr. **2**, 561–578 (2005)

37. U. Ledzewicz, H. Schättler, Drug resistance in cancer chemotherapy as an optimal control problem, Discr. Cont. Dyn. Syst. Ser. B, **6**, 129–150 (2006)
38. U. Ledzewicz, H. Schättler, Anti-angiogenic therapy in cancer treatment as an optimal control problem. SIAM J. Contr. Optim. **46**, 1052–1079 (2007)
39. U. Ledzewicz, H. Schättler, Optimal and suboptimal protocols for a class of mathematical models of tumor anti-angiogenesis. J. of Theo. Biol. **252**, 295–312, (2008)
40. U. Ledzewicz, H. Schättler, Multi-input optimal control problems for combined tumor anti-angiogenic and radiotherapy treatments. J. of Optim. Th. Appl. **153**, 195–224 (2012)
41. U. Ledzewicz, H. Schättler, M. Reisi Gahrooi, S. Mahmoudian Dehkordi, On the MTD paradigm and optimal control for combination cancer chemotherapy. Math. Biosci. Engr. **10**, 803–819 (2013)
42. L.A. Loeb, A mutator phenotype in cancer. Canc. Res. **61**, 3230–3239 (2001)
43. R. Martin, K.L. Teo, *Optimal Control of Drug Administration in Cancer Chemotherapy* (World Scientific Publishers, Singapore 1994)
44. L. Norton, R. Simon, The Norton-Simon hypothesis revisited. Canc. Treat. Rep. **70**, 163–169 (1986)
45. A. d'Onofrio, A general framework for modelling tumor-immune system competition and immunotherapy: Mathematical analysis and biomedial inferences. Phys. D **208**, 202–235, (2005)
46. A. d'Onofrio, Rapidly acting antitumoral antiangiogenic therapies. Phys. Rev. E Stat. Nonlin. Soft Matter Phys. **76**, 031920 (2007)
47. A. d'Onofrio, A. Gandolfi, Tumour eradication by antiangiogenic therapy: analysis and extensions of the model by Hahnfeldt et al., Math. Biosci. **191**, 159–184 (2004)
48. A. d'Onofrio, A. Gandolfi, The response to antiangiogenic anticancer drugs that inhibit endothelial cell proliferation. Appl. Math. and Comp. **181**, 1155–1162 (2006)
49. A. d'Onofrio, A. Gandolfi, A family of models of angiogenesis and anti-angiogenesis anti-cancer therapy. Math. Med. Biol., **26**, 63–95 (2009)
50. A. d'Onofrio, A. Gandolfi, Chemotherapy of vascularised tumours: role of vessel density and the effect of vascular "pruning". J. Theo. Biol. **264**, 253–265, (2010)
51. A. d'Onofrio, A. Gandolfi, A. Rocca, The dynamics of tumour-vasculature interaction suggests low-dose, time-dense antiangiogenic schedulings. Cell Prolif., **42**, 317–329, (2009)
52. A. d'Onofrio, U. Ledzewicz, H. Maurer, H. Schättler, On optimal delivery of combination therapy for tumors. Math. Biosci., **222**, 13–26 (2009)
53. A. d'Onofrio, U. Ledzewicz, H. Schättler, On the dynamics of tumor immune system interactions and combined chemo- and immunotherapy, in: New Challenges for Cancer Systems Biomedicine eds. by A. d'Onofrio, P. Cerrai, A Gandolfi, vol. 1 (SIMAI Springer series, 2012). pp. 249–266
54. D. Pardoll, Does the immune system see tumors as foreign or self? Ann. Rev. Immun. **21**, 807–839 (2003)
55. E. Pasquier, U. Ledzewicz, Perspective on "More is not necessarily better": Metronomic Chemotherapy. Newslet. Soc. Math. Biol. **26**(2), 9–10, (2013)
56. E. Pasquier, M. Kavallaris, N. André, Metronomic chemotherapy: new rationale for new directions. Nat. Rev.|Clin. Onc. **7**, 455–465 (2010)
57. K. Pietras, D. Hanahan, A multi-targeted, metronomic and maximum tolerated dose "chemo-switch" regimen is antiangiogenic, producing objective responses and survival benefit in a mouse model of cancer. J. Clin. Onc. **23**, 939–952 (2005)
58. J. Poleszczuk, U. Forys, Derivation of the Hahnfeldt et al. model (1999) revisited, Proceedings of the 16th Nat. Conf. on Applications of Mathematics in Biology and Medicine, Krynica, Poland 87–92 (2010)
59. L.S. Pontryagin, V.G. Boltyanskii, R.V. Gamkrelidze, E.F. Mishchenko, *The Mathematical Theory of Optimal Processes* (MacMillan, New York 1964)
60. H. Schättler, U. Ledzewicz, B. Cardwell, Robustness of optimal controls for a class of mathematical models for tumor anti-angiogenesis. Math. Biosci. Engr. **8**, 355–369 (2011)
61. H. Schättler, U. Ledzewicz: *Geometric Optimal Control* (Springer, NewYork 2012)

62. H. Schättler, U. Ledzewicz, S. Mahmoudian Dehkordi, M. Reisi Gahrooi, A geometric analysis of bang-bang extremals in optimal control problems for combination cancer chemotherapy, Proc. of the 51st IEEE Conf. on Decision and Control, Maui, Hawaii, 7691–7696, (2012)
63. N.V. Stepanova, Course of the immune reaction during the development of a malignant tumour. Biophys. **24**, 917–923 (1980)
64. G.W. Swan, Role of optimal control in cancer chemotherapy. Math. Biosci., **101**, 237–284 (1990)
65. J.B. Swann, M.J. Smyth, Immune surveillance of tumors. J. Clin. Invest. **117** 1137–1146, (2007)
66. A. Swierniak, Optimal treatment protocols in leukemia-modelling the proliferation cycle, Proc. of the 12th IMACS World Congress, Paris, vol. 4, 170–172 (1988)
67. A. Swierniak, Cell cycle as an object of control, J. Biol. Syst. **3**, 41–54 (1995)
68. A. Swierniak, Direct and indirect control of cancer populations. Bul. Pol. Acad. Sci. Techn. Sci. **56**, 367–378 (2008)
69. A. Swierniak, U. Ledzewicz, H. Schättler, Optimal control for a class of compartmental models in cancer chemotherapy. Int. J. Appl. Math. Comp. Sci. **13**, 357–368 (2003)
70. A. Swierniak, A. d'Onofrio, A. Gandolfi, Optimal control problems related to tumor angiogenesis. Proc. IEEE-IECON'2006, 667–681 (2006)
71. A. Swierniak, J. Smieja, Cancer chemotherapy optimization under evolving drug resistance. Nonlin. Ana. **47**, 375–386 (2000)
72. H.P. de Vladar, J.A. González, Dynamic response of cancer under the influence of immunological activity and therapy. J. Theo. Biol. **227**, 335–348 (2004)

MIX
Papier aus verantwortungsvollen Quellen
Paper from responsible sources
FSC® C105338

If you have any concerns about our products,
you can contact us on
ProductSafety@springernature.com

In case Publisher is established outside the EU,
the EU authorized representative is:
Springer Nature Customer Service Center GmbH
Europaplatz 3, 69115 Heidelberg, Germany

Printed by Libri Plureos GmbH
in Hamburg, Germany

150770398 8

UNIVERSITY OF BRISTOL
Geographical Sciences Library
School of Geographical Sciences, University Road
Bristol BS8 1SS

STANDARD LOAN

Date due for return
UNLESS RECALLED EARLIER

WITHDRAWN
University of Bristol